Experiencing Illness and the Sick Body
in Early Modern Europe

Experiencing Illness and the Sick Body in Early Modern Europe

Michael Stolberg

Chair for the History of Medicine,
Director of the Institut für Geschichte der Medizin, University of Würzburg, Germany

Translated from the German original by Leonhard Unglaub and Logan Kennedy

Softcover reprint of the hardcover 1st edition 2011 978-0-230-24343-9

Orig. published in 2003 as Homo patiens. Krankheits- und Körpererfahrung in der Frühen Neuzeit, by Böhlau-Verlag, Cologne-Weimar, Germany.

Research for this book and the translation of the German original were made possible by the generous support of the Deutsche Forschungsgemeinschaft and Wilhelm H. Ruchti Stiftung, Würzburg.

First published 2011 by
PALGRAVE MACMILLAN

Palgrave Macmillan in the UK is an imprint of Macmillan Publishers Limited, registered in England, company number 785998, of Houndmills, Basingstoke, Hampshire RG21 6XS.

Palgrave Macmillan in the US is a division of St Martin's Press LLC, 175 Fifth Avenue, New York, NY 10010.

Palgrave Macmillan is the global academic imprint of the above companies and has companies and representatives throughout the world.

Palgrave® and Macmillan® are registered trademarks in the United States, the United Kingdom, Europe and other countries

ISBN 978-1-349-31837-7   ISBN 978-0-230-35584-2 (eBook)
DOI 10.1057/9780230355842

A catalogue record for this book is available from the British Library.

A catalog record for this book is available from the Library of Congress.

10 9 8 7 6 5 4 3 2 1
20 19 18 17 16 15 14 13 12 11

Contents

Part III   Dominant Discourse and the Experience of Disease

Conclusion: A New Bourgeois Habitus 213

Introduction

This book explores how ordinary people in the early modern period perceived, experienced and interpreted illness and how they dealt and coped with it in everyday life. In this sense, it is an attempt to write a history of medicine from the patient's or layperson's point of view.[1] This approach to the writing of medical history is a comparatively recent undertaking.[2] For a long time, medical historians were primarily interested in the 'great physicians' and their contributions to medical 'progress'. Patients usually appeared as little more than faceless inmates of medical institutions or as the collective target of public health policies. The sole major exceptions were famous patients – rulers and artists above all – whose diseases and causes of death frequently gave rise to lively debates. Since the mid-1980s, however, in the wake of renewed interest in the social history of medicine,[3] historians widened the view to include patients' experience of disease and medical care. A number of studies have since produced valuable new insights.[4] Even studies on the history of public health and medical institutions have come to accept the need to pay attention to the needs and experiences of patients.[5]

Thanks to this work, we know much more today about how people dealt with diseases in the past, about the importance of self-treatment, for instance, about the wide range of curative approaches patients could choose from, about the relationship between patients and healers, and about people's attempts to find meaning and orientation in the face of their suffering. These issues have, in particular, been studied for the early modern period. Yet even in recent work about this period, the perception and subjective experience of the sick body and the interpretation and assessment of different symptoms and diseases by patients and their relatives are – if they are addressed at all – usually characterized only briefly and in a sweeping manner.[6] Some recent studies have opened valuable insights into the cultural meanings of the early modern body and its diseases from literary works of the period.[7] But we still know very little about how laypeople of those days perceived and interpreted the most important and common diseases such as, in contemporary terms, fever, fluxes, cancer, consumption and scurvy. Frequently and very misleadingly the lay understanding of all kinds of diseases is boiled down to the notion of a disturbed humoral balance.

Largely uncharted are the images, sensations and experiences that lay people associated with the disease terms commonly used at the time.

One major exception must be named: in her book *The Woman Beneath the Skin*, Barbara Duden, drawing on the Eisenach physician Johann Storch's case histories, has created a vivid picture of the female body in the 18th century; it is the fascinating image of a body shaped by a permanent flow of humors within and across its boundaries, a body that was always endangered by obstructions and congestions. Her study has rightly been acclaimed. But Storch's case histories deal specifically with 'women's complaints'. As a result, menstrual irregularities and the many complaints which contemporaries associated with them play a paramount role, while many other images and concepts which were crucial for the contemporary lay understanding of the sick body find little mention, because they did not in some way relate to the female reproductive system. Further, it is questionable to what extent Duden is able to access the personal experiences of the afflicted women relying solely on a physician's descriptions. A closer look at Duden's sources shows that, though the case histories often extend over several pages, the women's voices are usually expressed only in an occasional sentence or half-sentence; and even then, we do not know how truthfully Storch relayed their words and whether he perhaps gave preference to statements that supported his own interpretations.[8]

In what follows I will pursue a different path. I will rely primarily on personal testimonies in patient letters, personal correspondence, autobiographies and similar sources. As I hope to show they provide a very rich and nuanced account, even though it would be naive to assume that they offer an immediate, truly authentic picture of lay experiences of the body and its diseases.

This book is divided into three parts. The first part will give a primarily descriptive overview of how laypeople experienced and interpreted the body and its diseases and how they dealt with illness in everyday life. It will begin with the question of how assessment of health, illness and pain were described, assessed and given meaning. From there it will move to the fears that became associated with certain diseases and purportedly pathogenic influences from the environment and lifestyle, and it will close with issues surrounding nursing, medical care and the doctor–patient relationship.

The second part will focus on the perceptions and interpretations of individual types of disease. It will present the most important explanatory elements which framed contemporary lay perceptions and the ways different kinds of disease were dealt with. These explanatory elements, it will be argued, not only were decisive for the interpretation and choice of treatment; they also determined the vocabulary which patients used to describe their complaints and indeed shaped their physical, bodily perceptions, spawning descriptions of symptoms which patients in modern Western societies no longer experience. In particular, the doctrine of 'morbid matter' will be given considerable space, but also the fear of obstructions and the fear of a disruption of the flow of humors past the body's boundaries. A more detailed study of the most widespread and/or most feared diseases, such as fever, cancer and consumption, will show the concrete application and interplay of these explanatory concepts.

At the center of the third part will be the complex relation between the subjective body and disease experience on the one hand and the 'dominant' medical discourse and the values and interests that found expression in it, on the other. I will trace this relation by looking at the impact of two highly influential medical concepts of the 18th century. The rapid rise of 'nervous complaints' to the status of a fashionable disease shows how quickly – and creatively – medical laypeople adopted the new medical concepts of 'nervous sensibility' and 'nervous complaints'. Physicians combined the new model with a contemporary critique of civilization and also brought it to bear on the intense debate around the nature and social position of women. Looking at the reception and adaptation of the new model among the population, however, reveals that nervous complaints could also serve as a medium of self-fashioning or somatic protest. A look at the great campaign against the health hazards of sexual self-gratification, which began in the late 17th century once more shows clearly that the sole analysis of dominant medical discourse provides only limited access to the ways in which ordinary people actually experienced their bodies. At the same time, the confessions of patients who ascribed a wide range of diseases to the sins of their youth offer impressive empirical evidence for Michel Foucault's notion of a 'regimen of truth' which colonizes the individual not by repression but by offering a welcome source of meaning.

Since the original German edition of this book was published, in 2003, under the title Homo patiens. Krankheits- und Körpererfahrung in der Frühen Neuzeit a number of works on various aspects of the early modern experience and interpretation of illness have appeared, most of them based on a specific source or body of sources. Their findings have not substantially altered the picture at which I arrived in 2003. I will include references to the these works in the notes and will in some cases discuss their results in greater detail. I have resisted the temptation of rewriting the book, however, and have limited myself to minor revisions otherwise. Most changes are designed to make my argument still clearer, not least in view of the problems which a translation into another language inevitably brings with it, especially in the case of a study in which words and their changing meanings play an important role. Less frequently, changes reflect findings from new sources which I have analyzed in the meantime and results of my more recent work on various aspects of early modern medicine, in particular on the history of uroscopy, of ordinary medical practice and of palliative care.

Some Thoughts on Theory

For a long time, historians barely considered the body. They viewed the body basically as a stable, unchanging biological substrate in a world of historical change. The situation has become quite different over recent decades. Historiography has discovered the 'body'.[9] Some thirty years after Jacques Revel and Jean-Pierre Peter's programmatic plea for a historicization of the body[10] one might actually say, in analogy to the 'linguistic turn', that a 'somatic turn' occurred in many of the historical disciplines, and not only there: other disciplines in the social sciences and the humanities have also focused increasingly on the body.[11] The body – the seemingly natural, unchangeable Other – became itself the object of historical and cultural analysis.

The generic term 'history of the body', however, somewhat obscures the co-existence of quite different and at times even contradictory perspectives and methodological approaches in this field. For my purposes, roughly three levels can be discerned.[12] In the area of historical anthropology[13] and related approaches such as the German 'Alltagsgeschichte' (history of mentalities and micro-history), historians have aimed for the most part at a descriptive reconstruction or historical ethnography of somatic experiences and body-related practices. They have studied, for instance, the ideas which educated Parisian women of the 18th century had of the significance of digestion or of the heart; or what it meant to a 16th-century farm woman or maidservant in the Saar region to be pregnant; or what it 'felt' like, to an urban merchant or craftsman in 19th-century Berlin, to be in a skin that seemed highly permeable, from the outside as well as from the inside. The goal of this type of approach is to assemble a repertory of prominent body-related conceptions and practices and make them comprehensible to today's readers by locating them in the more general systems of ideas about man and nature to which they largely owed their importance and meaning.

When this largely descriptive plane is left behind and body concepts and practices are seen in a causal relation to the economic, social or political conditions from which they spring or which they reflect, the descriptive reconstruction passes onto a second level, which can be characterized roughly as social or cultural constructivist. It is not the subjectively experienced and lived body which is the center of interest here. What is explored instead is the influence of symbolic systems, power relations, societal structures, and political institutions on the production and development of dominant body conceptions and practices. 'Social' and 'cultural' construction cannot neatly be separated. The two terms do, however, refer to different levels of analysis, and it seems heuristically fruitful to make the distinction. The concept of 'cultural' construction is predominantly used in ethnology, primarily to investigate how dominant body images and body practices in a given society express or symbolize central collective values or fears peculiar to the culture in question. Very much at the center here is the search for analogies and metaphorical counterparts. The influential work of the English cultural anthropologist Mary Douglas has given crucial momentum to these approaches.[14] Douglas pointed out the close connection between body images and societal structure in many cultures. A society possessed by an extreme aversion to bodily defilement, for example, will frequently also tend to be particularly afraid of foreigners and invaders, and vice versa. From our own history, to take a different example, we are familiar with the manifold correlations between Western theories of the state and the interpretation of the body as a hierarchically structured entity within which the individual organs are attributed specific functions in the service of the structure as a whole.

More specifically, in the area of medicine, the so-called 'culture-bound syndromes' are prime examples of a cultural construction of illness. The term refers to disease patterns that are characterized by a typical concurrence of certain complaints and/or deviant behaviors that can be observed in this specific combination almost exclusively in one particular cultural area.[15] Some well known 'exotic' examples are 'koro', 'latah', 'el calor', 'susto', and 'nervios'. In Western societies as well, there have been attempts to interpret, for example, 'anorexia nervosa' and 'premenstrual syndrome' as culture-bound syndromes. I will take this subject up again later.[16]

The term 'social' construction, by contrast, tends to be used to describe the genesis of meaning, of discourses and practices as 'framed' or indeed brought forth by specific power structures, ideologies and special group interests, as we find them predominantly in more complex societies.[17]

Insofar as dominant body images in a given culture usually also express the hegemonic interests of a culture's elite, there are overlaps between 'cultural' and 'social' construction. However, with 'social' construction, the emphasis is decidedly on the 'dominant discourse' expressed in elite culture's written products and on the ways in which theories and images of the body are linked to the interests and world views of those who produce or support them or indeed to the power structures of a given society. Hence the most important tool of social-constructivist research is discourse analysis in the sense of a systematic and sometimes even serial examination of texts and bodies of text which are taken to reflect the 'dominant' discourse.

One of the most influential figures in the development of social-constructivist and discourse-analytical approaches in the area of body history – and not only there – was Michel Foucault. In his early works on the hospital, the prison, and the madhouse, he pointed out the paramount significance of the body as a target of social and political interests, as a place where leverage can be gained and power relations come into play.[18] Leading proponents of the more recent 'body history' have based their work on Foucault's analyses.[19] However, in many of these works we encounter a strangely disembodied 'body history'. The body appears as nothing more than an unchanging, anonymous, passive, suffering target of 'medical power' with its institutions and practices, while the body as a material entity that is experienced and lived frequently drops from view entirely.[20]

Foucault's later works on the history of sexuality are more rewarding when it comes to undertaking a body history that focuses on the lived body, on a body that is perceived and acts within a lived context. In these writings, Foucault focused on positive rather than repressive power, on forms of power which act by offering meaning and contribute to the very constitution of the individual. From this perspective, the history of sexuality in the Western world was not a history of silence and suppression. On the contrary, according to Foucault, it was through incessant talk about sex that even the most private and intimate emotions and thoughts were exposed and made accessible to 'power'.[21]

Social-constructivist approaches – including variants of discourse analysis à la Foucault – have been propagated widely in the sociology of knowledge and science. Starting from there, these approaches have also asserted their influence in more recent research on the history of science and medicine.[22] The science and medicine of the past are no longer regarded as a quest for an 'objective' truth. They are studied – like other ways of interpreting the world, nature and man – as social undertakings in which specific premises, values and interests are brought to bear. Instead of tracing a process of ever-increasing knowledge of and control over nature, scholars now pose the question: which historical context, which dominant interests or ideologies helped some approaches, theories or technical developments to prevail, while others – among them sometimes those that are later considered true – remained without significance or effect? Similar consideration is given to scientists' and physicians' strategies of self-promotion, rhetorical devices and 'networking' that initially secured attention and support for new findings or theories.

In this sense, social-constructivist studies have profoundly transformed traditional research in the history of medicine and science. In contrast to the naive glorification of 'great physicians' and revolutionary discoveries, they have shifted the focus towards the crucial impact of power relations and professional interests in the development of medicine and science. When such analysis is limited – as is often the case – to identifying a 'dominant' discourse and its context, its value for a history of the body remains rather limited, however. Looking at the discourse of the ruling classes and their 'organic intellectuals' ('intellettuali organici')[23] yields little insight into the actual impact of this discourse on the general public and on different parts of a population. Such an impact is frequently taken as a given although no evidence is provided that 'dominant' discourse was indeed communicated efficiently to and accepted by the wider population.[24] 'Dominant' discourse may well be rejected or ignored, however, or reinterpreted. In the history of the body, in particular, it is left entirely up to the speculative imagination of the historian, in the absence of other sources, to determine how and to what degree the 'dominant' discourse became 'inscribed' in body perception and experience, how it shaped and transformed them.

Insufficient consideration for the lived and experienced body as opposed to its mere linguistic 'representation' is a central problem shared by the third variety or analytic level of body history. For lack of an exact name and given its heterogeneity I will refer to it in a very simplified manner as postmodern body history. Its representatives can be found most of all in the study of literature and philosophy and, overlapping with the abovementioned approaches, in some parts of cultural and gender studies. A well known example is Judith Butler with her widely discussed thesis that not only 'gender' as a socially assigned role but also the binary conception of a male and female 'sex', in the sense of a biologically intended sexuality, is socially constructed.[25] These approaches have as their primary basis the insights of the linguistic turn and of the poststructuralist philosophy and literature studies which followed in its wake. Here, the body in its materiality drops from view almost entirely. Hardly any space is granted to an antecedent biological condition. Ultimately, everything appears to be the product of linguistic representations, implying that only that which can be expressed through language is 'real'.

This predominance of language and discourse as opposed to the lived and experienced body has met with growing opposition in recent years.[26] Today, hardly anyone will seriously assert that man's natural, biological condition can be described independently of existing cultural and linguistic categories or deny that even elementary, seemingly natural bodily phenomena, such as pain or affects, are culturally framed to a high degree. However, a look at the body in the past – just as in different cultures today – also reveals fundamental commonalities. Moreover, somatic experience and lived subjectivity are only incompletely rendered in linguistic discourse.[27] For some time, cultural anthropologists have therefore demanded that the biological aspects of human corporality be taken into consideration appropriately.[28] Along similar lines, Bryan S. Turner, one of the most influential representatives of the sociology of the body, has called the claim 'bizarre' that human activity has no biological basis; instead he has assigned to the sciences the task of exploring precisely those connections that exist between the body's biological condition and its social construction.[29] Representatives of feminist philosophy have argued for a 're-naturalization' of the body.[30]

Among historians, there is similarly growing skepticism toward a one-sided preference for the linguistic-discursive. Already some 15 years ago, Lyndal Roper raised her voice against 'a flight from the body and a retreat to the rational reaches of discourse', against a gender history that is all about language and 'leaves out the bodies'.[31] Among medical historians, the concept of 'framing'[32] diseases has emerged. It takes the body seriously as a historical actor in its own right but underlines, at the same time, that we can grasp the – presumably universal – biological processes in the body only through the language, images and notions of our respective culture.

Along these lines, this book takes, in epistemological and methodological terms, an intermediate position. It accepts without reservation that we can understand the body and bodily experience only from within our respective, historically/culturally shaped categories. But it aims for an understanding of the body conception which transcends the dichotomy of, on the one side, a neo-idealist, postmodernist cultural relativism fixated on the linguistic and, on the other side, static, biological essentialism. It takes as its starting point that corporality and body experience are fundamentally tied to a cultural and historical context, but also recognizes crucial aspects of the human bodily condition as naturally given with relatively little variation across cultures. It takes the body seriously as a discrete and resisting agent in its own right and, with it, the natural and man-made environment that influences it, an environment to which the body reacts and which it (co-)creates.

This type of approach in no way renders discourse analysis and deconstructivist approaches obsolete. In particular, the discourse of academic physicians about the body and its diseases merits a great deal of attention as a major source of meaning. However, we always need to ask whether or not this discourse had any effect among the population, whether it was accepted or rejected, or simply ignored, and we need to see to what degree the notions of physicians became changed or reinterpreted when they were appropriated and put into practice among the lay public. As we will see, the lay experience of the body and its diseases, especially among the upper classes, was in many cases framed by the medical discourse of the time, but the impact of this discourse was by no means pervasive or irresistible.

A second major methodological premise of this study has already been indicated: the perception and experience of the body are by no means identical with or indeed a mere product of their expression in language. To clarify the precise relation between experience and discourse is itself a crucial task of historical analysis. Resorting to concepts like 'embodiment', phenomenologist philosophers as well as ethnologists studying other cultures have pointed to the central significance of pre-linguistic elements in bodily experience.[33] This experience, they have argued, precedes its objectification in linguistic expression. Language or discourse is only one means by which experience of and knowledge about the body are communicated and subjected to the influence of culture and society. Beyond the realm of words, 'body techniques', as Marcel Mauss put it – seemingly natural, self-evident and yet culturally learned body practices – often play a much more important role. A small child learns through simple imitation how to walk 'correctly' in his/her respective culture, how to run or swim, how to move fingers and hands, in which situations the head should be bowed, when it is appropriate to show tears, and so forth.[34] Such body practices may indeed be a more effective, less easily resisted vehicle for the communication of values, norms and interests than the spoken or written word of the 'dominant discourse'. After all, it seems to the embodied individual as if the body itself speaks.[35]

Sources

This book relies above all on my analysis of many hundreds of patient letters, a source which has only recently found the appreciation it deserves among cultural historians of illness and the body.[36] Patient letters owe their existence above all to the early modern practice of consulting by letter. Among the educated classes it was fairly common to ask a renowned physician for advice, especially in the case of chronic diseases when the therapy proposed by local physicians did not have the desired effect.[37] Because the distant physician was, as a rule, expected to identify the nature of the disease and prescribe a suitable treatment without ever seeing the patient in person – that is, just on the basis of a written account – the letters are often quite thorough and detailed. In the 18th century, in the heyday of epistolary culture, patient letters could be over 30 pages long. And many patients not only described their current condition; they also sketched out the complete course of the disease in a more or less thorough manner – indeed their life stories, with all the circumstances and events that, in their view, might have had an effect on the genesis of their present suffering. In some cases, a series of letters documents a long-distance therapeutic consultation that went on for years, allowing us, to some extent, to trace the doctor–patient relationship in these particular circumstances from the patient's perspective.[38] More often, the letter exchange ceases after only a few letters or indeed after the physician's first response. Fortunately, from the historian's point of view, the first letters are usually particularly illuminating, since it was here that the patients or their relatives attempted to describe in great detail their histories and current complaints in order to provide as comprehensive an account of the disease as possible.

In many cases patients wrote personally or, if need be, dictated them to a scribe. Quite often, letters were also written by relatives, friends, the town's pastor or superiors, usually on the patient's request but sometimes secretly, without the patient's knowledge.[39] For the purposes of this book, such letters are hardly less valuable. They often convey what the patient communicated of his or her subjective perceptions in his or her own words. And even where they reflect the view of the writer, they provide direct access to lay conceptions of the body and its diseases and about the way in which those surrounding the patient dealt with disease.

Apart from the widespread practice of consulting famous but distant physicians, patients sometimes wrote to their ordinary physician, to call him to their bedside or to ask him for advice in times of absence or when patient and physician did not live in the same place. Some patients also wrote memoranda, reports or even diary-like chronicles of their complaints for the physician to read in addition to a personal consultation, trying to give the physician a clearer and more comprehensive idea than an oral account could achieve. In this book, I will, for brevity's sake, refer to all such letters, including those written by family members and friends, somewhat losely as 'patient letters'.

At the center of my analysis are three particularly large collections comprising a total of around 2,000 patient letters. The oldest collection dates to the years around 1580 and comes from the extensive correspondence of the Paracelsian Leonhard Thurneisser, the personal physician of elector Johann Georg of Brandenburg in Berlin.[40] Among contemporary physicians, Thurneisser had a rather dubious reputation. He enjoyed great esteem among laypeople, however, and even in the highest courtly circles, for his nostrums and his then new procedure of urine distillation.[41] The second large collection consists of letters to the professors of the faculty of medicine in Paris, and particularly to Étienne-François Geoffroy, between 1715 and 1735. The third collection has been preserved among the papers of Swiss physician Samuel Auguste Tissot (1728–97) in Lausanne and covers primarily the years between 1765 and 1795.[42] Tissot was one of the most famous physicians of his time and one of the most read authors of educational medical texts. His Advice to People was a bestseller all over Europe and his writings on the diseases of nobles and scholars and on the dangers of masturbation went through numerous editions.[43] Thus he was an obvious choice for a consultation by letter. All three collections of letters have been known to medical-historical research for decades.[44] But their value for a medical history from the patient's perspective and for the reconstruction of the body and disease experience of medical laypeople was not sufficiently appreciated until the advent of histories of medicine from the patient's point of view.[45]

I make use of a further, still more extensive, collection only to a limited extent, namely the letters to Samuel Hahnemann, the founder of homeopathy, and his wife Melanie, from the 1830s.[46] I have done a complete survey of all letters from Hahnemann's French patients[47] but my analysis of the thousands of letters he received from German patients has focused on initial letters.

In addition, I draw on several smaller letter collections. I should mention in particular the letters to Felix Platter in Basel at the end of the 16th century, to Daniel Horst in Frankfurt and to Sebastian Schobinger in St Gallen from the 17th century and, from the 18th century, the patient letters that Friedrich Hoffmann published in his 12-volume 'Medicina consultatoria',[48] as well as the patient letters to Albrecht von Haller in Bern.[49]

The choice of these various letter collections sets the geographical and chronological limits of this study. The study primarily treats the time from the middle of the 16th century to the early 19th century, and it concentrates chiefly on the German and French language areas; sources from England, Italy, and the Netherlands are used only occasionally and without attempting to be exhaustive.[50]

As a valuable supplement to the wide range of evidence from patient letters, I make use of other types of personal testimony or 'ego documents', in particular of more than a hundred autobiographies, diaries and printed editions of private correspondence.[51] The disadvantages of such sources are evident. Episodes of illness are mentioned only occasionally, scattered between accounts of many other topics and usually in a more succinct manner than in patient letters. In contrast to patient letters, women's voices are heard only exceptionally. Autobiographies were also often authored a long time after the unfolding of events and were usually written to be read by others. Elements of self-fashioning and retrospective restyling vis-à-vis the author's own life story are very much to be expected. On the other hand, we tend to learn more about the context – about the author's whole life and about the circumstances in which the disease in question took place.[52]

The broad scope of source material used for this book also provides a fairly solid basis on which to pin down national differences. Work on the sources revealed surprisingly little evidence for such differences between the various countries, however. The medical culture of the educated classes in German- and French-speaking areas proved by and large to transcend state borders. I shall point explicitly to exceptions, such as the role of 'scurvy' and the reception of the new model of 'nervous diseases' in different areas.[53] Long-term changes in the lay perception and interpretation of diseases also can be found only to a limited extent in the period I examined. This contrasts markedly with the profound changes in the explanatory models and dominant theories of learned physicians. For this reason, I will frequently place statements of patients made in the 16th and 17th centuries alongside those of the 18th or even early 19th centuries. It is only on individual points that some striking changes in the perception and interpretation of illness and the body come to light. Two of them are at the center of the third part of the book, namely the rise of nervous complaints and the increasing concern about the loss of semen.

A major disadvantage of using sources like patient letters is, of course, that they reflect predominantly the conceptions and experiences of those educated middle and upper classes that could read and write. Some patient letters, particularly from of the 18th century, were written by ordinary crafts-people or farmers and sometimes the choice of words and a highly idiosyncratic orthography suggest a rather limited degree of formal education.[54] The lowest social stratum represented in these texts is composed by and large of students, local clergymen, soldiers, merchants and better-off craftspeople, however. For the perceptions and experiences of the great majority of the population, the rural population in particular, we have to resort to other sources. Some useful clues are provided by records of interrogations designed to confirm or assess 'miraculous' cures.[55] The healed patients, their acquaintances and relatives, or the surgeons or physicians who had previously treated them, thoroughly reported upon the beginning of the disease, the long and usually difficult course it took, and the diverse, ultimately futile therapeutic efforts, until finally the prayers uttered at the grave of Monsieur de Paris, for instance, brought the longed-for cure, as even the surgeons and physicians confirmed. In such cases, the voices of craftspeople, maidservants, paid sickbed attendants and other members of the lower classes are recorded. Naturally this type of source comes with its own set of problems. The witnesses were questioned in a very specific context with a very specific intent. Their responses may often have reflected a desire to stress the patient's suffering in order to underline the miraculous nature of such cures which, at the time, were the subject of heavy debate.[56] But because, as a rule, the nature of the disease and the resulting changes in the body were not in question, the interrogations nevertheless provide valuable information about the perception and experience of the body and its diseases among the lower classes.

How the body and bodily phenomena, including diseases, pregnancy and fetal life, were perceived among the wider population also, at times, emerges from proceedings in criminal courts resulting from inquiries into the practice of unauthorized healers, for example, [57] or into cases of suspected abortion or infanticide.[58]

Aside from this, much of what we know about the views and experiences of ordinary people comes from accounts of educated contemporaries. Physicians' case histories and letter consultations, in particular, are rich sources. It is the genre that Barbara Duden used as the basis for her aforementioned study on the experience of the body and disease of the female patients of the Eisenach physician Johann Storch.[59] For the period from 1800 onwards, I will also draw on almost 300 mostly handwritten German medical 'topographies' or 'ethnographies' which local physicians, usually on the request of the government, wrote about their respective area. They had often come to know the region and its people very well over the years and were thus able to give detailed descriptions of the medical conceptions and practices which prevailed among the general population.[60] In view of the comparatively slow process of change within rural medical culture – something these ethnographies unanimously evidence – it can be assumed that the ideas and practices described in the ethnographies essentially hold for the rural experience of disease and the body in the 18th century as well, if not earlier.

All of this is not meant to conceal the fact that patients from the bourgeoisie and nobility are significantly better represented in this study. There simply are no similarly rich sources for the rest of the population. In a sense, this limitation may not be quite as serious as it would seem at first sight, however. Historians have largely come to agree that the line between 'elite culture' and 'popular culture' – to the extent that we can know about 'popular culture' in the first place – was far more blurred in early modern times than was long assumed. At times, it may make sense to distinguish 'popular' and 'learned' or 'elite' culture for analytic purposes: for example, when we study the dissemination and reception of medical advice literature. In the small world of early modern towns and villages, however, knowledge, practices, and norms were constantly communicated and exchanged between different social groups.[61] This is also – and particularly – true of the sphere of medicine.[62] To date there is no conclusive evidence that the medical culture of the lower classes in the early modern period was fundamentally different from that of the upper classes.[63] Differences – and quite important ones – can be found only in some specific areas and mostly towards the end of the early modern period. For example, astrological, sympathetic and magical healing as well as healing practices originating in folk piety, once a part of everyday medical culture in all ranks of society,[64] markedly lost significance among members of the 18th-century educated classes. They only rarely find mention in their personal testimonies, while they continued to pervade the medical world of the rural population far into the 19th century.

A final word about the challenges of working with this kind of source may be appropriate at this point. For the historian, letters and other personal testimonies are fascinating. People of the past seem to allow us direct access to their experiences and perceptions, to their hopes and fears, to their lives. This fascination is welcome in view of the considerable amount of time, effort, and patience which an analysis of such a mass of mostly manuscript – and sometimes quite badly written – material requires. But the easy accessibility, and the seeming clarity and comprehensibility of personal accounts from earlier times can prove deceptive. There is a considerable risk of a naive, anachronistic reading. After all, the way in which people perceived and described their bodies and diseases was inevitably and profoundly shaped by their culture and society. This is best illustrated by cases where patients of the past described bodily sensations that we are no longer familiar with in today's Western culture, such as the rising of a spherical mass from the abdomen in the case of 'hysteria', the sensation of 'vapors' ascending to the head, or the 'trickling' of coarse impurities through the blood vessels. As sociological and cultural-anthropological studies show, this cultural framing of bodily perception includes elementary phenomena such as pain, itching, sexual desire, and the sense of one's own bodily boundaries. But almost inevitably modern readers – embodied human beings themselves – will try to understand past accounts of physical phenomena through a comparison with their own familiar and seemingly natural bodily sensations. Since this is largely an unconscious process, the danger of anachronistic misunderstanding is great. When patients of times past complained of diarrhea, pains or itchiness, the physical sensations they referred to seem self-evident and easy to grasp to modern readers. We 'know' from our own experience what painful abdominal cramps are like, what it means to have a throbbing head, or to be unable to resist scratching an itching spot. It is quite possible, however, and in fact very likely, that the diarrhea, pains or itching experienced by people in the past 'felt' quite different in a very elementary, bodily sense and beyond all individual differences. All we can do is try to be as attentive as possible to the distortions and misconceptions which this framing of bodily sensation by our own culture may bring about in our encounters with accounts of the bodily experience in the past.

In my own case, my original training and practical work as a physician are bound to have left some traces. The diagnostic and therapeutic knowledge and skills which I acquired many years ago in my medical studies and which I have in part long since lost and forgotten play a negligible role in this; at best, they allow me to judge how doctors today might name comparable medical symptoms. But certainly my professional encounter with severely ill or injured patients, with people in pain and sometimes on the brink of death, has sharpened my eye to the paramount importance of pre-linguistic, pre-objective aspects of bodily experience to which the fashionable metaphor of 'the body as text' does not do justice. In addition, the several months I spent at Indian and

South African hospitals have given me experience of the fundamental differences between cultures in terms of the perception and interpretation of disease and the body and the manifold syncretisms as well as conflicts which can arise in the encounter between medical cultures. I think of, for instance, an Indian man with tuberculosis, who proudly showed me his chest x-ray, confident that his illness had been captured on it and that he would now be healthy, or the aghast look of a Zulu mother from the neck of whose sick child the nurse had just removed a protective amulet because it was 'in the way', leaving the child defenseless, in the eyes of the mother, against evil powers.

Another methodological problem has more specifically to do with work on patient letters. Even within one and the same culture, the seemingly most spontaneous private utterances do not offer direct, unadulterated access to the experience of others but only to the way they are expressed, in words and action. In speech as in writing, we constantly choose – consciously or unconsciously – what to tell and what not. And generally we only communicate what makes sense to ourselves and what we find worthy of being told to others. This is especially true when it comes to writing about the body and its diseases. Out of the many physical sensations and changes which arise in our bodies in times of illness, we consciously perceive and acknowledge only a few, namely those which, on the basis of our understanding of human physiology and pathology, seem to bear some relevance to the disease. For the purpose of this study, the selective nature of the resulting illness narratives is not just a limitation. It is often also highly illuminating. What patients consider worthy of sharing and what not provides critical information about their subjective, culturally framed experience and understanding of the body and its diseases.

A more serious limitation is that letter writers tend to anticipate and respond to the addressee's presumable expectations. Patients who consulted a renowned physician by letter clearly had a certain idea about how such a letter should be written. And they had good reason to conform to these expectations. Especially when they approached the physician as a stranger, they were in the position of a supplicant. They knew that the physician might easily refuse to respond to their letter or to make the requested visit. And in case he accepted, they wanted him to read their descriptions with particular care, to think thoroughly about their case, and, in spite of his presumed excessive workload, to answer them as quickly as possible, a request that repeatedly draws their letters to a close. Already in the letters to Thurneisser from the late 16th century and far more so in the longer letters from the end of the

18th century, the correspondents would frequently indulge in bounteous avowals of their appreciation. Tissot, in particular, was regularly praised as a man of Europe-wide fame, rightly renowned for his outstanding knowledge as well as for his humanity – which, as one sometimes reads between the lines, would doubtless ensure that he would attend to the personal request of the suffering patient in question.

A couple of patients worked through a list of questions which Tissot had annexed to his famous *Advice to People* and which was designed to help the physician come more easily to a valid diagnostic and therapeutic conclusion.[65] Some letter writers even tried to imitate the style and structure of professional medical consultations. Apparently it was a genre people were familiar with from the collections some physicians published of their consultations or, more likely, from experience in their own circle of family and friends. As they were quick to admit themselves, they rarely succeeded, however, and frequently ended up apologizing for their muddled style or post-scripting their letters with some forgotten yet, in their eyes, indispensable additional information such as that the patient's diarrhea had increased following intake of the medication or that he had not tolerated the goat's milk very well.

Since patients and relatives had a fairly clear idea what could be written to a physician and what not, their silence on certain matters is illuminating but sometimes also open to different interpretations. Patient letters are, for example, not very helpful in shedding light on the role of magical or pious healing practices or on the use of nostrums. Both are rarely mentioned but this could be simply because patients knew that most learned physicians thought very little of these procedures and 'remedies'. Only by comparing the letters with other sources can we confirm whether such practices had actually ceased to be widespread among those classes.

Similarly, contemporary ideas about the acceptable expression of emotions have to be taken into account. Men on average exercised much more restraint than women in giving affective, emotive accounts. While some women described, for instance, their cruel, unbearable dragging pain in the abdomen in dazzling colors, most men – with some exceptions – maintained a much more sober tone. Again it is difficult to decide in retrospect whether this affective restraint on the part of male patients (superseded at times by a detailed, even pedantic, listing of symptoms) reflected their different perception and experience of pain or whether men were simply more hesitant to violate the recognized norms of masculine composure and fortitude in a letter exchange with a famous male authority.

Occasionally even the patients' identities remain doubtful because some patients recounted their own case in third person as a 'memorandum', without a name or using pseudonyms such as 'Titius', which frequently appeared in writings by physicians. Sometimes writers changed in the course of the letter – at times in the very first sentence – from a third-person to a first-person narrative and began to write explicitly about their own stomachs, their own pains, just as the great majority of patients did from the start. Cases in which letters were written throughout in the third person and in which the physician to whom they were addressed did not, as they often did, indicate the patient's name on the top or in the margin create considerable problems for historical analysis. The historian may be sure that what she or he is looking at is a personal testimony because of the profuse style and loose structuring, a strong emotional undertone or highly unconventional spellings of technical terms, apart from the fact that physicians usually explicitly signed their letters. But ultimately there is no proof.

Part I
Illness in Everyday Life

The Concern for Oneself

It is widely believed that people in the past were far less preoccupied with their health than people are today and that they were guided in their attitudes by either faith in God or fatalism. Only in modern times, with the Enlightenment, is health said to have assumed its current position at the very top of most people's scale of personal values. Behind such beliefs is the valid and fruitful insight that health is a historically contingent and changeable concept. The idea that fatalism dominated matters of health among earlier generations is, however, a fiction, spawned and nourished not least by an anachronistic assessment of pre-modern medicine and its place in individual lives and in society as a whole.

Without a doubt, the time horizon of most people in early modern Europe differed markedly from that of today. Faith in life after death was more prevalent than in today's Western societies. This faith made it much easier to cope with disease, suffering and death with a degree of serenity. After all, a better life awaited one in the hereafter. At the same time, medical endeavors proved futile more often than today even in patients who were still in their prime, not to mention in infants and children, whose death rate in many places was between 20 and 25 per cent with peaks of up to 50 per cent. This was not a good basis for trusting that disease, in general, could be controlled. Still, the conviction that, as Martin Pansa wrote in 1618, 'in this world, nothing is dearer to us, than a healthy body'[1] is not an invention of modernity. If the importance of health as a political subject and economic factor has gained considerable ground in modern industrialized nations,[2] this has much more to do with the development of the modern state than with a changing private and collective appreciation of health. Even medieval monks, puritan clergymen and German Pietists combined deeply felt religious devotion with intense worldly efforts to protect themselves from disease and its consequences.[3] Medical advice books were one of the most successful literary genres from the beginning of letterpress printing. Works such as Luigi Cornaro's *Treatise of Temperance and Sobrietie* or Leonardus Lessius' *Hygiasticon* were bestsellers, and made their authors famous throughout Europe.[4] Numerous health manuals and plague pamphlets disseminated the basic rules of a traditional dietetics, preaching moderation as a cardinal virtue. In addition, there were many works which focused on the special health risks and needs of individual groups like scholars, courtiers and pregnant women.

Contrary to what historians like Philipp Sarasin have claimed, the conviction that the individual has considerable control over health, disease and even the time of death[5] was thus not at all new in the 18th century. It would be even more misguided to interpret the widespread trust in bloodletting, amulets, and healing charms as an expression of fatalism. The assumption that people must have recognized such cures as ineffective or even harmful[6] is wrong and anachronistic. As will become clear in the course of this study, people were, in fact, convinced that such remedies and procedures could and did prevent and defeat disease. And they were confirmed in this conviction by daily experience. After all, countless patients reported that they felt better after a treatment to which we today would attribute no possible beneficial effect. Excruciating pains vanished almost entirely after only a few baths. A single bloodletting produced the long-awaited menstrual period that had remained 'suppressed' for months. In retrospect, we would attribute the improvement to the placebo effect or to the natural course of the disease: at all times and in all cultures, most diseases – and the common acute, feverish diseases in particular – have a favorable outcome no matter how the patient is treated. The patients and the people around them, however, – and this is crucial in this context – will almost inevitably attribute this outcome to the medical treatment they happened to receive. They will consider it as proof that this treatment is indeed effective. Daily experience thus demonstrated that human beings were indeed capable of influencing and controlling the disease process. Of course, especially in chronic diseases, therapeutic efforts quite frequently failed in pre-modern times (and they still do today). In contemporary eyes, however, this did not by any means imply that medicine per se was powerless. Therapeutic failure could always be attributed to the individual healer who had arrived at the wrong diagnosis or prescribed an inefficient treatment.[7]

The basic trust in human power over diseases included beliefs in 'magic' and in 'sympathetic healing'. Recourse to protective charms and necklaces, to amulets, to sympathetic rituals, and to pilgrimages and vows remained common among the rural populations of France and Germany in the 19th century and beyond.[8] This must not be misunderstood as an expression of fatalism and resignation, as it usually was by physicians at the time, along with generations of medical historians thereafter. Since they were followed, in many cases, by considerable improvement or even complete recovery, magic and sympathetic healing, like bloodletting and other types of treatment, seemed to have proven their worth many times over.

To sum up: the desire for good health already ranked among the highest personal and collective aims in the 16th century and spawned numerous types of healing activities and medical publications. At most, the declining importance of transcendental meanings of illness as a test or punishment which God imposed upon the faithful, may have further strengthened the significance of the healthy body as the foundation of self-certainty and ontological security. The 'approaching crisis of metaphysics' in the 18th century, as Rudolf Behrens and Roland Galle put it, 'found its clearest and strongest antithesis in the supremacy of the body'.[9]

Disease and the Self

Across national and cultural boundaries, disease, and particularly chronic or life-threatening disease, ranks among the greatest challenges which most men and women face in the course of their lives. As proponents of phenomenological philosophy as well as medical sociologists have emphasized, disease profoundly calls into question what is usually taken for granted. The physical self that we experience as an undivided whole in everyday life seems to break apart. A split seems to appear between the body (or the painful, suffering body part) and the experiencing I, which finds itself facing that body or a part of it as a foreign, even hostile, antagonist.[10] Strange and sometimes agonizing physical sensations arise. Familiar activities cannot be performed as usual. Capacities and skills which were taken for granted are obliterated or begin to fade. In some cases even one's own thinking appears to be determined externally, dominated by moods and ideas that force their way into the mind as if from the outside and refuse to be pushed away.

Serious illness also shatters our deep-seated sense of our own invulnerability, which allows most of us to live from day to day without constantly fearing the many dangers that could bring our life to a sudden end. And by making us acutely aware of our own mortality, serious illness can also cause our subjective time horizon to shift. Our hopes and fears no longer circle around the realization of private or professional desires and ambitions. Our frail bodies and the threat of an imminent death take center stage. In this way, serious illness introduces an element of separation into our relationships with fellow human beings. Everybody else seems to live in a different, better, world. The sufferers feel closed off from others as if by an invisible wall. Patients with severe chronic pain, in particular, frequently find that they cannot adequately convey what they feel because no one, unless he or she has actually

been in such pain, can imagine how chronic pain penetrates the sufferer's entire person, entire existence.[11]

The experience of a shattering of that sense of physical unity which
most people take for granted also surfaces repeatedly in the personal
testimonials of patients from the early modern period. Many patients
described themselves as literally 'not themselves' anymore. They felt
battered, wan, weary, spiritless, 'sick as a dog'[12] or, as was a popular expression in 16th-and 17th-century German, 'baufällig', that is
'decrepit' or 'in disrepair'.[13] Their appetite waned. They needed to lie
down. Sleep was no longer reinvigorating. 'I didn't know if I still had a
body', wrote Mme de Guyon, weakened by fever and cramps.[14] In cases
of long-lasting, chronic suffering, the disease is frequently described
as overshadowing the patient's entire existence. It became a 'torment',
the 'misfortune' of their life.[15] Hopelessness and despair gain the upper
hand. The Hessian colonel von Jungken, for example, said that he loved
life but conceded that he would much rather die a swift death than
'see myself wasting away day by day, dying slowly and miserably'.[16] A
dyspeptic and childless Strasbourg cavalry captain explained that he
would give anything to be liberated from his suffering because 'it poi-
sons my existence'.[17] A 36-year old patient of Tissot with a hardened
uterine tumor was shaken by crying fits.[18] 'The entire time, I was like a
drowning person who cannot find a rope strong enough to hang onto',
said another patient in retrospect about the hopelessness she had felt.
She had become 'a burden' to herself.[19] Laconically the hypochondriac
Sig. Piazza remarked, 'I cannot live on'.[20]

Illness led people to experience not only their body as an adversary.
The illness itself was described by many patients as an entity apart, as
something foreign which had a separate existence from the physical
self. In modern medical theory, this is called an 'ontological' understanding of illness, as opposed to a 'physiological' understanding of
illness as a gradual departure from a healthy ideal state, such as the
notion of health as an equilibrium between hot and cold.[21] The terms
frequently used by patients and their relatives at the time (and sometimes still today) are revealing. In the language of patient letters, disease was something that was 'loaded' onto you or that 'weighed' on
you, something that 'had' you, 'grabbed' you, 'clutched' at you, 'over-
came', 'pinched' or 'enfeebled' you. Indeed, the disease or the 'morbid
matter' was an 'enemy' that 'struck' you, 'fell upon' you, 'assailed' you
or 'attacked' you, as it was put innumerable times. It 'invaded' you or
'snuck into' you and 'exposed itself' like a secret agent after it had established itself in the body, or it proved to be 'rebellious'. In accordance

with this martial, aggressive language, patients asked physicians for suitable weapons to 'victoriously attack' the illness, to 'wage war' on it, to 'exterminate' it.[22]

This predominant understanding of disease as a foreign entity with a life of its own is reflected in the widespread and, as it seems, deeply rooted desire to give illnesses as specific name. Even ordinary laypeople were familiar with a fairly sophisticated and nuanced disease terminology. Quite a number of medical terms for specific diseases were part of ordinary lay language: apoplexy, nervous complaints, epilepsy, mania, hysteria, hypochondria, smallpox, whooping cough, measles, scabies, three-day fever, jaundice, bilious fever, nervous fever, four-day fever, asthma, scurvy, thrush, chlorosis, consumption, catarrh, gonorrhea, syphilis, pleurisy, swine erysipelas, paronychia, scirrhus, cancer, rheumatism, gout, sciatica, and cataracts. Further terms were used occasionally by individual patients, including Latin/Greek terms such as 'arthritis vaga' ('migrating' or 'wandering arthritis'), which probably came to be known through physicians. From a modern perspective, it should be noted that some terms, like 'dropsy' or 'jaundice', do not refer to diseases but to symptoms which can appear in connection with very different diseases. In the understanding of the day, however, such complaints were, like 'cancer' and 'tertian fever', seen as specific diseases with a characteristic set of symptoms. Accordingly some patients underlined the absence in their specific case of certain symptoms which 'those ill in this way usually experience',[23] so as to rule out a particular diagnosis.

The importance of an ontological conception of disease in pre-modern medicine has been denied time and again. Pre-modern medicine, it has been claimed, was based on an individualizing, basically physiological understanding of disease as a state of imbalance between the natural humors and their associated elementary qualities. For those 16th- and early 17th-century physicians who held strictly to the Galenic tradition, this is true to a certain extent. Their treatment occasionally still aimed primarily at eliminating an excess of mucus or gall or of reducing an excess of heat, cold, moisture, or dryness. But medieval textbooks on medical practice were already to a great extent organized in terms of individual, separate diseases. Numerous dispensatories recommended a wide range of medicines, each to treat a particular disease. In any case, as far as early modern lay culture is concerned, an ontological conception of disease is virtually ubiquitous in patient letters and personal testimonies. People suffered from cancer, gonorrhea, scurvy, or putrid fever, or the disease was traceable to a specific morbid substance which could not normally be found in the body. Consequently, one could have the 'same illness' as others in one's family or community. The tendency to objectify disease, to focus the medical gaze on the disease rather than on the patient as a whole, which is often lamented today as one of modern medicine's central shortcomings, may thus well have much older roots. Perhaps it reflects a deep-seated emotional or psychological need to preserve the integrity of the self by separating the disease from the body and turning it into an independent entity which can be attacked head-on or driven out.[24] In fact, according to physicians' descriptions, country folks in the early 19th century wanted to know, above all, the name of their disease, as if simply naming the illness would already produce some magical power over it.[25]

Frequently disease was associated concretely with a specific, identifiable matter or entity that did its misdeeds inside the body. Sometimes it was attributed to an animal-like creature, as when patients excreted worms, or even, according to some spectacular reports, frogs, toads, snakes, and the like.[26] Old remedy books gave advice on what to do 'against the heart-worm' or against 'tooth worms', for example, or 'when a man has a worm in his ear' or 'a snake has stolen into the body'.[27] Much more frequently disease was associated with some kind of 'morbid matter'. I will deal with these ideas in detail later on, but this much can be said in advance: numerous patients described 'humors', 'fluxes', 'acrimonies', 'morbid matter', or 'pains', 'gouts', and 'rheumatisms' as active, harmful agents which seemed to have a life and will of their own and had to be killed to achieve a cure.[28] The poison must be 'completely killed', one of Thurneisser's patients pleaded, speaking of his 'great complaints and days of torment'.[29] According to one remedy book, when cancerous growths were moistened with menstrual blood, one would suffer great pain for half a day 'until the cancer died'.[30]

The popularity of 'evacuative' treatments such as bloodletting or giving laxatives and emetics should probably also be seen in light of this close association of disease with a morbid agent or substance.[31] These procedures were not only held to be empirically tried, tested, and true; they also accorded with the widespread experience of disease as something foreign, if not 'inflicted', which had to be driven out. In these terms, the boundary between the purgative, purifying effect of laxatives and the exorcism of evil demons was not as clear as it might appear at first sight.

The prevailing ontological understanding of disease also guided the use of medicines. German phrases like 'für jede Krankheit ist ein Kraut gewachsen' ('a herb grows for every illness') or 'dagegen ist kein Kraut gewachsen' ('no herb has grown for this one') are still in frequent use. In the early modern period, such phrases could be taken quite literally. There were, people believed, not only various diseases, distinguishable from one another, but also various medicines which God or Nature had endowed with a specific healing power against individual diseases. Sometimes these healing powers, according to the so-called doctrine of signatures, could be perceived from the outside, from their shape, color, smell, or taste. A flower whose blossoms resembled the shape of liver thereby suggested that it might be good for healing liver diseases. In other cases, the specific healing power of a plant was considered empirically proven, confirmed through experience. The best-known examples are guaiacum wood, which created a sensation in the 16th century as a specific remedy for syphilis, and Jesuit's bark – a precursor of quinine – as a cure for fever. Popular also were the laboratory-produced 'specifics' against certain diseases. They came from the alchemical and Paracelsian tradition of isolating 'quintessences' as well as from local traditions of distilling spirits from herbs and other vegetable matter.[32]

A significant difference from modern conceptions of disease was the widespread belief that one disease could 'transform' itself into another. The idea was directly related to the central role assigned to morbid matter in the genesis of disease. Since morbid matter could have different effects depending on the part of the body where it settled and since its nature could be changed by the body's vital heat or by Nature or by medical treatment, a transformation from one disease to another could easily be imagined. Abbot Erhardus' illness, for instance, began in 1574 with bloody urine. It then 'transformed' into a 'bleeding from the nose [...] through which we suffered great weakness and dizziness in the head'. As this came to an end, his 'weakness' 'transformed' into a 'hard bulge' so that, for the past eight weeks, his body was quite swollen and puffed up.[33] Numerous 18th-century patients reported along similar lines how certain complaints disappeared the moment other complaints emerged in another area, or how diseases 'threw themselves' from one body part to another.

The Experience of Pain

Pains, or 'Wehtage' ('days of torment') as they were often called in German, were in many cases a central and distinctive feature of the early modern experience of disease.[34] Some patients were content to describe pain as 'terrible', 'horrendous', 'unbearable', or 'unspeakable'. Most, however, tried to give their physicians a more precise idea, an impression of the specific qualities and the intensity of their suffering. Valten von Schaplo, for example, complained of 'such great cutting and tearing in my body [...] and my body is contracting in such a way that I can hardly stand up straight'. For weeks his wife and servants had been forced to watch 'how I must suffer'.[35] A woman with hip pain reported that she cried and screamed day and night.[36] In many patients' accounts, pain appears as something foreign penetrating the body – indeed, just like an independent morbid agent endowed with a will of its own: pain 'threw' itself into one body part or another or 'climbed' up or down or 'ran' into the back. This ability of pain to move by itself, analogous to that ascribed to morbid matter and disease itself, sometimes also implies a causal relation. Physicians and laypeople alike believed that the most important cause of (non-traumatic) pain were mobile disease substances, especially the so-called 'fluxes', to which I will turn in greater detail later on. Occasionally, patients perceived their pain as if it were an animal-like creature being hurting them inside. For a clergyman in Geoffroy's care, for example, it was as if someone were tearing at his diaphragm.[37] For the wife of one Irish clergyman, it was as if a living thing were running around inside and underneath her chest. As her pain diminished, she compared the sensation to a mouse which was now running around less than it had been. She also likened occasional pains in other body parts to, among other things, the movements of a small living thing about the size of a fly.[38] In the case of the 'tooth-worm' – a popular explanation of tooth-aches – the animal origin of pain was implicit in the term itself. Maybe the notion was supported by the common observation that something worm-like – which we would consider the nerve – could be seen coming out of dental roots when teeth were extracted. Similarly widespread, at least in Germany, was the notion of a 'Gebärmutterkolik', literally a 'uterine colic', also known into the 20th century as 'Bärmutterbeißen'. 'Bärmutter' was a widely accepted variant of the word 'Gebärmutter' (uterus). But 'Bär' also means 'bear' in German and 'Bärmutterbeißen' thus evoked an image of a mother bear biting the innards – which explains why boys and men could also suffer from the disease.[39]

To communicate the specific qualities of their pain, some patients invented a language and imagery of their own. A certain pain passed 'like clouds' through the chest of one count who suffered from hypochondria.[40] For another man, a rural parish priest, his pain usually made itself felt with a 'whimpering', 'prickling', or 'tensing' in the skin, as if he were being rubbed with salt.[41] Most patients, however, used more familiar images and comparisons from everyday life. In some accounts, sensations of heat or burning took center stage. One 78-year-old monk's urethra, for instance, smarted as if it had been touched by a red-hot iron.[42] One 34-year-old nun's head felt as if it were full of glowing embers.[43] The knots in Gertraudt Hake's chest burned 'just like a fire'.[44] Another patient described her pain by saying it was as if her limbs were filled with 'aqua fortis', that is with sulfuric acid.[45] In other cases, patients used images which expressed a state of tension. Thus a middle-aged patient with aching temples wrote: 'it is as if there are ropes attached to them, pulling me'.[46] For others, it was as if a bar were painfully locked across their chest,[47] as if someone had wrapped a bandage tightly around their skull,[48] or as if a rope had been pulled too tightly around their head.[49] Still others compared their pain to wounds from weapons or tools. One vice rector suffering from kidney stones wrote of a 'cruel pain in the small of the back' which felt 'no different than if my sacrum had been shattered or a stake driven through it'.[50] Friedrike Lutze complained of stabbing pains in the chest.[51] Monsieur Feger, who also suffered from chest pain, felt as if he had been transfixed by a lance.[52] An older patient of Hoffmann said that the fierce pain in his shins was 'as if a whole set of knives were whittling away in there'.[53] Of his 'weak wife', Johannes Hancke reported that she felt 'great tearing inside her body, as if being cut with knives and scratched and stabbed with awls'.[54] About another patient we read that there was 'sometimes such pain and pulling in every limb and in her entire body that it was as if they were being shattered by sticks that had been thrown at them'.[55] Scores of patients compared their pain to needle pricks.[56] One patient felt as if her stomach were being cut with razors – a particularly appropriate image considering that she was also producing bloody vomit.[57]

Pain on the surface of the body tended to be described accordingly in terms of influences that were more superficial, limited to the skin. This did not necessarily mean that the intensity of the torment was any less, however. One older woman felt as if she were being whipped with stinging nettles.[58] Others compared their pain to flea bites,[59] or to the sensation a dry sponge might cause when wiped across a fresh wound.[60]

Many patients were thus by no means speechless in the face of their pain.[61] On the contrary, they had a rich vocabulary at their disposal. It should be added, however, that pain had a more familiar place in many people's lives than it does today. Many among the chronically ill in particular suffered much more severe pain, often over longer periods of time, than most modern patients. After all, in hindsight, early modern medicine had in most cases no powerful drugs which could cure the disease. And surgery, if it was an option at all, was, without proper anesthesia, not only painful in itself but also very dangerous. In early modern medical literature, 'palliative' medicine, the 'cura palliativa' as it was then called, was therefore hailed as one of the physician's foremost duties.[62] At the same time, the means for fighting pain directly were very limited. Until the late 19th century, intense pain was treated primarily with opiates.[63] They were difficult to dose, however, often caused nausea, vomiting, and constipation, and their application sometimes proved deadly.

Alongside podagra or gout, which I will be going into more closely, kidney and bladder stones were described by contemporaries as a source of particularly intense pain. The frequency with which so-called 'stone diseases' are mentioned in the personal testimonials of the day suggests that they may have been more prevalent then, perhaps due to nutritional factors. The suffering often went on for years and the pain could be virtually unbearable. In 1596 Karl Utenhoven bewailed, for example, that 'there was no greater agony and martyrdom than the blocking of [my] urinary tract'. Even prayer was to no avail. He had 'often cried for the Lord's help and wished for it, but in vain'.[64] According to Michel de Montaigne, the last seven years of his father's life had been 'extraordinarily rife with pain' due to a large bladder stone until he finally, at the age of 9, died in 'horrific pain'. Montaigne declared that he would come to the defense of anyone who railed and roared over severe colic, even if he himself only moaned and groaned.[65] Patients who, after excruciating colic, passed their stones could consider themselves lucky. So it was for Hans Khevenhüller, who recounted how he 'had been hard hit with a stone and gravel and urinary retention' and suffered five days 'in grievous circumstances' until he 'threw out a big stone'. Subsequently his pain, 'praise be to God, ceased'. The next day he was already able to quit his bed.[66] Graf von Zimmern knew of the son of a gatekeeper who, after two or three days of suffering, passed several handfuls of pebbles, the size of hazelnuts. According to von Zimmern, many 'honorable and respectable people saw it with their own eyes and what was more astounding, as soon as the boy had passed the pebbles, he no longer felt any pain or harm. Many reasonable people thought urinating like this was not natural but witchcraft or a maleficium, which I now believe too.'[67]

At least bladder stones counted among the few internal illnesses for which surgical treatment was generally possible. Enthusiastic reports tell of 'adept stonecutters' who, 'with the help of God successfully cut the stones from the bodies of the sick', after which 'the sick recovered almost miraculously and became perfectly healthy'.[68] The operation itself was an ordeal, however. Vincentz, a goldsmith from Breslau, vividly described what was endured by Hans Schaller, another goldsmith's son, who to no avail had used concoctions of bear's garlic, holy thistle, strawberries, and ground ivy. Eventually his suffering was so great that he was 'shaking in agony' and paced back and forth because 'a fire was burning in his kidneys'. A 'master stone-cutter' from Nuremberg happened to be in Breslau at the time and Schaller decided to be operated on:

> He received the holy sacrament and everyone prayed to God to help this righteous man. Then the man from Nuremberg spread out his grizzly instruments, knife and pliers, and had the patient bound and fettered. After much digging and searching, the stone could not be found and [he] had to be dangerously ripped a second time. Watching all this was a great misery. When now his eyes broke, he was once more ripped open with might and main and the stone taken from him, and the danger of death overcome with God's great benevolence. But I will never ever forget the horrid work that I witnessed.[69]

Others were less fortunate. Zimmern recounted how Graf Hanns von Lupfen, fearing a bladder stone, asked for an operation 'before the latter increased and gained weight'. But 'the cutting or the cure failed'. The patient died and, according to Zimmern had thus 'willfully', so to speak, shortened his life 'by a number of years'.[70] Understandably, most of those who suffered from stones preferred to try 'stone-dissolving' medicines as long as possible, and some died without ever undergoing an attempt at surgical removal. Erasmus von Schenk, for instance, who suffered from stones for years, got advice from 'the most educated and experienced physicians' and then, when they were unable to help him, consulted various Jewish healers from Frankfurt and Worms who were renowned for their medical skills. But he did not undergo surgery and died in the end 'in his prime and in glowing youth, since he was not older than 40 years'.[71]

The experience of pain, as we know from comparative cultural studies, cannot be reduced simply to the laws of biology. It is to a substantial degree a cultural product. Individual societies (and sometimes social groups as well) do not only differ considerably from one another in the degree to which they allow or even encourage a dramatic expression of pain. Cross-cultural physiological studies show that even the ostensibly natural, immediate, bodily perception and sensation of pain vary markedly from culture to culture.[72] As historians we have no comparative physiological studies to rely on, and the words with which patients described their experience of pain allow only limited conclusions about the underlying bodily sensation of pain. But the phrasing used by patients from different social classes as well as physicians' descriptions of the practical means of dealing with pain among the population suggest that, within the individual European societies of the day, there was considerable variation and change in the sensation of and sensitivity to pain.

In their medical ethnographies, 19th-century physicians time and again voiced their surprise and indignation at the 'indolence' of the uneducated country folks. Their accounts have to be taken with more than a grain of salt. Undoubtedly, they also reflected the frustrating experience that many country folks did not consult them when they were sick or in pain. The fact that many people did not consult a physician did not by any means imply that they stoically accepted their pain. On the contrary, 'the soothing of pain' constituted one of the most important tasks of faith healers. People also trusted patron saints like Bibiana, Blaise, Ottilia and Aurelia, who could be turned to depending on the site of the pain.[73] Pain was also one of the major occasions for using sympathetic and pious healing practices. 'For earaches, the right index finger is inserted into the ear and a prayer is said', reported one rural physician from Bavaria.[74] A particularly elaborate healing ritual was practiced in Bavaria against the so-called 'Hauptschein' ('head shine'), a type of headache which felt as if the skull was coming apart.[75] The treatment consisted of 'measuring' or 'sizing' the head. The healer would lay a string or band around the patient's head and measure its circumference. Three candles were then lit and the 'measuring' was done again. If the treatment was a success, the circumference of the head would now be noticeably smaller.

Such rituals were still common in the 19th century. Among the upper classes, however, pain tolerance seems to have sunk considerably from around 1750, the time when the 'sensibility' of the upper-class body and nervous system became an important mark of distinction through which the elites could set themselves apart from the lower classes with their allegedly coarse, bodily nature.[76] Accounts of physical pain now sometimes took on highly dramatic tones in patient letters of the time, and some even complained of acute pain in the hair which made any contact with it unbearable.

The Search for Meaning: Religion, Witchcraft and Astrology

Few patients accept diseases as a given fact. The experience of illness – and of chronic or even fatal illness in particular – is often character-ized by an intense and sometimes desperate search for meaning, for an explanation. On a concrete level, the nature of the illness and the bodily processes that cause or accompany it have to be identified. It was this 'medical', 'pathophysiological', type of explanation which patients in former times expected above all from their physicians. Knowing, for example, that a 'flux' or an 'acrimony' of the blood was the problem not only provided an indispensable basis for treating the disease and for preventing its future recurrence, but also made the disease less threat-ening and countered a sometimes pervasive sense of powerlessness, of loss of control. The mysterious, uncanny changes taking place in one's own body and the strange, perhaps painful, sensations that went with them were rendered comprehensible.

Beyond meeting the need for control over and guidance on how to fight the disease, explaining an illness had (and still has) another important function for patients, especially in cases of severe and pro-longed suffering. It answers the timeless question, 'Why me?'[77] It gives the illness a subjective, personal meaning, thus making it more bear-able. In the early modern period, two major sources of meaning can be discerned, which sometimes surfaced simultaneously and became intermixed. Patients looked for a transcendental, religious meaning and they explained illness as resulting from their own biography and behav-ior. In the following, I will first look at the realm the 'supernatural'.

Today, religious belief and a trust in God are an important source of meaning and orientation even in largely secularized Western societies. Belief in God and an after-life can be of great help in coping with ill-ness, pain, or impending death. The religious dimension of illness was even more important in the past, when church and religion exerted a much stronger influence on most people's lives and outlooks.[78] Well into the 19th century religious belief acted as a major support and source of hope for most people. There are signs of marked national and social differences, however.

In 16th-century German and Swiss patient letters (a comparable body of sources is unfortunately not available for France at that time) God is constantly mentioned. 'With God's help' or 'God willing' were common turns of phrase. Some people used more eloquent wordings: 'may God our Lord turn everything to our favor'[79] or 'God the almighty father of our Lord Jesus Christ, who is the best helper, may he bestow his mercy on me and help me become healthy again.'[80] Or, as Caspar von Hobergk put it, 'especially and above all, the will of the Almighty' be done, 'since everything lies in his strength and power', which incidentally did not keep Hobergk from eagerly seeking a physician's advice.[81] Many other contemporary personal testimonies similarly referred to God's help or mercy, which would hopefully bring healing or had already granted a happy recovery.[82]

Only exceptionally was an individual disease – as opposed to an epidemic – attributed directly to divine wrath and seen as a chastisement for the patient's sins.[83] Many people saw the almighty and all-knowing God in broader terms as the ultimate cause of their illness, however. The Swabian pastor Johann Valentin Andreä, for example, wrote that through his son's serious illness, which he called a 'domestic afflic-tion', he had learned to revere God's will even more.[84] Some patients wrote explicitly that the 'almighty God' had 'attacked' or 'afflicted' them or claimed that their ailments stemmed from 'God's promulga-tion' or were 'a warning of God', who wanted to bring their lives 'to a blessed end'.[85]

At the same time pious belief opened prospects of healing. 'We have all seen what an earnest prayer can do in such unexpected and horrid cases. Thank God!' commented Caspar Hirsch on the swift recovery of his son, who had suffered a kind of convulsive seizure that had left him without speech and comprehension for two days.[86] Jakob Andreae simi-larly aired his conviction that his consumptive wife had been 'saved by the prayers of her husband, children and other pious people against all expectations and hope by an almost singular miracle'. They had long given up on physicians.[87] Sometimes people trusted in the healing powers of miracle pictures, relics and other holy objects, or engaged in pilgrimages or processions. The Hessen-Darmstadt envoy, Passer, for example, mentioned that Viennese Catholics kissed the statue of Saint Blaise on his feast day, 'which they superstitiously believed would heal their throat pain'.[88] In Paris in 1572, Lucas Geizkofler noticed old women selling Paternosters at an inflated price 'because they had sup-posedly been brought into contact with the great Genoveva during the procession and therefore offered protection from all kinds of illnesses and accidents'.[89] Until the 19th century, thousands of people traveled to the well known pilgrimage sites hoping to be healed. Even today, innu-merable votive gifts testify to this practice. Healing practices rooted in religion or pious folk beliefs such as 'incantations' and 'blessings' like-wise remained an essential part of everyday medical culture.[90]

Among the bourgeois and noble elites of the 18th century, the situation was somewhat different. Atheism and agnosticism remained a rare exception but with the rise of rationalism and the natural sciences religious faith, especially in France, lost considerable ground among the elites. People still went to mass and asked for the last sacraments, but signs of an increasingly secular, worldly orientation in life abound, particularly during the second half of the century.[91] Images of a rather distant God came to prevail who no longer involved Himself much in daily affairs and natural processes, and therefore no longer offered the same spiritual support for those struck with illness. To put it somewhat pointedly: rather than traveling to the Christian sanctuaries, the gentry were now taking pilgrimages to the famous spas.

In 18th-century patient letters from France and francophone Switzerland, references to God and religion are virtually absent. Only here and there can short standard phrases be found, such as 'Thank God' or 'with God's help;' or exclamations such as 'Mon Dieu!' or the wish for a 'favorable providence' – and it is telling that even these are predominantly found in the letters of clergymen. Very few patient letters discussed religious matters in any detail and sometimes in rather ambivalent terms. Some praised explicitly the beneficial, physical effect of their faith. They experienced pious thoughts as calming or even credited them with a favorable influence on their state of health. They found solace in books and religion, as was the case with 23-year-old Goret, who believed he might have otherwise have lost his mind when a serious ear problem had left him nearly deaf and robbed him, most of the time, of the pleasure of listening to birdsong and the joy of human company.[92] Other letters, in contrast, accorded religion a pathogenic role. This was especially the case with people who were suffering from the so-called 'vapors' or were of a rather 'melancholy' or 'hypochondriacal' constitution.[93] A melancholy clergyman in St Malo, for instance, was in a cheerful, even frolicsome, mood whenever he was, in his own words, in a 'materialist' state. When he had undergone bloodletting, he recognized his previous state as pathological and returned to his belief in God. But he was obsessed again with the constant fear of making a mistake during mass or of having a nervous fit.[94] Abbé Tinseau – who attributed his suffering to repeated acts of sexual self-gratification since the age of 13 – described vividly how 'thinking about certain frightening religious truths' had seriously affected not only his inner life but his body as well.[95] The physician of a 58-year-old patient saw the principal cause of her peculiar spells of dizziness in anguish due to 'a piety pushed to abnormal excess'. With her eyes fixed on the sky she would repeatedly raise her arms and hit her knees and finally lose consciousness.[96] Another female patient, who, according to her brother's descriptions, suffered from serious convulsive seizures and presentiments of death, would lose command of her voice when she intended to pray aloud. If she persisted, she would have a seizure. Her brother had even seen her grow pale simply because a child was saying a short prayer in her presence.[97]

For some patients, their mere presence in churches or at funerals was enough to produce a negative effect. The 58-year old female patient with dizziness just mentioned suffered one of her first serious seizures at one such occasion. And Mme de Chastenay, who, in her own words, served God 'like a slave who fears her master', was overcome by a morbid fear when she found herself beneath the high vaults of a church. She feared the ceiling might come crashing down on her at any moment.[98] The daughter of the royal counselor Nicolas Diacre was said to have lost her mind entirely when she was once locked inside her parish church for the whole night.[99]

One reason why religious belief played a minor and ambivalent role in patient letters written by the 18th-century French upper classes may lie in the specific function and context of patient letters. After all, as the patients and their relatives knew, the physician expected an exact case history, not a profession of faith. But in French autobiographical writing about disease episodes religious belief was likewise mentioned only occasionally.[100] This suggests an important difference between national cultures. In Germany, 18th-century patient letters, like other personal testimonies, far more frequently expressed the hope that healing would come 'from the God-given power' of the medication or from 'God's blessing'.[101] At the same time, illness was understood much more commonly as a 'domestic affliction' that the 'highest Lord has sent me'.[102] In Pietist circles, the religious interpretation of illness as a divine trial, warning, or punishment, seems to have played a prominent role at times.[103] And in the early 19th century, numerous members of the upper classes were still flocking to the public prayer healings of a high-ranking clergyman, the Geistliche Rat von Hohenlohe, who became famous for, among other things, the purported healing of the 17-year-old niece of the Austrian field marshal Karl Philipp von Schwarzenberg.[104] Although in comparison with earlier times, religious belief played a lesser role in German patient letters of the 18th century as well, it remained much more present than in those from France.

The belief that 'evil people' were able to 'injure' other people, or 'cast' or 'inflict' a disease upon them played a key role in the numerous witch hunts of the late 16th to mid-17th centuries. Inquisitors sought to expose all kinds of suspicious cases of illness and death as works of witchcraft through 'sharp interrogation'. But the belief in evil powers as a major cause of sickness was also part of everyday culture in the 16th and 17th centuries, across all social classes.[105] The vexing 'heaviness' of her head was thought by Frau von Closter, in 1571, to have been received 'from evil people'.[106] 'It seems indeed a strange illness to me', commented Ludolf von Closter about his wife's affliction.[107] Illnesses with a quick and dramatic progression or that evidenced peculiar symptoms and resisted all kinds of therapy easily aroused suspicion. Quite often people also had an idea about who might be the culprit. When 30-year-old Anna Vetter became seriously ill and lost weight dramatically, she initially suspected her neighbor, 'who seemed associated with sorcery and often said that she was able to make people buckled and lame'. The neighbor had also mocked her for going to church so frequently.[108] At the Himmelthal abbey in 1567, the abbess's niece began limping after climbing some stairs on which an old sorceress, it was said, had laid something harmful. She suffered pain in her legs, her ulcers opened up and she ultimately died.[109] In the case of Alexander Bösch, the physician suspected that his disease was the result of a love charm. According to the physician, a beautiful maidservant who 'had an eye for him', had 'given him a love potion'. The maidservant was dismissed.[110] Maria Elisabeth Stampfer gave a particularly detailed account of her gouty brother, an imperial forester, who suffered the most severe pain over the course of five years before he died: he had suspected a woman, who, he thought, had also crippled his dog. Shortly before his death, the woman had confessed that she had inflicted the illness upon him and offered to give him a herb with potent healing properties and to perform an incantation against the disease. But the afflicted man had declared that he preferred to die in God's name rather than accept the prayers and incantations of a sorceress.[111]

Some healers, in turn, were thought to possess a special gift for recognizing and treating 'inflicted', supernatural illnesses.[112] In the 16th century, for instance, a shepherd from Kreuzlingen explained to a paralyzed woman that she had been 'attacked' and asked if she had 'any suspicions'. The woman named a nurse and the shepherd confirmed that it was she who had done it. He then instructed his patient to address the suspect with the words 'For God's sake, I ask you to heal me'. The paralyzed woman did so and 'from this praying' the problem improved daily and she was able to walk again.[113]

Madness was also quite frequently attributed to supernatural forces. In this case, however, the fairly widespread belief was that the mad were possessed by demons or the Devil himself. Thus Gangolf Hartung affirmed that, in 1634, he had heard with his own ears and seen with his own eyes how the 'evil spirit' had spoken out of the mouth of the possessed daughter of a saddler. Succumbing to the relentless persuasion of Catholic priests, it then 'crept out of the ear, as big and black as a large black bumblebee' and the girl became well again.[114]

In the 18th century, however, the idea of 'inflicted' or 'cast-upon' illnesses was no longer widespread among the educated classes of France and Germany. This constitutes one of the major changes in lay medical culture in that period.[115] While even some academic physicians still admitted that peculiar symptoms such as convulsive fits, madness, and sudden impotence could be ascribed to a harmful spell, such beliefs, it seems, were now of marginal importance for the educated classes in their daily dealings with illness. One of the rare exceptions I have come across during this period was, of all people, a physician. His whole family had become ill after eating from a side of beef and his son had died. But before his death he had told a woman to her face that his illness was her fault. This woman had demanded to see the son and called out his name, upon which he got a terrible headache. After all this, the father became 'very suspicious that the whole affair had been wrought by magic arts'.[116]

Among the less educated, the situation was very different. In 1748, a few weeks after returning from harvest work in Alsace, Anna Maria Schittenhelm became languid, refused to eat and developed a swelling in her arms and legs. A local hangman known for his therapeutic skills gave her various medicines. When these brought no improvement, he explained to the patient's daughter, according to an official report, that 'it was simply an evil disease and there was nothing else one could do to help'. He was not the only one of this opinion. Everybody in the town called it 'an evil disease'. So the daughter went to a female healer who had helped her former master when his livestock had been 'attacked by evil people'. The healer inspected the urine of the sick woman and said that it was 'something evil', coming from 'evil people'. She gave the sick woman several remedies and boiled her urine and excrement in a pot, which she buried in the barn, promising the daughter 'that the person who had attacked her mother would now die'. When, soon after, the woman succumbed to her illness, her daughter was 'wholly convinced' that her mother's death 'was caused by sorcery'.[117]

At least among the rural population of southern Germany, judging from the unanimous reports and complaints of local physicians, belief in harmful spells and inflicted illness was still widespread in the 19th century. Some healers of the day were particularly in demand for their skill in dealing with people afflicted in this way, and exorcism thrived among Catholic and Lutheran clergymen alike.[118]

Placed literally between the natural and the supernatural was the realm of the stars and planets. Astrology had a long tradition in scholarly medical theory and practice[119] and was diffused widely by popular bloodletting calendars and health advice books.[120] At times astrology served divinatory purposes. According to Vincentz, for example, the devastating plague of 1585 had been heralded by the seventh great conjunction of the seven planets, which happened only every 792 years.[121] Kings and princes asked astrologers to calculate their natal charts, a task which, like the production of astrological calendars, was usually performed by learned physicians.

Far into the 17th century, however, the moon and the planets were not only seen as legible signs but also associated with a direct influence on earthly events and on occurrences in the human body. Epidemics were frequently ascribed to planetary constellations, which could explain, without the need to resort to ideas of contagion, why whole towns or countries were affected at the same time. Some of the most eminent 15th- and 16th-century medical and philosophical authors, like Marsilio Ficino and Jean Fernel, attributed diseases to celestial influences or described how these could be harnessed in the interest of human health and longevity.[122]

The significance of astrology as a diagnostic and prognostic tool in ordinary medical practice across the early modern period is much more difficult to gage. One of Thurneisser's patients, a 'battlefield secretary' in Upper Hungary, sent him not only his urine but also three different natal charts, 'along with the calculation and verification of my proper conception and nativity'.[123] It was exceptional in Germany, however, for patients to request a horoscope – even of Thurneisser, one of the most renowned astrologers of his time – for primarily diagnostic reasons. In England, in contrast, various medical-astrological practitioners flourished around 1600.[124] Michael MacDonald has convincingly shown the sophisticated way in which these astrologers not only gave medical advice but created meaning by embedding the patient and his disease in an all-encompassing cosmic order.[125]

Gradually, from the 17th century onwards, astrology lost credibility among the learned elites in general and among physicians and natural philosophers in particular. This was partly due to philosophical and theological concerns,[126] but it probably also reflected a growing disenchantment with the validity of astrological predictions, which were consistently belied by actual events.[127] Around 1600, medical astrologers in England were frequented by some high-ranking patients, but their fees were significantly lower than those of other physicians suggesting that their clientele was already predominantly among the less affluent. In early modern patient letters and other personal testimonies, astrological notions are rarely mentioned. Occasionally patients thought they had experienced the influence of the planets, and especially the moon, in their bodies rather as they sensed the effects of weather and climate. Her son was quite well, wrote Benigna von Lubbersdorff in 1579, 'except that his natural complexion and shape is very much on the decline at the end of the month and he becomes rather lean, as if his flesh were dropping off him'. But this went away again as 'the month waxes and gains'.[128] Another patient was hopeful for improvement because the air was going to be 'milder' and 'the constellations milder' as well.[129] Only in rare exceptions, however, do patients indicate a firm belief in and a fairly sophisticated degree of knowledge of astrology. Hieronymus Wolf was unusual in thinking that 'the Saturn square Moon weakened the vision in my left eye and the opposition of Moon and Mercury turned it, as I suspect, so that disfigurement was added to the damage.'[130] An unnamed patient of the famous Paracelsian Joseph Duchesne attributed his serious ailment to a 'celestial impression' caused by the malignant influence of Saturn passing through the starting point of its ascendant.[131]

In spite of massive criticism and its growing marginalization as 'superstition', medical astrology did not disappear entirely. In the early 18th century, a 45-year-old man suffering from 'red murrain' (erysipelas) reported that an astrologer had told him long ago that he would suffer from the disease, assuring him, however, that he would get away with his life. The patient had disagreed: not because he doubted the planetary influence but because he had arrived at a different interpretation. The astrologer, according to the patient, did not know what he was talking about, since the 'direction', the 'orbits', and the 'transits' of the planets unanimously indicated that he would perish the following March.[132] In early 18th-century Amsterdam, the German physician Johann Christoph Ludeman (1685–1757), an outsider to the medical establishment, was still able to establish a reputation among the populace as a medical astrologer, finding his patients mostly among the less educated. Using the place and date of birth of those seeking advice, he calculated their nature, illnesses, and prospects for recovery. In one of his consultations, for example, he found that the disposition of a 34-year-old patient was determined by Jupiter: around the lungs and dia-phragm an old acrimony and volatile gall had accumulated. The result was bloating and irritation around the liver and gall bladder; also, his blood and humors were turning scorbutic and it was to be feared that, against his nature, he would fall into a melancholy condition. At 40 years of age he would be struck down by a serious illness, which he would hopefully be able to shake, however, thanks to his prudence.[133]

The Search for Meaning: Illness, Way of Life and Biography

The idea of illness as something ordained by fate, or commanded or inflicted by God or another higher power, gradually began to be less widely accepted among the early modern educated classes. Increasingly educated patients and their families came to find the key to understand-ing their ailments and the answer to the question 'Why me?' in their own conduct and their individual life stories. In doing so, they relied in large part on the traditional principles of medical dietetics, that is, on the teaching of a healthy conduct of life as set out – ever anew and in many variations – in countless early modern booklets on health and plagues.[134] Knowledge of these principles, as numerous patient letters show, was widespread among the educated classes. Four dimensions of individual lifestyle and life circumstances were at the center: eating and drinking, physical and sexual activity, the quality of the air, and pas-sions.[135] Many patients referred to at least one of them in their letters and, especially in the more detailed letters of the 18th century, it was often two, three or even all four.

Physical exercise was deemed important because it aided in expel-ling harmful substances and waste products through perspiration and enhanced the body's consumption of food, viz. of the blood into which it was transformed. Alexander Bösch, for instance, regarded physical work as the best medicine. It helped him sweat and promoted sleep and appetite.[136] Lucas Geizkofler, after recovering from tertian fever, was advised to exercise in order 'to sweat out the rest of the fever', whereupon he walked from Strasbourg to Augsburg.[137] Johann Valentin Andreä believed that his long life was owed to his moderation and his physical activity. He bought himself a garden expressly to refresh his mind and enable him to go for walks.[138] Too much exercise or excessive sexual activity, on the other hand, exhausted the body and led to an excessive loss of 'spirits'. Along similar lines, educated authors of patient letters and other personal testimonies, following a topos of humanist self-portrayal, described excessive intellectual work – which involved the work of the animal spirits – as conducive to illness.[139]

The quality of the air was important for a number of reasons. First, air possibly contained morbid impurities, above all the unclean miasms, which were traditionally held responsible for malaria ('mala aria' = 'bad air') and other plagues. In 1591, Gideo von Boetzelaar, for example, suspected that his quartan fever was the result of 'infected air' in Zeeland.[140] In Silesia, people believed that the major epidemic of the plague in 1523 'came from corrupted deep wells, from which foul fumes rose, or from the large strong earthquakes in other countries, where the evil, poisonous fumes inside the mountains had become free, wafting over many hundreds of miles'.[141] Air was also experienced as irritating, and as capable of affecting the body via the sense of smell. He was 'unable to stand any strong smell and [grew] sick from it in an instant', explained one of Hoffmann's patients.[142] Not least, air affected the body through its warmth or coldness, which acted particularly on the fibers surrounding the vessels and pores. Warmth caused the fibers to become limp and widened the pores. This aided the release of volatile and liquid morbid matter but could also lead to a loss of vital substances. When the fibers around the pores contracted due to coldness, on the other hand, the release of morbid, impure matter was prevented and that matter threatened to undergo further corruption within the body. The coldness and limpness of his feet, thought Simon Roter, came without a doubt 'from this influence of the sky', especially because he had gone outside into the fresh air and his feet had become a little cold.[143] The rheumatism of her son, wrote one worried mother, appeared the morning after a warm day when the nine-year-old had been running around. He had become hot and sweaty and then unfortunately exposed himself to the cool evening air. Later, after he had recovered, he again stayed outside too long and the following day his left arm was numb.[144]

The effect of food and drink on the body was thought to be very much like that of medication and both were often consciously used in the treatment of illnesses. Especially when sick, too hefty a meal was to be avoided. It could easily overburden the stomach and the whole body. Easily digestible food on the other hand was conducive to health, and wine was often regarded as a welcome fortifier. The sick Joachim Brandis thought he could feel in his body how the wine 'restored' him and brought back his appetite, just as a lamp whose oil had run out was replenished.[145] More dangerous still were foods that were cold, uncooked, and hard to digest. They overtaxed the stomach, weakening its digestive heat. As a result, raw matter accumulated in the stomach and in the rest of the body. In addition, there was always the danger that food was spoiled. The means of preserving food were limited in those days.[146] Also, individual ingredients in food were associated with certain effects on the body. For example, it was suspected that very spicy food caused acrimonies. In this context, French patients and physicians were particularly wary of the popular spicy ragouts. Similar dangers were seen in smoked or very salty meat. Individual wines and even different kinds of water, it was known from experience, also affected the body in peculiar ways.

Intense passions, finally, according to a widespread belief, were among the most powerful causes of disease.[147] Sometimes, a single event that was followed by particularly intense emotions was at fault. For instance, Simon Roter related how an 'unexpected death had caused him great grief and sorrow, so that on the night of November 28th, at around 1 a.m., I was struck in my left thigh'.[148] Gottschalk Weinsberg fell seriously ill and was bedridden for several months after being scared out of his wits by a spider that had fallen from the spout of a pitcher into his glass.[149] A lady from Paris suffered a stroke followed by signs of palsy and intermittent speech loss an hour after a dog had terrified her on her walk.[150] Another patient dated the beginning of her ailment to the day when a soldier, fleeing after an illicit duel, almost ran her down.[151]

Others, by contrast, attributed their illness to repeated negative passions or long-lasting negative emotional states. '[I] incline very much to anger and zeal, which I feel instantaneously in my entire body', wrote a man who suffered from stomach problems and headaches to Friedrich Hoffmann.[152] Family tensions turned the life of an unmarried patient of Tissot's into a living hell. The woman was forced, wrote an acquaintance, to live with her family. And they treated her badly, especially one member of the family – probably her brother – 'who, for selfish reasons, did not content himself with abusing her with words and threats'. The resulting intense emotions had spoiled her blood. Consequently, she experienced a burning inner heat, and had dry skin and rashes; she felt weak, sad, and crestfallen, had a fever and toothaches, and suffered from bad dreams; and when she drank warm coffee, the space 'between the skins of the face' felt like dead.[153]

Anticipating in some measure modern psychoanalytical notions, individual patients also described the negative effects of suppressed emotions. Willibald Pirckheimer, for example, who attributed his podagra above all to 'the passionate nature of his spirit', said that he had taken injustices and hostility with lordliness but was unable to prevent 'their incessant and bitter stings' from boring deeply into his heart.[154] In 1714, a patient ascribed his flatulence, toothache, and intense rising heats to the fact 'that I have been living now for 20 months with a secret anger and have never been able to vent my passions'.[155] And in the late 18th century, Mme de Moncharle was convinced that the violent restraint she had imposed on herself at the time of her wedding 30 years previously was largely responsible for her present breathing difficulties, flatulence, and cramps.

These principles of a healthy lifestyle were disseminated through numerous advice books and pamphlets as well as in personal encounters and were widely known to laypeople. Historians like to see health-advice as a powerful means to spread and reinforce bourgeois norms of moderation and temperance. Its impact on the conduct of everyday life must not be overrated, however. Knowing the principles of dietetics and applying them on a day-to-day basis were (and are) not the same. There is in fact little reason to believe that in times of health laypeople actually carried on their lives in accordance with the prescribed rules. Patient letters and autobiographies suggest that dietetics became an important issue only when someone became ill, when a disease was to be treated or its cause established. In such cases, numerous patients asked their physicians to tell them explicitly how to conduct their lives properly or even asked detailed questions about the beneficial qualities of certain foods or drinks. Dietetics not only served prophylactic ends but also was an indispensable part of the therapy, and it seems that patients often heeded the medical recommendations. Much more rarely, on the other hand, did patients ascribe their ailments to dietary errors. A patient of Felix Platter's who attributed his chest ailment to 'a lot of drinking in company' was exceptional.[156] Somewhat more readily the unhealthy, immoderate lifestyle of one's fellows was imputed as the cause of their illness, for example when this or that person seemed too devoted to food or wine.[157] The judgment of Elisabeth Charlotte d'Orléans (Liselotte von der Pfalz), a German outsider at the French court, was particularly harsh, even scathing. The Duke of Berri, she said, had 'killed himself with his disgusting guzzling and swigging'. His wife also debauched and killed herself with 'her foolish bathing and gorging, indeed as efficiently as if she had shot herself in the head with a pistol, because she secretly ate melons, figs and milk; she confessed that to me herself'.[158] She even ascribed Anna Maria of Austria's breast cancer to her 'abominable' feeding four times a day.[159]

The four basic unhealthy influences presented here cover much of the spectrum of possible causes of prolonged diseases that dominated the etiological thinking of patients and their relatives in the early modern period. But two more causes of disease with likewise long-term effects, but both standing outside the traditional canon of the non-naturals, should be mentioned in addition, namely trauma and heredity.

From the perspective of the patients and their relatives sometimes a mere blow to a sensitive body part was all that was needed to bring about prolonged illnesses. For example, some patients ascribed tumors and cancerous ulcers, particularly in the female breast, to a blow. This might have happened years or decades earlier, but patients were convinced that they had retained a certain hardening or clogging.[160] Henriette de la Tour du Pin, for example, had no doubt that her ailment dated back to a cruise, when sailors had helped her disembark from a small boat, heaving her upward. She had felt an intense pain in her right side and believed ever since that she had suffered an internal injury in the liver area. Physicians, she wrote, were never willing to recognize this, 'but it is nonetheless true that since that day I have not ceased to suffer from this ailment and am still today, at 63 years of age, suffering from it'.[161]

Hereditary influences were also frequently mentioned. It was assumed that many diseases, or at least a predisposition to them, could be passed on from parents to children. In such cases, the medical condition transcended the limits of the individual biography. Thinking in this vein, J. G. Bövingh bluntly stated that his wife had 'congenital consumption'.[162] And a 44-year-old scurvy patient of Haller began her account remarking that she 'stemmed from scorbutic parents' and that both her younger siblings were also afflicted with the disease.[163] The father of Mlle Herbolin thought it worth mentioning that both of her grandparents had suffered from a similar 'gouty rheumatism'.[164] And the husband of Glückel von Hameln wanted to keep his hernia a secret, above all to protect his children because he thought people would say such an ailment was hereditary.[165] Some patients, in turn, explicitly mentioned hereditary influences only to exclude their possible role in the given case; in the 1670s, for example, Johannes Heinrich Hummel made a point of stating: 'Among my ancestors, as far as I know, no one was afflicted with the disease.'[166]

As these examples show, hereditary factors were taken into consideration for a wide range of ailments. But there were certain illnesses for which a hereditary connection was frequently presumed. The most serious ones were falling sickness and consumption. When it came to choosing a suitable spouse, these illnesses were thus particularly watched out for, and parents whose children suffered from them worried a great deal about their children's marriage prospects. In addition, there were some rarer diseases in which hereditary transmission seemed particularly likely. Deaf-muteness was the most obvious case, and physicians and authorities occasionally even demanded that those affected should not be allowed to marry.[167]

The Narrative Reconstruction of Personal History

Innumerable letters and case histories attest to an avid search for meaning – a search for harmful influences, for traumatic experiences, or violations of a healthy conduct of life which might explain the present ailment. Patients and relatives looked back over years or decades at life-changing events such as the death of a beloved child, but also at seemingly minor incidents, such as going for a ride in the cool evening air. Often they invoked a whole series of such morbific influences and incidents which had supposedly combined in bringing about the disease.

The immediate reason behind these searches for disease-triggering and disease- promoting factors, which often lay far in the past, was to identify the nature of the disease and to choose the best treatment. However, many of these stories were much more than mere recounting of memories of supposed disease-triggering or disease-promoting factors. In their search for the cause of their illness, the patients, in a sense, re-wrote their entire biographies. Their biography was reconfigured into a story leading up to the present ailment and it gained in itself new significance in light of the ailment. Conversely, a patient history that was 're-written' in this way gave the illness a personal meaning and importance. In this retrospective view, a patient's entire personal history almost inevitably led up to his or her present state of illness. Medical sociologists have termed this a 'narrative reconstruction' of personal history. By re-writing their personal histories in light of their illness, patients bridge the threatening chasm between their healthy, vibrant past self and their sick, ailing present state. In the process, past and present alike take on new meaning.[168] From this perspective, narrating the history of one's life and illness proves to be an act of creating meaning and identity.[169]

Four partially overlapping narrative patterns can be distinguished in early modern illness accounts and in patient letters, in particular. First, there is self-accusation. Its role in the process of 'narrative reconstruction' is particularly striking in 18th-century patient letters written by men, who ascribed their illnesses to sexual self-gratification during their youth and, in some cases, in adulthood as well. I will come back to this in more detail later in the book. To these men, a reading of Tissot's *Onanism*[170] or similar fruits of the Europe-wide medical campaign against masturbation became almost a Damascus road experience, the critical turning point in their lives. At last they understood that the cross they had been bearing, often for years and years, was a direct and deserved consequence of their earlier vice. Their letters turned into confessions rife with self-accusation, yet at the same time they also expressed a deep gratitude because their eyes had finally been opened. Their present illness was no longer a twist of blind fate. Now they finally understood its deeper meaning: it was self-inflicted, a punishment for their misdeeds which sometimes, as in the case of sexual impotence or an uncontrolled 'seminal flux', hit the very organ with which they had sinned.[171] Such interpretations of illness as the result of earlier misdemeanors can likewise be found in connection with other kinds of sexual 'excess' or with misguided eating and drinking habits. In a sense they constitute a secularized version of the old motif of penitence and repentance: the (dietary) 'sin' was followed by the just punishment and the return to God's or Nature's commandments.

According to a related narrative pattern, the present ailment was similarly an endpoint of sorts; it was the latest in a series of diseases which had accompanied the afflicted person throughout much of his or her life. Here, however, the complaints ultimately pointed not to 'unhealthy' or 'sinful' behavior but to a constitution that was fundamentally predisposed to illness. These people were virtually doomed to be ill. Illness was a central element of their identity. 'I was very sickly in my childhood', wrote J. V. Andreä, 'such that I only learned to stand on my feet at the age of two years, and this weakly body constitution I have felt throughout my life.'[172] It was especially those who thought themselves 'melancholic' or 'hypochondriacal' who structured their case histories according to this pattern.

A third narrative pattern also relied on the idea of a fundamentally weakened physical constitution predisposed to various diseases, but it saw a specific traumatic experience, such as a difficult birth or the loss of a close relative, as the starting point of this constitution. This made the illness appear as the result of something arriving from outside, as a matter of fate or as the consequence of the mistakes or misdeeds of others. The story of 49-year-old Mme de Merande's chest disease, for example, began immediately following her birth, when she was handed to a wet nurse whose milk was so bad that she almost died. Her state of health had been delicate ever since and smallpox, eye ailments, rheumatic fever, white discharge, and consumption were but some of the illnesses marking her biography.[173]

The fourth and final narrative pattern to be noted here combined, in a sense, the second and the third patterns. It reconstructed the patient's life story as a sequence of different traumatic incidents and pathogenic occurrences which had all left their indelible marks on the body – a harsh father, insufficient food as a student, a hard life as a soldier, an injury from a fall. The body became an archive of insalubrious influences which ultimately led to the current disease.

Similar processes of a retrospective search for causation and meaning can be shown to be at work in many cases of severe and/or chronic diseases today. Sometimes, in a process of shared 'mythopoesis' physicians and patients agree on the same story.[174] More frequently, the assumptions of patients differ considerably from those of their physicians.[175] Yet the patient's quest for meaning can prove helpful in coping with the illness even if the results are at variance with the medical view. In general, disease that can be linked to a specific cause even if it is one's own failure, rather than blind faith, appears less terrifying. Identifying a cause or meaning makes it easier for most patients to deal with their disease or even enables them to see something positive in it.[176] By ascribing their ailment to a cause for which they themselves were responsible or which they could have avoided, patients, in looking back, gain for themselves a sense of control over an illness which seems at first sight only a whim of fate. In this way disintegration, a threatening fissure through one's personal history, is prevented. Linking the present disease to one's life story is also helpful in coping with possible future disease episodes. The subjective experience of a continuous, coherent personal history – and with it of one's identity – remains intact.[177] On the other hand, patients who ascribe their ailments to an erratic, unfortunate, meaningless accident or to the inattention of another person, for instance, usually struggle more with their fate.[178] Embitterment and dolefulness dominated many patient narratives of this type in the 18th century. In these cases, the agonizing question, 'Why me?' remained unanswered.

Anxieties

Disease is often frightening. The character and intensity of the fear depends on several factors, however: on previous experience with similar diseases among family or friends, on one's individual anxiety level, on the hope for improvement through medical treatment and, not least, on the nature of the disease in question, the symptoms and images associated with it, its presumed course and potential negative effects.

In the early modern period, some ailments were feared above all because of the virtually unbearable pain typically associated with them.

Kidney and bladder stones, podagra or gout, and toothaches were at the top of the list. Other diseases were characterized by a loss of highly valued capabilities such as mobility and perception. The aforementioned Monsieur Goret, son of a public servant, for instance, narrated sorrowfully how his deafness had overshadowed his entire life since he was a child. He understood what people said only when they spoke loudly into his left ear. From a distance, he did not understand a word and at times he was completely deaf. Only occasionally, but never for longer than three days, was he suddenly able to enjoy birdsong and the company of others – just long enough, he said, to make him even more aware of the sad state he was in.[179] Others described their impending blindness with similar desperation. One 43-year-old clergyman was no longer able to fulfill his duties of pastoral care sufficiently due to his waning – or, as he expressed it, 'groping' – eyesight. At mass, he needed to read the liturgy word by word using a candle held close to the book, and this cost him much effort. Glasses brought little improvement.[180]

Other kinds of illness were most fearsome due to the swift, sudden, and frequently fatal course they took. Typical examples of such acute, life-threatening conditions were the plague and 'fevers' such as 'bilious fever', 'foul fever' and 'nervous fever'. In the 19th century, cholera, the 'Asiatic hydra', spread fear and horror. Its victims were said to die sometimes within hours and in gruesome agony, when just the night before they had been happily dining with their families.[181] Another common disease, 'apoplexy' or 'stroke' was feared because it lashed out suddenly and often unexpectedly. Within an hour, it could throw its victims paralyzed onto their sickbeds or even kill them. Understandably, its potential harbingers were taken note of anxiously.[182]

Illnesses of this kind were a stark demonstration of death's omnipresence. Even healthy, strong people in the prime of life could be attacked out of the blue, from one day to the next. Often they did not even have enough time to prepare themselves appropriately for death and to receive the last sacraments, let alone to suitably stage their own act of dying according to the precepts of a the centuries-old *ars moriendi*, the 'art of dying': looking death in the eye bravely and with composure, being at once a model and a comfort to one's next of kin. One had therefore good reason to 'courteously thank the dear Lord' when a sick woman, during her last days, following a stroke, was at least allowed – although she 'lay pitiably until the end' – to 'keep her wits about her'.[183]

With the declining force of religious faith, patients did, however, find themselves increasingly torn between the fear of a sudden, unexpected death which left no time for spiritual preparation and a farewell to one's family and friends, and the wish to escape the horrors of a long-lingering illness. There are signs that in the 18th century a sudden, unexpected death without any prior suffering started to be judged more positively.[184] Three chronic diseases particularly inspired fear of long-lasting suffering: cancer, consumption, and dropsy. I will discuss all three in greater detail later on. They were characterized by a progressive physical deterioration and, in the long-run, by a dramatic change of bodily identity – a change in one's appearance. With consumption, the body burnt up its own substance. With dropsy, the bodily cavities and/or the face and extremities filled with water. The blood, the 'source of life',[185] lost its natural consistency and became watery; in a strange contrast to their growing bulk, the victims' strength dwindled. Particularly harrowing images were associated with cancer, the 'most horrid of diseases', as it was called even then.[186] People were familiar above all with breast and uterine cancer, because their consequences were obvious to the senses. Violent, unappeasable pains accompanied massive physical decline. In a later phase of the disease, fetid ulcers and putrid secretion added to the horror. Victims seemed to rot alive.

Other diseases jeopardized not so much the physical integrity as the reputation and sometimes the marriage prospects of the afflicted person. Diseases of the genitals were particularly stigmatizing and tainted with shame. They were for good reason frequently kept a secret, even from one's own servants, who otherwise knew just about everything. Very common were male gonorrhea – at the time a catch-all term for the discharge of semen and other fluids – and in women, 'the whites' (vaginal discharge). To the contemporary eye both suggested insufficient control over one's genitals and their respective openings. They were considered repugnant and disgusting and they called one's ability to procreate into question. One of Thurneisser's aristocratic patients had already spent close to 3,000 talers on treating his gonorrhea – at a time when many ordinary workers or even scribes earnt less than 50 talers per year. However, he was willing to pay even more, if he 'only could be helped', because he was planning to marry, hoping that his lineage would 'not perish entirely'.[187]

Even more of a threat and burden in this regard was male impotence.[188] The ability to perform well in the marital bed was a crucial element of male identity and status. There was, as the famous English physician John Hunter put it, 'perhaps no act in which a man feels himself more interested, or more anxious to perform well, his pride being engaged in some degree'.[189] A patient of Tissot saw 'all the happiness in my life' as lost because his impotence rendered him unable to marry.[190] Another patient could no longer enjoy social gatherings because his diminished virility and his weak genitalia made him feel 'inferior'.[191] And a third man was distraught because, on his wedding night, he remained impotent despite the 'liberties' his beloved bride 'allowed' him and despite their mutual 'caresses'. Subsequent to this he had to witness how his chagrin over his failure even seemed to increase his impotence. He still considered himself lucky insofar as his wife met with all of this with 'quite impressive and commendable patience and reason' and was worried only about his health.[192] But he had good reason to conduct his letter consultation anonymously via two middlemen. If impotence became publicly known, the consequences could be disastrous. The man's standing in society was called into question and, in the case of wedded couples, there was the specter of a scandalous marriage annulment trial.[193]

Two other diseases which were associated with a loss of fiber and control in and around the genitals were also experienced as particularly shameful: uterine prolapse and – somewhat surprisingly, from a modern perspective – hernias. Marie Jeanne Orget, for example, refused to have a hernial truss put on because only men had the necessary skills to do so. She would have had to disrobe herself in front of them, and she did not want to be seen 'in this state'. It was not so much the nudity itself that she found embarrassing, she indicated, but the sight of her hernia.[194] When the hernia of Glückel von Hameln's husband dramatically worsened after a fall, he too did not want anyone to call a physician and only tolerated people he could trust around him, saying he would rather die than reveal it. Finally, though ultimately in vain, he accepted medical treatment while still refusing to have any strangers around and he asked his relatives to keep everything secret.[195] Junker Hans Adam von Hohenfürst, who, according to Felix Platter's account, had been 'secretly ruptured for many a year', did not even tell his own wife. Without her knowledge, he finally went to Colmar for surgery and died from the intervention.[196] Similarly, Hermann von Weinsberg told his wife about his hernia, which was the size of a chicken egg, only when the pain became so acute one night that he saw death before his eyes.[197]

Rashes and skin changes could drastically – and visibly – affect the body's appearance. His rash had 'quite violently [...] erupted and flared up below his face', complained Christoph von Falckenberg in 1577, 'with a whole lot of evil redness'. He was so ashamed to show himself in public that he stayed in his inn for three days, unwilling to go outside.[198] Along the same lines, Johann Georg Bövingh in the early 18th century wrote, 'I felt almost like an outcast and was ashamed to see people.'[199] In the case of scarring diseases such as smallpox, women (and to some extent also men) had to fear lasting damage to their appearance and their value on the marriage market. Mme de Staal Delaunay, writing of the severe smallpox she had endured as an adolescent, reported in her memoirs that she had not dared for months on end even to look at her face in the mirror.[200] Her husband's family began treating her even worse than before, said Mme de Guyon, after smallpox disfigured her at the age of 22.[201]

Among other examples of diseases attended by particular fears, convulsive seizures and madness must finally be mentioned. There was general agreement about the typical symptoms of 'epilepsy' or 'falling sickness'. Tongue biting, which is considered typical today, was hardly ever mentioned, but many patients and their relatives did report foaming at the mouth and, as a particularly characteristic symptom, thumbs turned inward toward the palm.[202] She had not taken the fatal illness of the child 'for falling sickness', one old midwife put on record, 'because such children turn their hands violently inward and roll their eyes'.[203] It seems that epilepsy was especially feared on account of the extreme loss of control that characterized it. It meant a total breach of the norms of self-control current at the time. Epileptics seemed like wild animals.

The shattering, unsettling features of epileptic seizures become most clear in letters from patients who ascribed their own, quite different, illnesses to the effect of simply witnessing an epileptic's seizure. Thus Monsieur Baville, a 34-year-old secretary and tutor, dated the beginning of his ailment to the age of 17, when he was present at the epileptic seizure of a comrade. For a long time afterward, his whole body would tremble violently and uncontrollably whenever he encountered 'the unfortunate victim of this cruel illness'. In the end he attended two anatomy classes to rid himself of this fear. But he was convinced that his mind and health had been permanently damaged. Since that time, he had been suffering from, among other things, severe headaches and nocturnal choking, and would start awake at night.[204]

As this case shows, the mere presence of epileptics made them a hazard to others. Pregnant women were at a particular risk. According to a then still widely held belief, women's emotions affected the fruit of the womb directly through the imagination.[205] The child of a woman who had witnessed a seizure during pregnancy was likely to develop a convulsive disease as well. Epileptics – and the same presumably went for patients with what we would call spastic paralysis – thus not only suffered from seizures. Their whole existence in society was at stake. They were likely to be excluded and ostracized. Thus, the 68-year-old servant Marie-Anna Couronneau was barely able to assert her wish to go to church. 'Her disorderly and forced movements', according to the interrogation records, 'caused her to twist with every step and very often to have convulsions, which alarmed everyone who saw her and made pregnant women avoid her'.[206]

Epilepsy was feared to such an extent that some physicians and relatives tended to play the condition down as mere 'nervous complaints' even in the presence of massive attacks. One sick man who had had many severe seizures – once he fell face first into a fire – was told by his family that he was only suffering from the 'vapors'.[207] When certain characteristic symptoms were absent, patients and relatives also held on to their hope that perhaps it was not epilepsy after all. The mother of a 16-year-old girl, for instance, stressed that her daughter remained conscious during her seizures and did not turn her thumbs in, as was common with epileptics.[208]

Like epilepsy, madness was sometimes subject to ostracism. When a captain of Mölln had become 'stark raving mad', Brokes, the mayor of Lübeck, prayed that God might save any Christian from such a disease.[209] In the eyes of their contemporaries, those afflicted resembled animals more than humans. Physicians treated them with bloodletting and medicines or attempted to talk reason into them and return them to their human self by cajoling, scolding, or if necessary chastising them. Ordinary people often seem to have been remarkably tolerant toward the insane. But unruly or violent behavior and the fear that they would kill themselves or commit arson sometimes made them feel that there was no other choice, in spite of the often substantial costs, but to put them in chains or to lock them in a prison or another 'safe place', for their own protection as much as for that of their fellow human beings.[210]

The Physician's Audience: Illness and the Bedside Community

To a far greater extent than today, illness was lived as a collective, indeed public, event in the pre-modern world. Patients at home were often surrounded by people: by their own families, but also by friends, neighbors, and acquaintances. This was anathema to the physicians, who warned that this corrupted the air in the sickroom even more. But it seems that most people flocked to the sickbed not only out of compassion or curiosity. Certainly among the nobility visiting sick relatives was *de rigueur*, and one's absence had to be explicitly excused and accounted for.[211] Among craftsmen and the common rural population as well, patients could expect numerous visitors. Even in times of epidemics, a sense of duty and solidarity sometimes outweighed the fear of contagion. When cholera swept through Europe in the early 19th century, many people fled in panic, but at the same time physicians complained about friends and relatives gathering around the severely ill and even sitting on their beds in the evenings or on Sundays.[212]

Relatives and acquaintances did not come only to give words of encouragement. They also took a lively interest in the medical condition and shared their personal assumptions about the nature of the disease and the most promising treatment. They especially suggested using gifted healers[213] or remedies that had proven to be effective;[214] they recommended new diagnostic methods such as Thurneisser's urine distillation[215] and sometimes they advised the patient to stop treatment because the disease would go away by itself.[216] Educated people also shared information about tried and tested medications in their correspondence. Having heard 'that the evil toothache is still a burden to you', the countess of Solms, for example, wrote to her 'darling Bellchen' that she wanted to 'communicate a little remedy which often helps me and has recently helped me'. She also mentioned a remedy that her mother had learned about in a similar case from another woman.[217] Other people's positive experiences and recommendations were often decisive in the choice of a remedy. One parson, for example, had his son, who suffered from edema, drink lice in wine because the wife of a befriended clergyman had told him about the beneficial effects of this remedy in a similar case.[218]

Some patients found themselves virtually swamped with well meaning therapeutic advice. 'Now it being look'd upon as a slight infirmity, amongst my visitant neighbours', recounted 80-year-old John Evelyn, who suffered from painful hemorrhoids, 'everyone is ready to recommend their remedies'.[219] Many people were gracious in commiserating in the suffering of his son, who had a severe case of dropsy, said a clergyman. But impatient to see him recover they pressed him to have his skin cauterized with a hot iron or to have an opening made in the belly in order to let off some of the accumulated fluid.[220] Patients whose friends happened to be physicians found it particularly difficult to resist such recommendations. All his physician friends, one dyspeptic V. Ferguson lamented 'did so importune me to use many things [that] were indeed rationall, but many very disagreable to me on tryall, and my relations did so presse me to reiterated trialls of those receipts that

[it] was much against my grain. I was overpressed with too much advise and too manye medicins, and dayly declined till all men concluded me hopeless; tho in truth I never thought myselfe past cure.' Finally, he resorted to a treatment of his own devising and to everyone's astonishment got much better and was thinking about traveling abroad over Christmas.[221]

Nursing Care

Bedridden patients need good nursing, someone to look after them, to bring food and drink, to take care of their physical and emotional needs. Presumably the quality of nursing care played a crucial role in the subjective experience of sickness then, just as it does today. There is, however, hardly any aspect of the everyday medical life of past centuries about which we know less.[222] The sources are sporadic and fragmentary: hints in personal testimonials, household accounts and last wills, short references to nursing in medical case histories, court records when there was a dispute about a wage for paid nursing. A major reason for this lack of sources is that nursing was almost always done in private, in the patient's home. Although some hospitals served curative purposes long before 1800,[223] early modern hospitals were first and foremost asylums for the old and invalid and not for the medical care of the sick.

Often it is not even clear who did the nursing. Wills and autobiographical writings suggest that among the wealthier classes nursing was commonly done by maidservants, lackeys or other domestics. In 1636, for example, the eldest daughter of Christoph von Bismarck died in the lap of a maidservant.[224] The sickly husband of Mme de Guyon was nursed by a lady's maid among others.[225] Occasionally domestics and subordinates would even look after the genteel sick in their own, modest, homes.[226]

Attendants coming from outside, presumably for pay, are also frequently mentioned.[227] Vincentz, in Breslau, for example, complained about the young women who preferred to leave the night watch at the sickbed to the older women.[228] Caspar Questel in the late 17th century was even more critical of the elderly lower-class women who nursed the sick for money. Tired of their work or hoping to get the deceased patient's clothes they sometimes, he claimed, even tried to make the patients die faster by depriving them of their pillows – a widely reported practice at the time to which even the learned physician attributed fatal effects.[229] Occasionally, paid nursing can also be tracked in court records. In 16th-century

Nuremberg, for example, Margretha Flaschnerin was allowed 10 pfund for the 10 days and nights that she had 'attended and nursed' a Frau Heintzin in her illness.[230] The court allowed Margreth Weberin 40 pfennig for every week she had 'nursed' a now recently deceased widow who had taken ill.[231] Sometimes salaried attendants also provided food and lodging in their own homes. In Nuremberg a woman received 33 pfund for 'having nursed and housed' the unmarried Margaretha Behaim in her severe illness.[232] In 18th-century Nuremberg, the urban administration paid for nurses to lodge and nurse indigent patients.[233]

Neighbors, friends, acquaintances, landlords and landladies also appear to have helped out.[234] Bövingh praised the people of his 'lodgment' who provided him 'with all the care possible'. In 1595, exhausted by his colic, Abraham Scultetus was nursed by his student, Magister Müller, who led him by the hand wherever he needed to go.[235] In the 18th century, Mme de Graffigny wrote to her friend that she would be acting as a 'sickbed attendant' for Mme Eynaud; her choice of words indicates that she would not simply be there as a companion.[236] It seems that, among the common people, visitors took it as a matter of course that they would help out. It was nothing new in rural areas for visitors to offer to make the bed for a sick person, stated one 17th-century source in Württemberg.[237] Neighbors and other visitors would even assist with medical procedures such as bloodletting, if only by restraining the patient.[238]

But in most cases, it was the family – above all mothers, wives, sisters, and daughters – who were primarily responsible for nursing. Mme de Guyon, for example, nursed not only her father but initially also her husband, who was 22 years her senior.[239] God had allowed his wife to become healthy again after a severe fall, said the Swiss pastor Johannes Heinrich Hummel, 'so she could attend to me in this my last illness, which she did to the best of her abilities.'[240] The sick Hieronymus Birckholtz wanted his wife around even when he was traveling, 'for the purpose of nursing and attending'.[241] A burgher of Nuremburg who had been 'for such a long time burdened with the great and long-drawn French disease' left his 'dear housewife' Christina a special legacy because she had 'shown all her good and kindly will and work [...] with wiping, washing, lifting and laying down'.[242] A burgher of Cologne took similar action, noting that his wife had given him 'support and help in his troublesome illness'.[243] The sick Hieronymus Wolf was nursed by his sisters Anna and Maria.[244] When Gideo van Boetzelaar became ill in 1591, he came to depend on his mother for nursing.[245] She had hardly any time at her disposal, complained Mme Marnais two centuries later, because she had a very old mother who needed close attention and daily care.[246] Even noblemen took it for granted that it was their close relatives' duty to nurse them personally. After 20 years of happy marriage, the Marquise d'Agrain assured the physician that she was prepared to serve as her husband's sickbed attendant for the rest of her life.[247]

Nursing a chronically ill person was often demanding, both emotionally and physically. Added to this was sometimes the fear of contagion. Even years later, some patients ascribed their own diseases to the physical and psychological strain of nursing an aunt, mother, or beloved child. For three years, until his death, the Comtesse de Mouroux nursed her sick husband; according to medical opinion, he was suffering from consumption and had an ulcer in his lung. To the countess, the exhausting support she had given, the discomfort she had endured, and the worry and fear she had experienced, not only for herself but also for her children, whom her husband had insisted on seeing until the end, had ultimately ruined her.[248] Another patient dedicated more than four years – from her 16th to her 20th year – to nursing her consumptive aunt. Soon she contracted a chest ailment herself.[249] Mme de Möhn nursed her moribund brother for only eight days, together with her mother, but it was enough, in her eyes, to prompt the return of her severe coughing fits.[250]

Indications that men also engaged in nursing are much rarer, although there are a few.[251] When the mother of future superintendent Fabricius fell ill with consumption, her husband left her but the son, apprenticing as a cobbler at the time, took care of his bedridden mother.[252] Ulrich van Hutten even recounted approvingly how Georg Gros had spent much time in his sickroom when he was ill with syphilis and had diligently seen to it that he always had everything he needed, 'even though I stank horribly from the abominable disease'.[253] In the 18th century, Baronne de Staël-Holstein said of her father that he had devoted himself wholeheartedly to the care of her mother during her long illness, not budging for hours on end when she had finally found some sleep in his arms.[254] Devaux, in writing to Mme de Graffigny, said that he left his sick mother alone only for brief intervals.[255] And the male friends of Jean-Baptiste le Doulx took turns at his sickbed when he was seriously ill.[256]

The situation could become particularly hard to bear for the families and friends of the insane. Pleas for a relative's 'safe custody' in the nearest jail, tower, or madhouse bear ample testimony to this.[257] In the late 18th century, Mme Develay gave a detailed account of her domestic drama. She and her husband had seven children and the eighth was under way. Despite a considerable age difference between her and her husband, she wrote, her marriage had been happy and full of tender feelings. But over the past several years her husband's mental health had gone from bad to worse. He became upset over trifles and had fits of rage, was gripped by fear and presentiments of death, and broke into tears. His thoughts and language had become muddled. In the end, he did not want her to leave him alone, even during the night – until his mood eventually changed and he accused her of being indifferent to his suffering. He refused all food, drink, and medication from her or anyone else in the household; he was convinced that she actually sought his life and was conspiring with the physician, servants, and his best friends to kill him. And so she suddenly found herself cruelly refused by the 'one and only I love, I value, and I cherish'. Further delusions became intertwined with his suspicions of a planned murder. He believed he was due a large inheritance and thought he was one of the most distinguished gentlemen of France – and hence there was every reason for murderous intentions. Finally, out of his own will, as his wife stressed, he took refuge in the local hospital.[258] The Earl of Derby also had a difficult time when he took in mad John Getting, thinking falsely that the man's state had improved. The man soon proved 'ungovernable' and was of hardly any use for work. There was no getting along with him. In the end the Earl, at a loss as to what to do, had him admitted to Bedlam.[259]

The suicidal intentions of lunatics were similarly trying for relatives. A young Savoyard, for example, caused a great commotion in his community with his repeated attempts on his own life. Several times, he tried jumping into the nearest river; at other times he meant to cut his throat, shoot himself, or hang himself and once he refused all food for 17 days, intending to die of hunger; but in the end someone forced a drink down his throat. He was cared for by his siblings in the family home. In the end, he felt somewhat better after bloodletting, purgatives, words of comfort and cold baths, which he initially welcomed but later had to be forced to take. But he would still spend entire days in bed, not wanting to get dressed.[260] An engaging literary account based on a true occurrence can be found in the novella *Lenz* (1835) by Georg Büchner. It describes the fate of the insane writer Lenz, a historical figure, who took refuge with various people including the well known vicar Oberlin, whose family he put to the test with his irrational behavior and suicide attempts.

The Medical Marketplace

In modern Western societies, most patients immediately consult a university-trained physician when they fall ill. In comparison, the early modern medical marketplace was much more varied and pluralistic. Patients and relatives could resort to a fairly wide range of healing practices and healers.[261]

At the outset of a disease self-help was common. Diseases such as dropsy, fever, dizziness, diarrhea, rheumatism, gout, cramps, and discharge could be recognized by laypeople. Like the physicians, patients and relatives sometimes inspected the urine and other excretions or discharges, which gave important clues to processes inside the body.[262] If needed, the throat was inspected,[263] the pulse taken, or the belly palpated when constipation or a hardening was suspected. One Irish clergyman even examined his wife's breast, saying he was unable to detect any swelling or hardening, despite having felt for one at length. His wife, however, complained of 'a small hardening and a perpetual straining, as she calls it'.[264]

Quite often patients related diagnoses suggested by their acquaintances. In the Irish case just mentioned, for example, some 'good women' ascribed the patient's stomachaches to her insufficient menstrual evacuations, bad blood, flatulence, and stones.[265] 'The folks I talk to here about my case tell me it is nervous', related history professor Shallett Turner from Cambridge, who was plagued by dizzy spells. They advised him to go see a physician before he developed apoplexy or palsy.[266]

Many laypeople were apparently also familiar with home remedies. They were usually the first choice for lesser ailments and illnesses and they might be reverted to when a professional treatment proved ineffective. Numerous home remedies were known and recommendations were shared freely.[267] Some people relied on the same remedy for everything that ailed them. Monsieur Develay, for example, who mistrusted the medications prescribed by physicians on principle, found his panacea in the onion.[268] Generally speaking, however, particular home remedies were used in curing particular ailments. The Comtesse de Wedel, for example, treated her missing menstrual periods with hot moss – a peasant remedy, she said – after the physicians had come to their wits' end; and indeed, her periods returned.[269] Some remedies and modes of treatment were quite complicated. It was recommended to a man with stone disease, for instance, that he 'ingest a head of garlic pickled in brandy at the start of the new moon'.[270] Among the most widely used remedies we find laxatives and, for skin ailments and localized pain, all kinds of animal fat.[271] Some people also changed their diets and way of life without consulting a physician. Such changes could be drastic. Thus for his severe colic-like groin pains, Major Bouju simply put himself on a rigid milk diet.[272]

Somewhere in between home remedies and medicines from a pharmacy, numerous 'specifics', nostrums, and cure-alls were widely available. Apothecaries and distillers but also physicians – particularly in the 18th century – entered this flourishing market with ever new remedies. Some of these became known all over Europe and made their inventors wealthy.[273] Among the best known remedies, which were also frequently mentioned in patient letters, were various kinds of theriac and orvietan, Gräfinnenpuder or ('poudre de la comtesse'), Rufus Pills, Le Lièvre's balm of life, James's English powder, Hoffmann's anodyne (a mixture of alcohol and ether which is still known today by this name), Schauer'sche Balsam, from Augsburg, Pippel's animal oil, calming salt from Homberg and, particularly widespread in France, Ailhaud's powder.[274] The effect of many of these remedies was laxative or emetic, that is, they prompted abundant excretions through stools and vomiting. Additionally, specifics for various illnesses were on sale: for example, the 'balsamic syrup' made by the nuns of Ste Perrine de Chaillot, which was recommended to Monsieur Gringet for his asthma.[275]

Drinking mineral waters and bathing at spas were alternatives to the specifics and cure-alls. They gained additional appreciation with the rise of iatrochemistry and improvements in transport in the early modern period. Physicians and patients alike were convinced of their positive effects. Numerous patients traveled to famous baths such as Schwabach, Sedlitz, Ems, Spa, Vichy, Plombières, Forges, and Barèges to drink from or bathe in the curative water. Some patients described how their condition improved significantly after only a few sips of mineral water. In the 18th century, bottled mineral water became more and more widely available and its sale evolved into a flourishing and highly profitable business.[276] Its uses were, at first, purely medicinal. Only later would mineral waters become demedicalized and turn into the common drink we know today.

Depending on its nature or composition, water from different springs was considered beneficial in different kinds of diseases. Choosing the proper spring was therefore essential, much as was choosing the proper medication. In their letter consultations many patients asked their faraway physician which mineral water or bath was best for their particular case.

In a wider sense, bloodletting can also be counted among the home remedies. Rather than consulting a physician, some people seem to have gone immediately to a bloodletter when they were sick and deemed a bloodletting necessary.[277] Many healthy people underwent a prophylactic bloodletting, usually in the fall and in springtime, before and after the cold winter months which, as was commonly assumed, made corrupt and waste matter accumulate in the body. As late as the early 19th century, one of Hahnemann's female patients calculated that she had undergone 36 bloodlettings in 18 years.[278] Only in the course of the 19th century did physicians note a gradual decline in the popularity of prophylactic bloodletting. It became limited to pregnancy, for which healers, including many physicians, continued to recommend it as indispensable for fear that the accumulating menstrual blood would press towards its natural exit and take the fetus with it.[279]

When self-treatment practiced within the circle of family and acquaintances proved unsuccessful or when the nature of the disease was unclear, patients could usually call upon a range of professional healers – and they made ample use of this freedom of choice. The political and legal framework concerning medical practice differed quite markedly from one country to the next and changed over time, which makes it difficult to generalize. On the whole, however, three groups of healers can be distinguished.

Learned physicians were concentrated largely, though not exclusively, in the larger cities.[280] The great majority of them represented 'official' (even though in itself heterogeneous) medicine as taught at universities. A small minority followed their own paths and developed or adhered to heterodox theories and approaches. In the 16th and early 17th centuries, it was mainly Paracelsism that took the spotlight. The late 18th and early 19th centuries then saw a whole range of new doctrines and procedures enter the health market, such as homeopathy, Mesmerism, natural healing and Baunscheidtism.[281]

Even orthodox physicians differed widely in their interpretations and preferred modes of treatment, however, thus giving the medicine they practiced a characteristic flair. Patients' frequent experience of dissenting medical opinions jeopardized the authority of the medical profession as a whole. But individual physicians could in this way secure a niche for themselves in the medical marketplace. When other physicians' treatments failed, their own approach, for the very reason that it was different, held the promise of a radical cure. Théodore Tronchin[282] and to some extent also Samuel Tissot, for example, placed particular emphasis on 'natural' methods of healing like fresh air, milk, and exercise. Others built their reputation on one kind of treatment which they used for almost any illness. The famous French physician Pierre Pomme, for instance, prescribed hours of bathing to most of his patients.[283]

Much more numerous than the learned physicians were the barber-surgeons, who acquired their knowledge and skills as craftspeople in the course of an apprenticeship. Though ordinances sometimes limited their approved area of practice to bathing, bloodletting, cupping, and the external treatment of injuries, ulcers, skin lesions, and the like, they usually practiced all aspects of medicine and gave medication, too. Well into the 19th century, the medical care of the large majority of the population rested on their shoulders. Only from the 18th century, as learned physicians grew in numbers and successfully pushed for professionalization, did the barber-surgeons increasingly find themselves subject to police repression and limited in their field of activity. In the end, regulations and laws in some countries recast them as little more than assistants to the academically trained physicians.[284]

Finally, in towns and rural areas, there were innumerable unauthorized healers or lay practitioners of both sexes. The scope and extent of their medical practice varied widely. They ranged from bonesetters and herbalists to diviners and exorcists. Some covered all fields of medicine, others specialized in specific diseases, diagnostic procedures, or therapies. Some treated diseases only occasionally, in the family and among friends, while others earned their living from their medical work or even acquired considerable wealth.[285]

In most parts of Europe, patients could thus choose from a range of offers, guided by preference and the presumed nature of their disease. And if one healer failed, they could always try their luck with another one. People with serious, prolonged illnesses sometimes consulted a dozen healers, one after the other, in the hope of a cure.

There were, however, marked differences in the extent to which the different sectors of the population consulted different kinds of healer. As early as the 16th and 17th centuries, the upper classes generally preferred learned physicians,[286] and tended to consult them right from the start. Often in upper-class letters an 'ordinary' physician is mentioned, that is, a family physician whose services were 'usually' sought. When the disease was serious or prolonged, other physicians might be asked for their advice as well or the patient might even convene them all at once around his bed.

This upper-class preference for learned physicians cannot be ascribed simply to the superiority of their therapeutic results over those obtained by the use of herbs or prayers or over the natural outcome. From the viewpoint of modern medicine, the large majority of remedies and treatments which were available at the time were useless if not harmful, and it did not matter whether they were prescribed by a physician or by an illiterate unauthorized healer. If many educated people were willing to credit physicians with superior competence, this was due, it seems, above all to their esteem for academic learning in general. Physicians presented their medicine as 'learned' and 'rational' and thus successfully appealed to prevailing upper-class ideals.[287] Especially in the 18th century, consulting a learned physician, for bourgeois patients, could also be a means of social distinction, a way to acquire status.[288] There was prestige attached to having, like kings and princes, an academically trained family physician or personal physician. And the most 'symbolic capital', we may assume, could be acquired by laying claim to the services of a famous and highly praised medical authority – a Bordeu, a Petit, a Tronchin, or a Tissot – or even gathering all of them around one's sickbed, no matter what it cost.[289]

Even the upper classes did not rely exclusively on learned physicians, however. Especially when the physician's therapy did not seem to have the desired effect, wealthy and educated patients, including the royal family,[290] would like everyone else call for another healer or even travel to his home to see him or her.[291] When, around 1600, the serious 'podagra' of former Swabian bailiff Ritter von Trozberg swiftly improved following his visit to a 'miracle-man' who cured only with water and by applying 'blessed herbs', several other high-ranking aristocrats also set off to consult the man about the same ailment. The renowned Jesuit Petrus Canisius explicitly defended the healer against the allegation that his methods were all smoke and mirrors for the superstitious.[292] In the 18th century, Mme de Chastenay – though full of praise for her physicians – was not shy about going to uroscopists a couple of times, if only to find out that their diagnosis was in agreement with that of the most learned physicians.[293] One family from Lorraine which was known by the name of its village, Valdageoux, acquired fame throughout the country for their skill in setting broken bones. When one member of the family successfully treated the fractured arm of the Duchess of Luynes, which had mended badly following treatment by surgeons, this became a hot topic among the higher circles.[294] Members of high society all across Europe are known to have converged at the practice of yet another healer, Michel Schüppach in Langnau (1707–16) in the Swiss region of Emmental.[295] Even in their letters to Tissot, who railed at length against such 'quacks'[296] in his writings, patients and relatives sometimes dared mention Schüppach's name, as in the case of a sick friar who was urged by his fellow brethren to consult him.[297]

Historians of medicine once believed that in the early modern period only a small elite of wealthy bourgeois and aristocrats relied on the advice of learned physicians. A closer look at contemporary medical observations and case histories shows, however, that numerous ordinary craftspeople consulted physicians, too; and physicians' case books –

unfortunately a rare source – confirm this finding. In fact, in numerical terms, wealthy, educated patients often constituted only a small minority of a physician's patients.[298] Though for craftspeople and farmers a learned physician was often not the first choice, they were nevertheless able to afford his fees, and the very poor were, in many places, entitled to free treatment. In other words, although money was an important factor in the choice of healers, it was not the only one. With most physicians working in major towns, the rural population was generally much less likely to call one, as he would have to come a long way to reach their homes and charge accordingly. Usually barber-surgeons and unauthorized healers were much closer at hand and they frequently also had a competitive edge for other reasons. They were prepared to base their advice only on a close examination of the patient's urine – something learned physicians were increasingly loath to do.[299] Anyone – a maidservant, a relative, a friend – could bring the patient's urine. This spared patients expensive house calls; and the fee for a simple urine examination was modest. Though it was generally forbidden, many barber-surgeons and unauthorized healers also handed out medication rather than writing prescriptions to be taken to an expensive pharmacy. As learned physicians complained, many rural people in particular did not feel at ease with them. With their elegant attire, including capes, neck cloths, cuffs, and canes, physicians fashioned themselves as members of the affluent elite.[300] And their perceived close ties with those in power could be a source of distrust rather than respect.[301] Last but not least, learned physicians were considered outright incompetent in the treatment of certain diseases. As late as the 19th century, convulsions and madness, for example, were widely ascribed to supernatural, demonic powers. They called for sympathetic healing or exorcism. The physician's art was deemed powerless in such cases.

The Doctor–Patient Relationship

The history of the doctor–patient relationship has attracted considerable interest in recent years, frequently with a more or less explicit political agenda behind it: the fatherly, sensitive family physician of the past who took care of his patients from the cradle to the grave is contrasted with the hasty and impersonal encounters in modern medical practice increasingly dominated by financial considerations and time constraints.

Such findings do not withstand scrutiny, however. It is undoubtedly true that the structure and context of patient–physician encounters in early modern times were in many respects quite different from the way they are today. Since medical practice at the time – with the important exception of uroscopy – consisted almost exclusively in house calls there was generally much more opportunity for close, personal encounters. It was only gradually in the 19th century that physicians began to set times at which they received patients in their own residences. Before 1850, only a few renowned Parisian authorities like Tronchin and Corvisart obliged even rich patients to visit them. For the patients, this evidently took some getting used to. Marie Victorine de Chastenay, for example, called her experience with Corvisart 'bizarre', because she had to request an appointment on a certain day at a certain time.[302] In Germany, Samuel Hahnemann, the founder of homeopathy, was still trying in vain in the 1820s to dissuade his colleagues from the common practice of house calls – a practice in which, according to his judgment, the physician was always at risk of bowing too low.[303]

Physicians, then, came commonly as guests, as visitors. Having traveled, sometimes for a considerable amount of time, they were not in a hurry to leave again. On the contrary, they seem to have taken their time, making sure to convey the impression that they had carefully considered the individual patient's case. For example, the hypochondriacal bailiff Gringet was greatly impressed with the care taken by the renowned Villermoz, who felt his pulse for more than an hour before drawing the conclusion that he did not have a 'fever'.[304]

Preliminary results of an ongoing research project on pre-modern physicians' case books suggest that even a successful physician might see no more than four patients a day, leaving him ample time to devote to each.[305] In noble Parisian households, physicians could expect a place at the table and took part in games and music-making.[306] Physicians who were called to the castles or mansions of wealthy patients in more remote areas might even stay for several days, sometimes for weeks on end. The prerequisites were thus in place for a personal relationship that could easily transcend the bounds of professional interaction. They were 'first physicians and then friends', wrote Marie Victorine de Chastenay in the 18th century about the numerous physicians who frequented her paternal home.[307] When necessary, some physicians did much more than treat their illnesses – for example, helping their aristocratic patients to escape revolutionary riots.[308] Even in letters to distant medical authorities, patients expressed their desire for a close, personal relationship time and again. One of Tissot's patients explicitly voiced his wish to see him in person, 'more to profit from the pleasure of your company than to tell you about my ailments'. This was because 'I would be visiting Monsieur Tissot, and not my physician'. It would be very nice 'to have, in addition to the esteem and admiration I feel for you, Monsieur, other feelings that arise from a more special relationship, and to be rewarded with a certain reciprocation on your part.'[309] Another patient wrote that he would be happy if he had won a little of Tissot's friendship.[310]

Much more than today, patients could also expect a physician to tailor his treatment to their individual bodily constitution and lifestyle. Patient letters give a good impression of the wealth of detailed information which the patient had to provide for this purpose and which the physician had to gain by asking a series of questions – about the type of meat or wine the patient ate and drank, about his or her reactions to previous treatment, to various kinds of medicine or mineral water, and the like. The physician who managed to convince his patients that he had carefully investigated their individual case and their underlying constitution could forge strong patient loyalty. He knew his patient's body better than any other physician or healer and thus seemed better equipped than anyone else to treat him or her successfully. One patient of Thurneisser desired his advice 'because you know my complexion, nature and illness particularly well'.[311] Another patient said that because 'my physician, who knew my nature, has died' and she 'was as yet unable decide on another', she had for now taken treatment into her own hands with the help of books.[312]

Physicians did not always live up to these desires for individualized care, however, and sometimes left their patients disappointed. The Marquise d'Aglie urgently requested of Tissot that he proceed with her treatment carefully and test its success at every step, 'which the common physicians do not want to do'.[313] Another patient vividly described his visit to the renowned physician Petit. Before the sick man was even able to open his mouth, Petit told him he had nervous complaints. Because he was on the heavy side while his brother, who was with him, was thin and thus seemed more of the nervous type, the patient initially believed there had been a mistake. But Petit had indeed meant him, and did not want to listen to him properly, instead yelling out 'of course – nervous complaints!' He did at least briefly examine the patient's abdomen and assured him he was not suffering from constipation. But because Petit had not let him talk and had concluded merely from his appearance that he had a nervous affliction, the patient did not heed Petit's advice until the counsel of another physician had proven fruitless.[314] Another patient consulted his physician for a jabbing pain in his side. The physician, 'evidently distracted by other remarks', did not take his pain into account. The patient had barely left when his pains began to worsen drastically, leaving him tormented by anxiety.[315]

Contemporary descriptions of the relationship between physicians and their female patients portray it as especially close. Physicians who wanted to build a lucrative practice were well advised to foster such relationships. It was widely held in medical circles that among the upper-classes female patients were frequently the ones who decided which physician should be called to treat members of the family.[316] For some women the physician was the only male interlocutor outside the family to whom they could confide personal matters. In this respect physicians had taken over, to use Monsieur de Levis's words, the role of the father confessor, a role still current in their grandmothers' day at the end of the 17th century. Three-quarters of the time, de Levis said, physicians were called more out of 'luxury' than necessity.[317] Thus, professional success probably hinged as much on a physician's ability and readiness to ingratiate himself with female patients as it did on his reputation for treating patients successfully. Physicians – according to a slightly sarcastic de Levis – had to 'have a sensible heart or the ability to feign one'. They had to 'listen to their patients' lengthy stories while appearing to have the keenest interest in them'. They should neither take their patients' fears too seriously and reinforce them, nor curtly shrug them off as a figment of their imaginations and risk damaging the women's amour-propre or being regarded as unfeeling. 'The trick was to strengthen the courage of these effete souls, prescribing with apparent care some harmless directives that assuage the soul without harming the health; and then, with tactful, light jocoseness, to end the visit, which had initially been consecrated to sensibility.'[318] The famous physician, Anne-Charles Lorry is said to have been a master at consoling and cheering up his patients. He apparently immersed himself so deeply in their hardships that he seemed to share their suffering, describing it as precisely as if it were his own.[319] A deeply moved Mme d'Arblay recounted how 'the good Dr Larrey' had maintained during her long illness 'the warmest friendship'. His eyes, she said, had filled with tears when her cancer was diagnosed.[320] And even the famous Bouvart,[321] whom some described as sarcastic, was said to have cried when the child of the French author Marmontel died under his care.[322]

Important as the personal element in the relationship between physicians and their high-ranking patients was, the images we associate today with a private, personal consultation reflect the historical situation only in part. More often than not the physician was not alone with the patient when he made house-calls. He had to deal with the patient's family and servants, and often with visiting family members and acquaintances as well. For physicians, this had significant consequences. They had to convince not only the patient but everyone else, too, that their diagnosis was accurate, their interpretation of the underlying disease process correct and their treatment best suited for the patient and the disease in question. This could demand considerable rhetorical skills and the situation has aptly been compared to a theatrical stage, on which the physician had to put up a convincing 'performance' in order to win the trust and favors of his 'audience'.[323]

Research on the relationship between physicians and patients in premodern medicine has delved deeply into the question of power relations, of dominance and subordination. The discussion has largely been sparked and fueled by two contributions by British sociologist Nicholas Jewson. Drawing on English sources, Jewson came to the conclusion that the physician–patient relationship of the 18th century was characterized by 'patronage'. According to Jewson the physicians' clientele – predominantly, as he thought, aristocratic and very limited in number – generally ranked higher on the social ladder than the physicians themselves. Since the physician's economic and professional prospects hinged on these patients' favor rather than on his colleagues' esteem, Jewson argued, he was forced to accommodate the preferences and desires of his patients as much as possible. Because medical research was then still largely carried out by individual practitioners, this, according to Jewson, also had important consequences for the development of medical science. Only those innovations that were well received by rich patients, which were in line, that is, with their expectations and preferences, were able to gain recognition.[324]

Jewson's theses have met with some criticism. They have been corrected and amended, particularly as concerns the situation outside England. It has been shown that the relationship between early modern patients and physicians can be described as one of 'patronage' only to a very limited extent. As we have seen, in fact, the majority of patients, in particular on the Continent, were of the same or even a lower social status than the physicians. Jewson is certainly right, however, when he claims that the individual physician was in a rather precarious situation with respect to his patients and that their relationship was much more symmetrical than it is today. With the exception of the personal physicians of kings and princes – whose position was in certain respects similar to that of a lackey – this was not because the patients were in a superior social and economic position, however. The patients' wishes and desires carried great weight simply because there was usually a wide range of other healers and healing practices at hand which they could easily resort to if they were not happy with a physician's advice or treatment.

As the patient letters make abundantly clear, most patients were certainly not prepared to follow their physician's orders without question.[325] On the surface they usually signaled complete obedience to the 'divinely gifted' healer or to the 'oracle'[326] of the famous medical authority. They admitted that 'the patient should follow his physician'[327] and many – though not all – of Thurneisser's patients, and likewise Hahnemann's later on, were ready to take medication in blind faith without knowing what it contained. Yet most educated patients made it clear that they did not unreservedly accept the physician's authority. They rejected what did not suit them and clearly expressed their own wishes. She desired on this occasion to take only a purgative and no further medicine, wrote a nun from Maggenau in 1613 to the 'highly learned' Doctor Schobinger.[328] One clergyman was adamant that in his case 'a bloodletting would be necessary'. He only wanted to know how – that is, presumably, from which vein – and on which day this should best be done.[329] Hans Georg Reinach asked Felix Platter 'to prescribe a purgation after looking at my urine.'[330] Nearly 135 years later, a 51-year-old major wrote to his physician, 'I expect of you, Monsieur, a sure specific to strengthen my nerves.'[331] Alongside these explicit requests were questions which, if answered in the negative, demanded that the physician justify his standpoint. Wouldn't this mineral water or that spa be highly useful in their case? Might not 'for instance, a good bloodletting be helpful'[332] or was it not time to take a laxative?[333]

In their initial letters, most patients did not immediately express therapeutic preferences. But once the physician had given his advice, many started arguing and put forward reasons why they could not follow the physician's instructions. They refused to take a medicine because they feared it would be hard on their stomach. Or they inquired about a preparation before taking it. One of Thurneisser's patients, for instance, could not understand from the enclosed slip of paper 'what is in the two little bottles that you, Sir, have honored me with, what it is called, and to what end I should actually take it.'[334] Or they reduced the dosage because they held a lesser one to be sufficient.[335] Or they now made their own wishes clear, asking instead of, say, a sarsaparilla decoction, for 'a diet drinke brew'd with malt and what drugs you think proper'.[336] Or they briefly attempted the prescribed treatment only to report that, for instance, the baths had agitated their nervous system so unbearably that they were forced to cease the therapy.[337]

Those patients who had followed the physician's instructions for some time were even less apt to mince their words when the desired recovery was a long time coming. She had 'most diligently' taken the medications, and done 'what you prescribed to me but I have nonetheless found no improvement whatever', complained Anna von Bradowa to Thurneisser in 1574. She had not wanted to 'keep this from you as you are my very good friend', and she hoped the dear Lord would 'be merciful and help'.[338] 'I have adhered strictly to your orders', wrote Freifrau Louise von Werthern 135 years later to Tissot, but 'unfortunately they in no way produced the effect I was expecting'.[339] 'Your pills have done absolutely nothing to calm the cough', wrote one clergyman impatiently to Verdeil; and it had already been eight weeks since he had asked for 'the true medication' for it.[340]

If a patient's condition worsened rather than improved during treatment, it was almost inevitably the physician who got the blame. One older female patient complained that an overly vigorous bloodletting had enfeebled her and that all the organs of her body had been affected.[341] Similarly, one of Herman Boerhaave's female patients feared that too much blood had likely been taken from her, 'which is why I now risk having dropsy in the belly and limbs'.[342] One 20-year-old male student suffered for two months from a 'black melancholy' after his physician had diagnosed constipation and given him powerful purging agents; the treatment, he believed, had dried him out and made his blood viscous, as was typical of melancholy.[343] Another patient started having severe nervous attacks and even felt 'a kind of sand circulate in her vessels' after a physician had prescribed the – for her case – too powerful Seltz water. 'Consider, Monsieur, whether it is right to insist', her daughter added.[344] Others did not even ask such questions. Acting on their own authority, they discontinued the use of medications with which they found fault[345] or they left out certain ingredients – guaiacum wood, for instance, because it heated and dried out the body.[346]

It took a good degree of tenacity for a physician to insist on his orders being followed amidst such complaints and objections. A number of physicians asserted themselves over the declared desires of patients and their relatives, and their patients accepted this, at least for a time. Occasionally physicians would also find support among relatives and friends who pushed the patient to heed the physician's advice. Generally, however, the physician who ignored his patients' notions and requests was taking a considerable risk. Any cure that failed to deliver the desired results called his authority into question; but if, on top of that, the physician flouted the express wishes of his patients and their relatives, he carried full responsibility and could expect serious repercussions if things went badly. One patient of the renowned London physician Hans Sloane desired never to see him again after Sloane had prescribed laxatives that failed to produce a single bowel movement but caused intense cramps. 'In his ravings', reported the man's brother, 'he gives you very hard words, for attempting to purge him with the waters, which, he sayth, he told you would not do'.[347] Physicians' careers at the time could be built on a handful of sensational recoveries,[348] but a physician's reputation was ruined just as quickly. The most celebrated of authorities could become the target of vitriolic criticism. Even the renowned Cabanis was not spared accusations of ignorance and charlatanism,[349] and the fashionable physician Bouvart was described by Geneviève de Malboissière as a downright 'murderer' following the unsuccessful treatment of her fiancé; she would have liked to have him thrown out the window, she wrote.[350] Charlotte d'Orléans even held the famous royal physician Fagon responsible for Queen Marie Thérèse's death in 1683: 'Our Queen died from an ulcer under her arm; instead of pulling it out, Fagon let the Queen's blood (it was the greatest mis-fortune that he was her doctor at the time), which made the ulcer burst inside her, and everything went to the heart, and the emetic that he gave her in addition suffocated the Queen. The barber who let the Queen's blood on Fagon's demand said to him: "Monsieur! Consider it well, it will be my mistress's death."'[351] Presumably Charlotte, like the barber, was convinced that it was necessary in such cases to support Nature's effort to evacuate the morbid humor through the festering ulcer and that bloodletting, on the contrary, would pull it back inside, towards the vital parts.

Such extreme accusations throw light more generally on the com-plex mix of confidence and distrust that surrounded the early modern physician–patient relationship. In many cases the patient got better and trust in the physician thus seemed to be justified. As pointed out in the introduction to this book, we must not be misled by our modern, rather negative, assessment of early modern therapy. Of course, by the standards of modern medicine most of the medications and procedures of the period were without effect and some of them were harmful, like administering mercury compounds for skin alterations or powerful lax-atives for diarrhea, or bloodletting to treat heavy uterine bleeding. To most people at that time, however, not only did these treatments make sense, but they constantly experienced what they took as proof of their effectiveness: many patients got better – as indeed most patients do, especially those with acute or minor ailments, no matter what treat-ment is applied. It was only because they were convinced that the phy-sicians had the means to fight disease successfully that they were ready in many cases to undergo disagreeable and painful treatments. Their letters tell of dozens, at times hundreds, of bloodlettings, of powerful laxatives and emetics – which had them running to the toilet every half an hour – to drive out the presumed morbid matter, of artificial ulcers that would exude sanious liquid over months and years. Indeed, some explicitly called for a massive and drastic treatment. No 'paltry medications' but rather 'strong correctives' were desired by the husband of an English woman suffering from dropsy.[352] Because, for most lay-people, the primary goal of a successful treatment was the elimination of the presumed morbid matter, a good laxative was one that emptied the body powerfully and repeatedly. Thus the brother of the sick Lord Hatton praised Hans Sloane's 'purging physick', 'which hath wrought very well with him as they say here that is often'.[353] And contrariwise, another man fiercely reproached his physician for being overly timid. The physician had foregone a third bloodletting, thus permitting the formation of a 'deposit' in his legs, which had then dissolved, accompanied by severe general illness.[354]

Physicians also had to beware of hasty prognoses, especially negative ones. When J. H. Hummel fell very ill in his student years, his physician, on the basis of a uroscopy, came to the conclusion that he would not recover; Hummel lived to the age of 63.[355] Triumphantly, Monsieur Decheppe recounted how, unimpressed by his surgeon's warning that one jar of a certain medicine would kill him, he had taken nine jars and with good results.[356]

Even in what were seemingly desperate cases, patients and their families often maintained the hope that a disease might be cured or at least arrested if only the right treatment could be found. And so, when the desired results did not materialize, many patients and their relatives usually did not seek the cause in the severity of the illness, in divine will, or in the given limits of medical treatment. They suspected first of all that it was the physician's fault for not giving them the best possible treatment. They had reason for their doubts. Again and again patients and relatives encountered extreme discrepancies between the diagnoses and therapeutic recommendations put forward by the physicians they asked for advice. P. D. Steelant in 1614, for example, lamented that he did not know who to believe as the different physicians' recommendations for treatment contradicted each other.[357] The chronically ill, who often consulted a fair number of physicians in the course of time, related especially bitterly – indeed despairingly – how they were constantly confronted with new and contradictory diagnoses and prescriptions. For all the money they spent they only got worse. Physicians, complained the Augsburg merchant Lucas Rem, were unable to agree whether his disease derived from excessive fluxes, as he himself believed, from other natural causes, or from poison, or whether a woman had cast it upon him; one physician would 'say one thing, another something else'.[358] Similarly, one of Tissot's patients complained that physicians had him stumbling 'from one conjecture to the next'. 'At one point everything was taken for chest ailments, and then for digestive weakness, and then it was all supposed to be sanious sputum, then tubercles, then a 'virus' (meaning poison), and then hemorrhoids.'[359] One Dutch patient had already consulted at least 10 physicians and 'each one started from scratch with me'.[360]

In the end, some patients were no longer willing to accept guidance from any physician. According to one of Thurneisser's patients, one physician thought he had overheated his stomach with wine and fouled his brain with acrimonious phlegm, while another claimed his stomach was not overheated but rather the illness was caused more by cold moisture in the stomach. Yet another physician said it was a kidney stone, while a fourth said there was no stone. Since these 'humoristic' physicians in their 'judgments do not agree with one another' but rather 'utter contraria', he no longer wanted 'to submit himself to their uncertain art'. He himself believed he had 'severe tartar'.[361] Similarly, Guido von Boetzelaar related in 1593 how one physician thought he had intestinal colic while another suspected a diseased spleen. But the location of his acute pain and the palpable hardening above his navel, he thought, disproved both theories. He himself believed it was rather a stomach problem and requested an appropriate treatment.[362]

One reason for such contradictory medical judgments – as laypeople were well aware – was the growing heterogeneity of contemporary medical theories. Physicians resorted to different theories and therapeutic rationales and thus inevitably came up with different results. Already within traditional Hippocratic-Galenic medicine, there were manifold discrepancies. In the 16th century, the Paracelsians developed a radically new alternative to 'orthodox' medicine. This pluralism became all the greater in the late 17th and early 18th centuries.[363] At the bedside, traditional humoral notions of morbid matter and healthful evacuations retained paramount importance, but the theoretical framework surrounding such ideas became more and more diverse. Mechanistic and Cartesian physicians understood the body as a kind of hydraulic machine and focused above all on the interaction of fluids and fibers. Helmontians and iatrochemists saw everywhere acrimonies and alcali, fermentation and effervescences at work. In Germany, Georg Ernst Stahl and his followers elaborated on traditional notions of the body's natural healing powers and attributed all disease to a disturbance of the guiding influence of the soul. In the course of the 18th century, the new concepts of nervous sensibility and irritability took the limelight and many diseases were attributed to the 'nerves'.

For patients, this coexistence of different theories and practices could open up an array of options. When he broke out in brown spots all over his body, wrote Georg Stange of Magdeburg in 1571, he had asked for advice from 'doctors'. They claimed the problem came from drinking and a hot liver, and that the spots were called 'liver spots'. But he knew that 'the advice or the medicine of the Galenists is not to be heeded, and less than nothing has ever been achieved because they use no quintessence whatsoever, nor do they know how to prepare it.' For these reasons he was now turning to Thurneisser.[364] Most patients, however, seem to have found the coexistence of various schools of thought confusing and responded with skepticism. Some patients in the 18th century even stated explicitly that they preferred to write their medical histories themselves, because they were concerned that a physician might immediately fit the history into his own system of thought instead of providing a truthful account. A related worry was that a doctor might rashly reach for a proven standard medication without considering the precise nature of the complaint or the patient's individual constitution and way of life. Since the treatment prescribed by the physicians had 'not yet taken any effect, and as they have no sufficient reasons to offer', one of Hoffmann's patients doubted that 'they have adequate knowledge of illnesses and [I] instead suspect that all they do is cure what they know from experience, while not weighing all circumstances enough.'[365] Apart from this, the patients and their relatives took the individual skill of the physician, his experience and power of judgment, to be decisive for therapeutic success. For this reason, patients and relatives made a considerable effort to obtain reliable information about individual physicians and about the cures they had performed. Their skills were thought to differ so widely that it would have seemed naive to simply have faith in the next best physician. A fair measure of skepticism was in fact crucial for survival.

To sum up: the early modern doctor–patient relationship can neither be sweepingly characterized as a relationship of patronage nor be seen as the expression of an increasing disempowerment of patients by an omnipotent medical profession which medicalized every aspect of human life.[366] Rather, what emerges is a complex interplay of claims to validity and self-fashioning, of attempts to influence and sanction, of domination and subordination, of glorification and condemnation, which lent the doctor–patient relationship a different form in each case. Yet it can generally be said that patients of the time tended to be in a significantly stronger position than they are today. They were able to channel their physicians' decisions in a certain direction. Physicians for their part saw themselves obliged to take patients' wishes into account and to justify their own diagnoses and treatments. Otherwise, the patient was likely to consult another, more accommodating, physician or healer.

It was usually not power that patients sought, however, but a successful cure. Patient letters time and again document the longing for a medical luminary, for a genius who, thanks to divine inspiration or an infallible power of judgment, would finally guide them in the right direction. Even the triumphant words of Mme de Graffigny belie a precarious balance between confidence in her own judgment and a desire for a medical authority beyond all doubt and skepticism:

'To my satisfaction, everything I imagined for myself has so far gone well, since it was again I who asked for the whey. The physician discusses things for a while to keep his honor. I, who only listen to straight talk and reason, push him into a corner. He concurs – and the effect never fails to materialize. If only he had let me take the [mineral] waters earlier, I would have saved myself 15 days of stomach cramps. Oh, how I feel confirmed daily in my certainty of the foolishness of these beasts, and how unhappy one is in their hands, without light and without knowledge of the remedies. I know very well that we have to listen to their reasoning and that there are some cases in which we don't see any clearer than they, but at least we have to counter their reasoning with what we feel and make the choice that seems most appropriate to us.[367]

While the prevailing forms of interaction between learned physicians and their educated patients emerge quite clearly from the sources, thanks, above all, to the patient letters, our knowledge of the relationship between lower class urban and rural patients and their physicians or healers is only fragmentary. It is above all in cases of conflict that it comes to light in the archives: when there was a dispute over an unsuccessful attempt at healing, for example, or an unpaid bill. Otherwise we have to rely on the often one-sided polemical accounts of learned physicians. From these sources, it emerges that less-educated patients also met with medical professionals eye to eye, with self-confidence. This was much to the chagrin of academic physicians, who decried the 'arrogance' of common country folk, who failed to give them the unconditional respect which they felt they deserved in the light of their superior, erudite knowledge. Learned physicians were particularly enraged by the stubborn insistence of the illiterate majority that they diagnose diseases just from their urine without seeing the patient and indeed without any further information about the patient's complaints and history. Due to their precarious position in the medical market-place, many physicians found they had little choice, however. If they refused, the patients would simply consult a barber-surgeon or an unauthorized healer, who were renowned for their uroscopic skills.[368]

Financial considerations also had a considerable impact on the relationship between physicians and lower-class patients. Patients who spent the little money they had on a learned physician's advice expected something tangible in return. The degree to which patients and relatives perceived medical consultations in terms of a financial transaction or trading becomes clear in cases in which they refused to pay because the treatment had not been successful. For instance, the maidservant Lucia Heppenstreit brought an action against a barber-surgeon in Cologne in 1750 because he demanded payment in spite of a failed treatment.[369] And the sister of a patient who had died rejected the demands of another healer, saying the man did not deserve anything, since he had forced his assistance upon them and had wrongly assured the sick man 'that he would cure him'.[370]

Part II
Perceptions and Interpretations

Severe illness, as we have seen, is an existential challenge in many respects. The taken-for-granted vigor and vitality of one's body is lost. One's whole life appears in a new light. The familiar world seems to falls apart. Across cultures, the experience of severe illness spawns a deep desire for meaning and orientation. Philosophical and religious interpretations of illness, as a divine admonition or a test of piety, for example, or as a chance to rethink one's aims and way of life, can be helpful in such circumstances. At the same time, however, patients and their relatives wish to understand the nature of disease – or at least of the one that afflicts them – more concretely, in physical terms. The early moderns were no different in this respect. They wanted to grasp what was going on inside the body. They wanted to know what caused their complaints, what laws governed them and, of course, how they could be successfully fought. They sought, in other words, a 'natural', 'medical' explanation – which raises, first of all, the question as to which medical theories and explanatory models they could resort to.

Medical Popularization

In the early modern period, lay notions of the body and its diseases were tightly intertwined with the theories of learned physicians. In a concrete case of illness, patients and physicians might hold different opinions, and diagnosis and therapy were frequently a matter of negotiation between the physicians on the one side and the patients and their relatives on the other. But to a large extent, physicians and patients lived in a shared medical world. They spoke a common language and moved within the same humoral explanatory framework. This raises the questions as to where laypeople obtained their knowledge of medical theories and how they applied it.

In recent historiography, the popularization of scientific and medical knowledge has attracted considerable attention. This attention is due in part to the mounting interest in the various aspects of what has come to be called 'public understanding of science'. In the history of medicine, studies have so far concentrated almost exclusively on investigations of 'popular' advice literature and its authors.[1] Analyzing such published texts is undoubtedly appealing. They contain valuable information about the kind of reader their authors anticipated and hoped to attract and about the image they aimed to convey of the significance of medical knowledge and practice. Such writings also ranked among of the most popular literary genres on the book market in general. Some of them were bestsellers and went through numerous editions and translations, and their authors became famous throughout Europe.[2] But even a close reading of these texts provides little information about their impact on medical lay culture, on the ideas and health practices of the wider public, and about their significance in comparison with other sources of lay medical knowledge.[3] In this respect, personal testimonies and above all patient letters are a much richer source, especially for the wealthier and more educated classes to whom such works – despite titles like *Advice to People* – were primarily addressed. In addition, important clues to the way in which such works were read and used can be gained by a systematic scrutiny of extant copies of one and the same widely known work.

Dozens of patient letters, especially in the 18th century, mentioned 'popular' medical texts. In all likelihood, many other patients who consulted by letter had read health advice books, too. Patients usually only referred to works which had been been written by the doctor to whom their letter was addressed. Most references to such works suggest that they were used in a specific way. Rather than reading them from cover to cover, people seem to have browsed them for advice on how to diagnose and treat certain diseases as they occurred. 'I read', wrote one patient of Tissot, 'I checked, I went through all the diseases of the chest; I thought I had found my disease in the obstructions and took this as the starting point to becoming my own physician.'[4] A priest in St Malo concluded from his reading of Tissot that his fainting spells came 'from the nerves'.[5] And for another man who suffered from anxiety attacks, Tissot's work on the diseases of scholars[6] put an end to years of uncertainty about 'the state I was in, which was quite strange indeed'.[7] Repeatedly, patients had, on the basis of such texts, initially attempted to treat their illnesses themselves before they consulted a physician – usually the author – because their efforts had not been successful. Some of Tissot's patients even gave references from a numbered list of medications (or concoctions) that Tissot had included in his *Advice to People* to indicate which of them they had tried.

Patient letters may, of course, give us a distorted picture. In general they were ultimately motivated by a particular disease for which a diagnosis and therapy was sought. Without a doubt, there were individual readers who did not only read up on specific diseases but attained quite a broad range of medical knowledge from such texts. Some even gloated about their medical knowledge and skills, such as Mme de Maraise, who 'with my Tissot in hand' examined and treated patients and thoroughly discussed medical issues in her letters as well.[8] Others also reported how they had read a certain book for guidance on how to treat patients with various diseases. The Comtesse de Vougy, for instance, expressed her gratitude to Tissot because she had, with the help of his *Advice to People*, successfully treated sick people on her estates and how happy it made her when these poor people came to her and told her how she had saved their lives.[9] Most patient letters and the markings or marginal notes in extant copies of health advice books suggest, however, that they were often primarily read in relation to particular cases and for practical purposes rather than theoretical edification. In one French copy of Tissot's *Advice to People* in Munich, for example, only two notes in the margins are found, both in the section about diseases of the throat. One says 'this, I think, is the disease', and the other, shortly thereafter, 'and this the remedy'.[10] Similarly, in a copy of the German edition of the same work, the numerous notes are concentrated solely in the chapter about heat stroke.[11]

By all appearances, even these copies are exceptional. It is striking that most of the circa 80 copies of various editions of Tissot's works that I have seen[12] exhibit hardly any signs of use at all; they often seem new. This is no conclusive evidence but it supports Roy Porter's suggestion that ownership of such books may have had above all a ritual, a psychological, or indeed a talismanic significance.[13]

At any rate, even if we grant that such works were sometimes studied and put to use in a thorough and systematic manner, my limited survey of extant copies suggests that they were not by any means the only source of lay medical knowledge. Indeed, scholars have probably grossly overestimated their influence on and significance for popular medical culture. First, many members of the educated classes at whom 'popular' advice books were primarily aimed were quite able also to access academic medical literature. They could read and understand Latin without much difficulty; in their letters, some patients even cited passages from scholarly Latin treatises.[14] Even more importantly, medical knowledge was more efficiently passed on via channels other than books. In all likelihood, the most significant medium by which medical knowledge was disseminated was not the written but the spoken word. Each time attending physicians or surgeons proclaimed a diagnosis or recommended a treatment they also communicated, implicitly or explicitly, their understanding of the body and its diseases. And apparently they often explained in detail what they believed was going on inside the patient's body. Certainly physicians' written responses to patient letters often contained elaborate explanations of the disease in question, which showed at the same time why the prescribed treatment was necessary. This was an important strategy in the physicians' ongoing endeavor to showcase their skill and expertise and distinguish themselves from their less learned competitors.

The physicians' explanations were underscored by the way they diagnosed and treated diseases at the bedside. When, for instance, a physician prescribed a bloodletting or even showed the patient how his blood was 'inflamed' or covered with a layer of 'mucus' or 'acrimony', he lent a degree of self-evidence to his conceptions of pathological heat and morbid matter that made them almost unquestionable: patients could see the truth of the physicians' words with their own eyes.

The comparably 'public' character of illness – the presence of family, relatives, and friends in the sickroom during a medical practitioner's visit – created ideal conditions for the dissemination of medical ideas and practices far beyond the narrow circle of the physician–patient relationship. Disease was an important and popular topic of conversation across all social classes. As patient letters and physicians' case histories alike indicate, those who witnessed in person what physicians (or other healers) did and said would usually tell it to others, who would often spread such hearsay even further.

The Cultural Framing of Disease

Many of the notions and explanatory models which laypeople used at the time to grasp and understand their diseases corresponded, at least in their basic outline, to those of contemporary learned medicine. Physicians and patients, particularly those from the upper classes, lived in a shared medical world. This does not mean, however, that they held identical views. Physicians today are increasingly coming to acknowledge that medical 'lay theories' are often at odds with their own views and that this is a major reason for the widespread phenomenon of noncompliance. Patients will often not follow the physician's prescriptions or advice if they have a different view of their disease and do not accept the physician's rationale for prophylaxis or treatment.[15]

In the early modern period, the differences between learned and lay theories were more blurred than today. In most cases, physicians and laypeople did not hold entirely different, incommensurable views. The differences related more to the uses to which the theories were put. Learned physicians sought to understand down to the smallest detail the signs and symptoms of diseases and the underlying pathological changes in the body. To do so, they developed complicated and frequently contradictory theories and models and availed themselves of predominantly Latin terminology. Laypeople borrowed from this terminology and conceptual framework, but their approach was more pragmatic, more goal-oriented and more eclectic, particularly when they themselves were sick. For them it was often sufficient to identify the general nature of the disease and to arrive at a plausible image of the underlying changes in the body, without the complicated pathophysiological theories and manifold distinctions of types and subtypes of diseases that can be found in the medical literature of the day.

For the historian, this difference in perspective has important methodological consequences. Reading the medical literature of the day or even the relevant entries in 'general' encyclopedias – which were also, for the most part, written by physicians – provides us with only a partial and distorted picture of how patients perceived and explained illness on a daily basis. It is essential to look also, and as directly as possible, at what laypeople themselves had to say.

The terms, images, and ideas that early modern laypeople (and physicians) connected with the body and its diseases and which I will take a close look at in the following have in certain respects become foreign and even exotic to us today. The body as we encounter it in the patient letters and other personal testimonies, as well as in the writings of scholarly physicians, was subject to very different laws than those we take as given today. The body was, for instance, much more permeable and determined by the constant movement of fluids, spirits, and vapors through it rather than by processes and structural changes in its solid parts, in the very substance of the individual organs. This very different conception of the human body, which, in the eyes of contemporaries, was based on irrefutable evidence, makes historical analysis difficult and intriguing at the same. We must take the early modern theories and images of the body and its diseases seriously in their own right, if we want to understand how they worked and why they weathered social and cultural change for many centuries. We must do justice to the coherence, the inherent logic and rationality of these theories. But in order to make this strange world comprehensible to modern readers, we also have to translate those images and ideas into our own language. This is particularly difficult in this case because we tend to see the way we perceive and experience our bodies as something universal, naturally given – even if we are, in theory, aware of the profound influence of culture.

Approaches from cultural anthropology or ethnomedicine offer helpful methodological and conceptual tools here. The encounter with medical conceptions and practices that depart radically from those of modern Western societies is, of course, at the very core of cultural anthropology. Even what counts as a 'medical' problem and what does not varies markedly from culture to culture.[16] For the analysis of lay medical theories the concept of 'semantic networks', borrowed from cultural anthropology, proves particularly useful.[17] The concept reflects the insight that the terms, images, and ideas which members of a certain culture use to describe and understand the body and its diseases are often only loosely tied together, based on analogy and associated imagery rather than strict deductive reasoning. They constitute a medical cosmology which leaves ample space, according to the epistemological standards of modern science, for inconsistencies and contradictions. 'Semantic networks' as Thomas Lux described this approach are not necessarily built logically. Different, sometimes contradictory, concepts and explanatory models can be applied simultaneously to the same phenomenon with no urgent need to resolve the tension between them. Rather than seeing medical cultures as a closed, rigidly structured system of thought, the concept of 'semantic networks' understands the terms, images, and concepts at the core of a medical culture in terms of a collage.[18] They provide a reservoir from which individual elements can be more or less freely combined to make sense of diseases, bodily experiences, or medical practices.[19]

An important consequence of the 'fuzzy' structure of medical 'semantic networks' is that the resulting perceptions, interpretations, and explanations cannot conclusively be refuted in the same way in which scientific theories as a whole can be shaken by specific contradictions or anomalies. Empirical evidence or theoretical arguments which call into question individual elements of the 'semantic network' or fill them with new meaning can leave the validity of other elements largely untouched. The interpretational framework remains intact.[20]

It is for obvious reasons that the model of the 'semantic network' or 'collage' proves particularly valuable for the analysis of early modern lay medical theories of disease and ultimately for the theories of learned physicians as well. The medical ideas and practices in early modern Europe had melted together over a long period. They were a conglomerate of images, ideas, and practices which had formed over more than 2,000 years. Especially in the Middle Ages, Western medicine was on the receiving end of transfers from other medical cultures – most pervasively from the Persian and Arabic world. More often than not the new images, notions, or explanatory elements which enriched the existing framework of learned medicine – like 'contagion', 'diseases of the whole substance', and 'quintessence' – did not simply replace previous explanations but extended physicians' interpretative repertoire. In this sense, early modern medical culture can be best understood as a multi-layered collage or patchwork whose components originated in different historical periods but nevertheless, from the contemporary perspective, formed a coherent whole.

At the same time, the 'semantic network' approach does justice to the often considerable individual variation in the interpretation and experience of similar episodes of disease within one and the same culture. Even when the symptoms seem almost identical, different patients may select different images and fragments of knowledge to connect them to a specific, sometimes idiosyncratic, whole, in order to make personal sense of their experience, reflecting their individual preferences, their personal history, and the particular circumstances of their lives.[21] Like the cultural anthropologist, the historian thus cannot describe the early modern experience and interpretation of a disease like 'dropsy' or 'cancer'. His or her task is to reconstruct the rich vocabulary of disease, the wide repertoire of explanatory elements, images, analogies, and practices from which the patients and their relatives could choose in many different combinations in order to arrive at their specific understanding of the disease in question.

With this in mind, I will present in the following the more common and widely used explanatory concepts which framed the experience and interpretation of disease in early modern lay medical culture, and the manifold links between them. Obviously, I cannot claim to present all the medical images, terms, and concepts that were meaningful for the lay public at the time. Even some quite interesting but less common concepts like 'pica'[22] and some of the rare or regional diseases such as 'plica polonica' or 'nostalgia'[23] will hardly be mentioned. Nevertheless, an overview of the most important terms, images, and explanatory elements within early modern lay medical culture should be helpful not only for medical-historical studies in the stricter sense. Students of general social history, the history of literature and the arts, and even the history of philosophy frequently encounter early modern notions and concepts of disease in their sources – and will, without proper guidance, often misunderstand their specific contemporary meanings and connotations.

From Temperament to Character

Even very basic early modern notions of disease and the body have often been misunderstood and misrepresented. One of the most persistent and, it seems, almost ineradicable errors is the assumption that early modern medicine generally attributed disease to a disturbance of the natural equilibrium of the four humors, that is yellow and black bile, blood, and phlegm, and that medical treatment generally aimed at reestablishing this equilibrium. 'To become healthy again', as one recent – and otherwise very laudable – overview of early modern medicine puts it, 'it was necessary within the doctrine of humors to regain the lost balance of the humors'.[24] This description is not entirely incorrect, but it reduces the range and complexity of early modern conceptions of the body and its diseases to a single aspect – and in no way to the most important one. The notion that disease was above all the consequence of a disturbed equilibrium of qualities or humors was developed in ancient Greece and remained of great importance until the late Middle Ages. In early modern medicine, however, diseases were only rarely explained in this manner. Other disease concepts had come to prevail.

The notion of an equilibrium of the humors was now largely restricted to the notion of 'temperament'. Developed in antiquity, the term 'temperament', as well as its Greek counterpart '(eu)krasis', originally referred to a specific, healthy mix of humors or qualities.[25] Depending on the prevailing humor, a healthy person could have a choleric, melancholic, sanguine, or phlegmatic temperament. If, say, black bile and phlegm were predominant, then the temperament would be melancholic-phlegmatic and so on. Two qualities were assigned to each humor: blood was warm and wet, yellow bile warm and dry, black bile cold and dry, and phlegm cold and wet. Hence each temperament was linked to a specific combination of the respective qualities, resulting in the so-called 'complexio'. The choleric person was fervid and dry, the phlegmatic person cold and moist, and so forth. The material subject or carrier of 'temperament' and 'complexio' was initially the human body as a whole. Each person had his or her own temperament in which one or the other humor or one or two qualities preponderated. This preponderance – for instance of phlegm in the phlegmatic – was not considered pathological in itself. It only indicated the basic constitution of the person in question and at most signaled a predisposition to pathological deviation in a specific direction. Thus, according to ancient medicine, the phlegmatic was particularly susceptible to illnesses that entailed an excess of mucus or that were characterized by coldness and moistness. The choleric, on the other hand, was more threatened by bilious, hot, and dry diseases.[26] Not only the whole body, but the individual body parts and organs as well each had a particular temperament befitting their respective task. Thus the 'slimy' brain tended to be determined by cold and moist phlegm. The liver, on the other hand, where blood was concocted from chyle, was rather warm and dry.

People in the early modern period were familiar with the doctrine of temperaments. Knowing one's own 'nature' or 'temperament' was considered by many to be an indispensable prerequisite to a well directed diagnosis, treatment, and/or preventative measure. Thus it is not uncommon for patient letters to begin with an indication of the temperament as, for instance, 'sanguine', 'choleric', or 'phlegmatic-melancholic', or as 'dry', 'heated', and/or 'fervid'. In this vein, one of Thurneisser's patients said he was 'around' 42 years old and that he was 'a phlegmatic and melancholic'.[27] Others expressly requested that Thurneisser determine their 'nature and complexion' from their urine,[28] or they trusted that he already knew their 'complexion, nature and characteristic', and would send 'a good preservative according to my nature and complexion' or would treat them accordingly.[29]

People knew how to identify the different types of temperament. A sanguine temperament revealed itself in a patient's lively facial appearance, for instance, or increased sexual desire. The hot, dry yellow bile of cholerics inclined them toward fits of rage. Her own defect, wrote one of Hoffmann's female patients, was that 'I easily get irritated, so it could be that bile has entered the blood.'[30] Like physicians, laypeople also accorded the individual organs a particular natural temperament, which was sometimes subject to pathological change. Isaac Keller, for example, explained to Thurneisser that he had a 'heated complexion' and, particularly, a large 'heated' liver, which was, in his opinion, why he had a red face.[31] Another patient related that a treatment had, thank God, brought his brain 'back to a good temperament' and that he no longer had 'such heaviness from the catarrhs'.[32]

As early as the 16th century, however, 'temperament' was increasingly used in a wider sense. It was no longer used only to describe a peculiar, idiosyncratic mix of humors or qualities but became largely synonymous with 'nature'.[33] 'Temperament' now referred more generally to characteristics and traits which were once ascribed to the idiosyncratic balance or mix of humors and qualities. And eventually the notion of 'temperament' as such lost importance. More and more letter-writers mentioned their temperament as one of several aspects of their individual constitution. 'I am, first of all, around 40 years old', wrote one person, 'am physically fat and stocky, of phlegmatic and sanguine nature, and have a large belly'.[34] Others described themselves as having a large frame or a wide chest, or as chubby, and sometimes even the color and thickness of their hair was considered worthy of mention in this context.[35] At times, the terms 'temperament' or 'complexio' were even used in a manner which was incompatible with the basic assumptions of the four-part system. Some patients, for instance, described their 'temperament' as sanguine-bilious-phlegmatic, which according to the traditional understanding made little sense, because the body could hardly be dominated by hot and dry (bile) and cold and moist (phlegm) at the same time.[36] Others used 'temperament' only to generally characterize their constitution or bodily build, with no reference to humors or qualities whatsoever. In this vein, a 'meager' temperament is mentioned,[37] or a 'quite constipated' temperament,[38] or a patient thought that her 'temperament' had once tended toward diarrhea.[39] Still others equated 'temperament' with physical verve or power. They considered their temperament to be 'powerful',[40] 'strong', or 'energetic', or 'weakly', 'exhausted', 'delicate', or even 'effeminate'.[41] As early as at the end of the 17th century, Johann Valentin Andreä described his temperament as 'weak, unable to bear the cold but able to endure work'.[42]

In the 18th century patients in their letters started to talk about 'temperament' in today's sense of the word, that is, primarily as an affective predisposition or as vivaciousness – characteristics which laypeople and physicians at the time still understood to be largely rooted in the body. Accordingly, patients and physicians mentioned 'earnest', 'glowing', 'lively', and 'very cheerful' temperaments.[43] In English, French, and Italian letters, talk of 'temperament' began to approach that of 'humor' (or 'humeur', 'umore'), a term which similarly referred back to the doctrine of humors but which already by the 17th century had begun to be used figuratively for 'disposition' or 'mood' in general.[44] Thus when the letters mention a 'humor' that is 'dreadful', 'bad', 'impatient', or 'fluctuating', it is analogous to a 'disposition' or 'mood' in the modern English sense, and was at most indirectly linked to the predominance of one humor.[45] One patient of Helvetius, for instance, described himself as follows: 'My temperament is in my judgment quite staid ['posé']. But everyone says that I am very vivacious and this is because I do things a little fast and catch fire like gunpowder.' His mood ('humeur'), he said, fluctuated in this, but tended more toward laughter than irritation.[46] Along similar lines, François de Becker, in 1803, reported of count Karl Anselm von Thurn und Taxis from Regensburg that his 'humor, gaiety and appetite' had returned.[47]

Understood in this way, temperament could be more or less pronounced. One could have 'much', 'more', or 'less' temperament. One Regensburg baroness said that it was a new experience for her 'to have as much temperament as at present'.[48] Sometimes 'temperament' was even used to indicate sexual desire. In allusion to her sexual needs one of Tissot's patients was said to have more 'temperament' than she thought.[49]

In this way, the notion of 'temperament' became increasingly disconnected from its traditional humoral-pathological frame of reference. Some patients, it seems, were even aware of this change to some extent. One man, for instance, mentioned his 'sanguine' and 'choleric' temperament, and then qualified this statement with 'as people say'.[50] Many others simply avoided the term. As early as 1593, a law student did not mention his temperament at all when asking Johannes Heurne for advice. He did, however, describe his 'constitution' in detail: his small, thin frame, his perpetual appetite, his sensitivity to cold, his bumbling speech and his often unmotivated tendency toward dejection and sighing.[51] By the 18th century, such detailed epistolary self-portrayals had become common. The habit of describing, in detail, the individual properties and characteristics of one's body, with its weaknesses and its susceptibility to certain diseases, continued for some time. But a particular mixture of humors and qualities was at most only one factor among many thought to determine the specific physical and, as we would say today, psychological characteristics of a person.

Plethora and Apoplexy

Though the notion of 'temperament' increasingly lost its original, specific meaning, the idea that a person's peculiar mix of humors and qualities determined, at least partly, his or her general bodily constitution and disposition to diseases remained alive throughout the early modern period. When it came to explaining concrete episodes of illness, however, the notion, as we have seen, was already of minor importance in the 16th century. Only in a few cases was a disease simply attributed to an excess of yellow or black bile or phlegm in the body and the treatment aimed at reestablishing their natural balance.

Excessive blood, 'plethora', as it was commonly called, was frequently mentioned as a presumed cause of disease. But this was not due to an imbalance between the four humors but to the restricted amount of space in the body and, in particular, in the blood vessels. The vessels would be congested, their walls stretched to the limit, with the result that the flow of blood would slow down and in the worst case the stagnating blood would putrefy. William Harvey's new theory of blood circulation of 1628[52] heightened such fears and 18th-century images of the body as a kind of hydraulic machine made them even sharper. On the one hand, the walls of the vessels might rupture. On the other hand, the abundance of blood threatened to slow the life-sustaining circulation or to block it altogether.

In the opinion of physicians, plethora was due above all to excessive eating. This is explained by their assumptions about the origin of blood in the body. Traditionally, blood was thought to be produced in two steps directly from food. The food was initially 'concocted' to chyle in the stomach and then, in a second step, turned into blood in the liver. From the liver, it passed into the heart and the rest of the body, slowly, in a one-directional movement towards the periphery through the veins, according to the traditional Galenic concept, or quite fast through the arteries according to the new theory of blood circulation. The individual body parts or organs took those components of the nutritious blood which they could literally assimilate to their own substance. Blood was, in other words, primarily conceived not as a means of transportation, but as food and nourishment.

Because the production of blood was thus closely connected with food intake, plethora became an important issue in the 18th-century medical critique of civilization. As physicians explained to their enlightened readers, plethora struck above all rich urban ladies and gentlemen, who constantly consumed overly rich food, which the liver then turned into more blood than the body needed. At the same time, the blood was inadequately consumed, due to the sedentary lifestyle typical of the rich, and of wealthy women in particular. Among farmers and farm women, as Tissot argued, for instance, plethora was much less widespread.[53]

The symptoms and consequences ascribed to plethora were clearly not due to a disturbed balance between the blood and the other bodily humors. They reflected the increasing congestion of the blood vessels and of the body as a whole. Some patients expressly characterized their temperament in this context not as 'sanguine' but as 'plethoric' or 'full'. Physicians of the time listed a range of typical consequences of plethora,[54] and many of these can be found in the letters of 'plethoric' patients. The limbs became heavy, the face reddened, the pulse felt hard and full. Her daughter sometimes seemed to have too much blood, wrote one concerned mother, because at times her face was quite red.[55] In the head, in the limited space of the skull, an abundance of blood made itself unpleasantly known. Headaches and dizziness, and at times a buzzing in the ears, were often connected to plethora.[56] Lower abdominal pains were also considered typical and interpreted as the result of an excessive tension in the vessels there. A prominent example were pains around the anus, which sometimes ceased almost at once when bleeding from hemorrhoids relieved the pressure.

In order for excessive blood to be got rid of, it had to be consumed by physical exercise or it had to be evacuated at regular intervals. For boys, nosebleeds were considered particularly helpful in this respect, and for men, bleeding from hemorrhoids. Hemorrhoids were often disagreeable and painful, but bleeding hemorrhoids were seen as healthy and expressly welcomed. If only, prayed one 46-year-old colonel, his earlier, regular spells of hemorrhoid bleeding would return, then, true, he would have one more complaint, but he would be freed of his other pains, namely of droning and a sensation of heat in his head, both of which had set in when his hemorrhoids ran dry and left him fearing an apoplexy.[57] In Germany patients even spoke of a 'golden' vein ('goldene' or 'güldene Ader'[58]) in this context. To imitate or support the body's efforts to rid itself of excessive blood was one of the main reasons for bloodletting.

In women, the menstrual period was the body's natural and most efficient way of eliminating excess blood.[59] From the late 16th century, the great majority of physicians even considered the evacuation of excessive blood the primary function of monthly bleeding. In their view, a sexually mature woman produced an excess of blood each month. During pregnancy this excess blood served to nourish the child inside her and after birth it was turned into milk – which was the reason why pregnant and nursing women did not have menstrual periods. When they were not pregnant, women had to eliminate the blood because otherwise excessive blood would rapidly accumulate in their body and blood vessels. After all, a woman's monthly bleeding was traditionally estimated to amount to around one pound or half a litre. From the perspective of the medical critique of civilization, a regular monthly elimination of excess blood was particularly important for inactive and well fed upper-class women. Their periods were described as markedly heavier than those of, say, farm women. Some physicians even claimed that menstruation resulted only from the unnatural lifestyle in modern European societies and pointed to travelogues recounting that women in other countries had fewer menstrual periods or even none whatsoever and nevertheless stayed healthy.[60]

There were numerous other bodily openings and pathways through which excess blood could be evacuated and which could take over when habitual bleeding from hemorrhoids ceased or menstruation was suppressed.[61] The newly wed daughter of Monsieur Cerrier, for instance, got a rash when her monthly period abated and her urine turned red in spite of a bloodletting.[62] In the case of Hans Khevenhüller's wife, the bleeding after she had given birth let up when she got a severe nosebleed.[63] One of Verdeil's female patients started having massive nosebleeds as she approached menopause and her periods began to cease.[64] A town clerk of Feldkirch by the name of Amberg claimed to have lost seven 'Maaß' – about seven liters or almost two gallons – of blood from his right nostril and, soon after, five more from his left. After a bloodletting he felt 'physically' better again.[65]

Sometimes, when such bleeding became uncontrollable, the excessive loss of valuable, nutritious blood was seen as a threat. A particular cause for concern was massive bleeding from the lungs and from the uterus – probably because it was often, in people's experience, connected with two of the most dreaded illnesses, consumption and uterine cancer. Even when a patient coughed blood, however, this was in many cases interpreted as a welcome effort undertaken by the body or by Nature to eliminate excess blood. Accordingly, the physician of one Berlin preacher attributed the man's pulmonary bleeding to the cessation of his previously frequent nosebleeds: 'Nature' had 'taken a different, more dangerous route of excretion instead'.[66] And according to the same logic, another patient whose legs and abdomen were swollen felt 'slightly relieved' when he coughed up two mouthfuls of blood in the evening. He also had nosebleeds, all of which, his father postulated, meant that the excess blood had risen upward.[67] Unpleasant as it sometimes was, in most cases bleeding thus offered welcome relief, and deciding whether to undertake anything to stop it was a matter of great care.

If all efforts undertaken by the body or Nature to channel the excess blood out of the body failed, serious consequences loomed. As health advice books explained to readers,[68] if the movement of the blood in the overstretched and overfilled vessels slowed, it became thicker and more viscous and began to stagnate.[69] It was particularly dangerous if the blood rushed toward the head or collected in the limited space of the skull and clogged the vessels of the brain. At this point there was a risk of 'stroke' or 'apoplexy',[70] frequently with a fatal outcome.[71]

The dangers of a stroke appear to have been widely known among the general population and the more technical synonym 'apoplexy' was commonly used. As we have seen, the disease was feared as a major cause of sudden death. The symptoms that educated laypeople connected with a 'stroke' still sound familiar today. Whoever was 'touched' by a stroke suffered primarily from loss of sensation and movement on one side of the body, but also from difficulty in speaking, and, in severe cases, was rendered unconscious. Joachim Brandes related how Arneke, the mayor of Hildesheim, died in 1601: 'his language and reason almost left him, and he was touched on the right side of his body'.[72] 'Language as well seems almost difficult to me', complained one of Thurneisser's patients after a stroke; when he spoke, 'it was as if my lips and palate were thick or filled with stuffing'.[73] One of Hoffmann's patients still complained of partial paralysis in the mouth, tongue, and arm three years after he 'had been afflicted by the Almighty with a stroke'.[74] Consequently, those in the company of a 68-year-old attendant at the Hôtel Dieu who was suddenly unable to speak for half an hour immediately suspected an apoplectic stroke. Without even waiting for the surgeon, a bloodletting was performed. Afterwards, she still had difficulty in speaking, and the corners of her mouth curled spasmodically.[75]

A sign of excessive amounts of blood flushing into the head and brain and/or accumulating there – the trigger of an apoplectic stroke – was extreme redness in the face or, worse, visibly protruding veins.[76] Sometimes a massive stroke was presaged by smaller disorders or signs of more limited paralysis: describing the facial palsy of the Princesse de Ligne, Mme de Graffigny said, 'her mouth is pulled back to her ear and one eye is open and motionless, without the ability to shut it'. If nothing were done about it, she feared, the patient would surely suffer an apoplectic stroke.[77] Even dizzy spells, as La Mettrie warned his readers, could degenerate into apoplexy.[78] And apoplexy was also feared by the relatives of a privy councilor and chancellor because he would enter strange states of confusion and impaired consciousness.[79] Even symptoms that were commonly attributed to 'hypochondria', 'vapors', or 'nervous complaints' could occasionally stir fears of ensuing apoplexy, for instance for a lady from Reims, who complained of, among other things, a persistent sensation of coldness in her head and neck, a sudden weakness in her legs, a disturbing ringing in her ears, and general listlessness.[80] And 43-year old Mme Faugeroux was convinced that she was suffering from apoplexy, although Tissot considered her symptoms purely nervous. She feared she would soon die, a concern shared by her husband. Her periods had been irregular for over a year – implying that she could no longer sufficiently eliminate excessive blood. She felt a burning heat in her body and had difficulties speaking, headaches, and heart palpitations. According to her husband, her arms and legs were so numb that four men could slap her hands without her feeling it.[81]

It seems that laypeople thought about apoplexy and its causes in much the same way as physicians. In their letters, patients and relatives resort to similar images of congestion, of a sluggish flow of corrupted, unhealthy blood, and of mechanical pressure on the brain in the narrow cranial cavity.[82] As a rule, however, they did not expressly articulate these conceptions or elaborate on them. Amberg, the above-mentioned town clerk of Feldkirch, for instance, had not only eliminated about three gallons of his apparently excessive blood through the nose. He also suffered from dizziness and copious fluxes from the head and when he had a bloodletting, the blood was 'burnt black'. He urged his physician to give him a preventative medicine 'for the apoplexy (about which I am most concerned)'.[83] Since getting married, complained one 39-year-old patient of Tissot, his previous bilious temperament had been increasingly replaced by an excess of blood, which, despite bloodlettings every six months, had become increasingly 'viscous'. Sometimes he even saw something like small blood particles before his eyes, and he experienced a feeling of heaviness. His relatives feared 'obstructions' or the onset of an apoplexy, which they wanted to see combated early.[84] Similarly, another patient with 'heavy and thick blood' feared palsy or apoplexy if his body were unable to rid itself of its humors.[85]

The most important procedure to counteract an excess of blood and to prevent 'apoplexy' and other dangerous effects of plethora, was to evacuate the superfluous blood. The conviction, shared by patients and physicians alike, that bloodletting was indispensable in such cases provides a particularly striking proof of the persuasive power and inner logic of the notion of plethora.[86] It was not uncommon for patients and their relatives to expressly demand a bloodletting when they suspected that plethora was at the root of their suffering, such as in cases of severe headaches or nosebleeds.[87] And afterwards they reported upon the beneficial effects. Nosebleeds or bloody coughs, for instance, stopped after a number of bloodlettings at the arm and foot, indicating that the body no longer needed to be freed of surplus blood.[88]

In cases of massive bleeding from the lungs, uterus, or other organs, bloodletting to today's reader must seem like pure foolishness, with patients probably already dying from blood loss. And sometimes patients and relatives did indeed complain that bloodletting had overly weakened or exhausted them. But according to the concept of plethora, bloodletting was, in principle, no less indicated in cases of massive bleeding – indeed it could seem particularly urgent. If the bleeding was not due to an injury, it showed that there was still too much blood in the body, which Nature was trying to get rid of. The logical and sensible thing to do in this situation was to support – not to impede – Nature's efforts.

Of course, the treatment might fail and the patient might continue bleeding. But even that did not necessarily mean that bloodletting had been the wrong choice. Failure could indicate, on the contrary, that the quantity of blood let had still not been sufficient. One severely ill nun, according to Klara Staiger's account, underwent 12 bloodlettings within 14 days, and still 'the excess blood wanted to smother her'.[89] Marguerite Françoise Duchesne, who later recovered miraculously, underwent countless bloodlettings over the course of many years, sometimes up to four per day. She had begun, after taking many falls, to spit up blood and to bleed from the nose and it was feared that she would suffocate if nothing were done. In the end, complained the sick woman, she hardly had any real blood anymore; it was 'like water'. Apparently she did not doubt, however, that the bloodlettings were necessary and justified. She only began to place her hopes in the miraculous healing powers of the tomb of Monsieur Paris when she became deaf and dumb and suffered from signs of apoplexy and when blood began to press forth from the corners of her eyes, making even the attending poor people's physician give up on her.[90]

Fluxes, Gout and Rheumatism

In the overwhelming majority of cases, disease was not explained by an excess of one of the four natural humors but by more or less specific, harmful, or corrupt morbid matter. Such morbid matter could enter the body from the outside or originate from processes of corruption or putrefaction within. Sometimes the nature of the morbid matter was manifest to the senses, as in the case of the sharp, biting 'acrimonies', which oozed from ulcers. Most of the time however, the nature of the morbid matter was quite simply defined by the disease or the symptoms it typically brought forth.

Among the most common diseases of this type were the 'fluxes'. They were attributed to mobile, fluid morbid matter and were mentioned in numerous patient letters and other personal testimonies. Almost any illness with local and/or painful symptoms could be considered in general terms as a 'flux', even more so if the location of the symptoms changed. This concept of 'flux' was closely connected to the notion of 'catarrh' (contemporary French: 'rhume', 'catarrh', German: 'Katarrh'). Etymologically, in fact, the term 'catarrh' is derived from the Greek words 'rrheo' meaning 'flow' and 'kata' meaning 'down'. In the early modern period, the word 'catarrh' still expressed the traditional idea that the liquid that flowed out of the nose during a cold came from the brain. The French term 'rhume de cerveau' (flux from the brain), which was commonly used by laypeople and physicians alike at that time, illustrates this idea neatly.[91] In the late 18th century, scores of patients still took it as a matter of course that the brain was the place of origin. But 'fluxes' elsewhere in the body also originated, according to the traditional doctrine, in the brain. The brain was held to be particularly moist and cold; superfluous fluid matter could easily accumulate in it. Such matter originated, in particular, from hot damp vapors which ascended from the warm stomach and abdomen. They condensed in the cool brain and from there pushed outward via the nose or 'flowed down' within the body to one or the other body part.

Georg Traupitz could feel the accumulating fluid. He complained that his head was heavy and water was forcing its way into his eyes. In the morning he had to cough heavily. Then the fluid fell from his head to his chest and lungs, making him 'so afraid that with great pain I vomit and can hardly breathe'.[92] Traupitz could also feel how 'phlegmatic matter' was pushing outward through his eyes and 'beclouding' them.[93] While the term 'cataract' today refers to pathological changes in the lens of the eye, it was at this time closely related to this notion of flux, which, in this case, was 'flowing' or 'falling' into the eyes, clouding the vision – a meaning which survives in the modern use of the term 'cataract' for 'waterfall'.

A particularly frequent and unpleasant type of flux manifested itself as toothache.[94] When Monsieur Budé de Boisy, who had suffered from coughing and expectoration for some time, began having toothache, he had little doubt that morbid humor had thrown itself from his throat into his tooth.[95] Mme Du Bouchage also considered her toothaches to be a 'flux' or a 'fluxion'; they had begun at the same time as her rash, another manifestation of morbid humor, first appeared.[96] And the Comtesse de Mouroux began to suffer from toothaches whenever she did not 'sufficiently clear her head through her nose'.[97]

Further down in the body, ailments of the chest and lungs were almost routinely interpreted as the result of a flux that had 'fallen' from above. Here, the liquid morbid matter seemed to become visible in the expectorated sputum. For Albrecht Sauerman in 1572, for example, a 'severe catarrh' had 'fallen into his chest'.[98] And Isaac Keller turned to Thurneisser because of his 'continually descending catarrh and flux from the head', as the cause of his constant coughing and 'humidity of the chest'.[99]

Fluxes epitomized mobile morbid matter. They could settle in any place within the body and could also move away from there again, invading other body parts. According to the sick Johann Christoff Amberg, for instance, a flux fell to his eye, ear, nose, and chest, then moved on to his stomach and finally settled in both kidneys, between his shoulders, and under his toes.[100]

As well as the chest, the joints were particularly susceptible to fluxes. Petrus Groß, for instance, complained that 'about a year ago, the flux fell into [my] right arm'.[101] It is here that we encounter two concepts which were closely related to that of 'fluxes': 'rheumatism' and 'gout'. In etymological terms, 'rheumatism' is, like 'catarrh', derived from the Greek word 'rrheo', 'to flow', and 'gout' goes back to the Latin word gutta, 'drop' and the French word for gout, 'goutte', also means 'drop'.[102] Especially among physicians the terms 'rheumatism' and 'gout' were used, as they are today, specifically for painful diseases of the limbs and joints. In ordinary language, however, the terms could be used almost synonymously. According to a 19th-century German physician from the Palatinate, 'flux', the way country people saw it, was 'any feeling in the body that gads about uncomfortably; now here, now there, appearing with severe pain. All rheumatisms are fluxes'.[103] In patient letters and other personal testimonies, pains assumed to be caused by mobile morbid matter in the extremities and joints were almost routinely attributed to 'rheumatism' or 'gout'. The Irish bishop Cary, for instance, wrote of his wife: 'We tend to think that the wandering pains in her arms, hips, shoulders and other body parts are rheumatic in nature.'[104] Gout manifested itself particularly frequently in the extremely painful form of podagra, as it is still called today, that is, in the big toe. But sometimes several or even all of the joints would swell simultaneously – a disease pattern that is more reminiscent of what physicians today call a 'rheumatic fever'. He had had 'podagra' again in both feet, 'and at times in the right hand as well', complained Georg Wagener.[105]

The consequences could be grave. He was so 'burdened with gout or podagra', wrote Lenhardt Stradell in 1574, that almost all of his strength had been taken from his limbs.[106] Monsieur de Schilden, during a bout of gout, could hardly move an inch for 15 consecutive days due to his painfully swollen feet, knees, and wrists. His illness had begun with an extreme shortage of breath and a feeling of pressure in the chest, which may have suggested to him that the flux had first settled there before moving on to the joints.[107] Three days after having had a purulent catarrh, the curate J. Gilbert felt 'sick inside', and his fingers, hands, and arms, as well as his hips, knuckles, and feet, were so painfully swollen that he could hardly walk.[108] Johannes Heinrich Hummel's 'podagra' was limited to his feet and hands, but nevertheless he, in his own words, lay 'very gravely and dangerously ill' with it. He had to have himself carried to his wife, who lay mortally ill, and even missed his own daughter's funeral 'because I lay very low with the podagra'.[109]

Typically gout and rheumatism were thought to be characterized by a swelling, warming, and reddening of the affected joints.[110] When the disease took a longer course, the morbid matter could in the end be seen and felt from the outside, in the form of hard knobs under the skin. For the sick mother of the Comtesse de Lucinge such nodules were enough to make her fear that she suffered from 'gout'.[111]

Like other fluxes, rheumatism and gout were not limited to the joints, however. They could also appear in very different body parts and could move from one place to another. Monsieur de Beaucourse, for example, suspected a 'gouty humor' in his bladder as the cause of his severe urge to urinate and the burning pain in his urethra during urination. Despite the sand-like sediment in his urine, he believed he could rule out a bladder stone as the cause because he was sometimes free of complaints for weeks on end and could even ride a horse, something that at other times immediately gave rise to bloody urine.[112] One sick woman from Ludwigslust wrote that she had to suffer dire toothaches due to 'flying gout' ('goutte volante'), which had thrown itself onto the nerves of her head and had settled there 'mercilessly', considerably weakening her vision as well.[113] The 30-year-old Mlle Herbolin was, according to her father's account, convinced that she suffered from a 'gouty rheumatism', which she could feel moving to and fro between her head, shoulders, arms, and legs and which gave rise to unbearable stomach pains.[114] Monsieur de Schilden's 'gout' of the joints ultimately entered his stomach, accompanied by a biliary fever and jaundice.[115] The 74-year-old Monsieur de Croix noticed that his gout moved, in the space of only a few days, from his foot to his chest and intestines and gave rise to stomach cramps and vomiting.[116] In Germany, children who would 'cry day and night, turn this way and that, draw in their little bellies, will neither eat nor drink and cannot find rest' were said to be suffering from 'intestinal gout'.[117]

Gout and rheumatism could be caused by anything that contributed to the formation and/or accumulation of a harmful 'rheumatic' or 'gouty' humor. Cold air and a suppressed evacuation of morbid matter via sweating were thought to play a prominent role in this. Thus one army adjutant was convinced that the coldness of the Jura Mountains, where he had had to serve as a soldier, had 'anchored' the pain-producing 'rheumatic fluid' below his shoulders and in the flesh of his legs.[118] Among the rural population of the 19th-century Palatinate, too, rheumatism was said to be linked above all to 'colds' or 'getting chilled', an explanation which sufferers of rheumatism still invoke today.[119] Hereditary influences were also often mentioned. Scores of patients made reference to similar illnesses in their families or, on the contrary, explicitly emphasized that they did not know of any. Monsieur Torchon de Lihu, for instance, suspected that he had inherited something from his maternal side, since he had been 'rhumatisant' from childhood on. But it was also possible, he thought, that his mother's misdirected care was responsible: wanting to protect him, she had made him sleep under several blankets, which hindered – it seems this was his point – the excretion of the morbid matter via the skin.[120]

Academic physicians sometimes attempted to grasp more precisely the 'chemical' nature of the morbid matter and its effects, describing it, for instance, as 'lime-like'.[121] But even among educated laypeople, such efforts met with little interest. Apparently they deemed it sufficient to characterize the morbid matter as 'gouty' or 'rheumatic'. In any case, its chemical nature could hardly explain why gout and rheumatism could appear in the form of sudden attacks and sometimes shift quickly from one place to another. This suggested that the morbid matter was an active, dynamic agent with the power of movement. For some patients, precisely the shifting site of their complaints constituted a prerequisite for the diagnosis of rheumatism. Lieutenant Roussany, for example, rejected Tissot's assumption that he suffered from rheumatism on the basis that his pains had been in the same spot for years.[122] Another man believed he could sense the physical movement of the rheumatic fluid inside his body: 'I can feel how it spreads out over my stomach and up one side and then rises into my chest; I was almost smothered in this way this past winter.' When the fluid then flowed into his feet, he promptly felt markedly relieved. A previous attack of rheumatism had quickly improved when a severe red rash with white blisters developed in the painful area, a rash – as is implied – into which the morbid matter emptied itself.[123]

When treating rheumatism, gout and other fluxes, it was essential to rid the body of the morbid matter in question with purgatives or enemas,[124] bloodletting, cupping, blistering, or a similar evacuative treatment. Some patients are said to have purchased and used 'secret' or 'universal' remedies, which were also generally characterized by their drastic purging effects.[125] Those with less money relied on local herbs like gentian.[126] If the morbid matter could not be dispelled entirely, at least it had to be lured to the extremities, where it was less dangerous; this could be achieved by bloodletting from the feet, for example.[127]

Evacuating morbid matter became particularly difficult when it was viscous or had hardened. When one nun suffered from shortness of breath and lost her voice, physicians were only temporarily able, through bloodlettings, evacuants, and broths, to loosen and dissolve the 'sticky', mucous fluid they suspected was in her lungs.[128] The morbid matter first had to be mobilized or made more fluid. In view of his son's 'rheumatic' pains, a Protestant clergyman expressly asked whether one should not give him dissolvent or melting juices ('jus fondants').[129] Certain foods, such as cherries and raisins, were said to have a 'softening' effect and patients were advised to add them to their diets.[130] Many patients and physicians believed in the dissolvent and purifying effect of mineral waters. Monsieur de Beaucourse imbibed them for two whole years even though they had little effect.[131] Of course, the appropriate spring had to be chosen, and if the patient's condition worsened, physicians might have to put up with accusations of having recommended the wrong water.[132] The best remedy in E. F. Geoffroy's eyes, however, was donkey milk, whose mildly evacuative effects were trusted by Tissot as well.[133] Meadow saffron (colchicum autumnale), by contrast, which was known in the Middle Ages and was recommended by Anton Stoerck around 1760 as a treatment for gout,[134] was not even mentioned in letters from the late 18th century. In the proper dosage, it is still today regarded as one of the most effective remedies for an acute attack of gout.

Otherwise, one could only try to keep the morbid matter from accumulating and hardening further through an appropriate way of life and diet. Gout in particular, like plethora, was regarded by early modern physicians as a consequence of a misdirected lifestyle, such as was maintained (although no longer exclusively) by the upper classes and characterized by immoderate eating and drinking (particularly wine), lack of exercise, and sexual excesses.[135] Gout and podagra came to be understood as diseases of civilization. In earlier times, said Elias Anhart, on the authority of Galen, the 'painful flux' had been less common but 'as soon as people abandoned moderation and indulged in crapulence and gluttony this foot disease became rampant. For this reason, it is not only called an illness of the masters but also one of the servants.'[136] Some patients appear to have taken such ideas to heart insofar as they changed their diets, avoided 'coarse' foods, and contended themselves with a smaller evening meal.[137]

'Gichter' (Convulsions)

As has been repeatedly emphasized, 'fluxes' and 'rheumatisms' were associated with images of a harmful, mobile entity in the body which, at its own will, could literally 'throw' itself on one body part or another and 'take hold' there. In German-speaking areas, such images of a hostile agent crystallized in the notion of 'Gichter', the plural of the German word for 'gout': 'Gicht'. Especially in rural areas, the notion was still quite familiar in the 19th century. 'Gichter' ranged from toddlers' convulsions during weaning and teething to febrile seizures and to the abovementioned 'intestinal gout' and even the dramatic attacks of epileptics.[138]

Characterized by strange and often highly dramatic symptoms, 'Gicht' or 'Gichter' sometimes also referred to pain in the joints but used in this sense it was much closer to the notion of demonic disease than to a simple flux. 'Gichter' were generally presumed to have supernatural causes, and treating them constituted one of the principal domains of magic and sympathetic healing.[139] Often the disease was addressed directly as one would address a person or a demon. 'Christ traveled the land', began one of the Biblical short stories which were typically used by soothsayers on such occasions; 'there he encountered a Gicht: "Gicht, where are you going?" "I want to go into people and rip and mangle them." "By the power of God I command you to go to the wild woods and the woods you shall rip and mangle, in the name of the Father, the Son and the Holy Ghost, Amen."'[140] Like demons each 'Gicht' had its particular ways, and certain healers specialized exclusively in the treatment of the '77' or '99' 'Gichter'.[141] Some of them were highly sought after.[142] So-called 'Gichtzettel', little pieces of paper on which were written curative sayings such as 'accursed Gicht, go creep into the deepest woods', were also popular. In fact, as one Franconian physician reported in 1860, no patient with 'Gicht' went without such a 'Gichtzettel'.[143] They were swallowed or worn as amulets around the neck, directly against the skin or sewn into a little sachet – the so-called 'Büscherl' – together with other objects such as nails or bits of lead or a frog's leg. They were handed out by healers who specialized in 'Gichter' but they could also be purchased like ordinary medicines.[144]

Acrimonies

Patients and their relatives were usually content to speak generally of 'fluxes', 'humors', or 'morbid matter' without specifying the nature of the fluid. One kind of 'flux' or morbid humor stands out in patient letters, however, namely 'acrimonies'.[145] 'Acrimonies' and 'acrimonious blood' were frequently mentioned and discussed and in the 19th century they were still said to play a prominent role in the medical world of the countryside.[146]

In learned medical writing, 'acrimonious' humors had a long tradition, reaching as far back as Hippocrates. Their significance, however, grew considerably in the medical literature of the 17th and 18th centuries. This was mostly due to the spreading of Paracelsian and iatrochemical ideas, which saw 'acrimonies' as closely related to chemically defined 'salts' and 'acids'.[147] In the 18th century, 'acrimonies' were a prominent feature in the medicine of Herman Boerhaave and his followers. Their opponents claimed that they attempted to explain almost every ailment in terms of their 'imaginary acrimonies'.[148] But in the face of certain diseases or symptoms many physicians, not only Boerhaavians, resorted to the concept of 'acrimonies'.

Patients only rarely alluded to the 'chemical' nature of 'acrimonies', as when a worried mother observed in her infant an expulsion of 'salts' due to 'acrimonies'.[149] Patient letters referred to 'acrimonies' primarily when the morbid matter gave rise to sensations similar to those which 'acrid' or 'acid' substances produced on the skin or mucous membranes. 'Acrimonies' offer another striking example of how perceptions of the body in health and disease which patients took to be immediate and naturally given were in fact shaped or framed by the medical culture of their time. Some patients, for instance, thought they were suffering from throat aches due to an 'abundant melting of an acrid and biting serum from the brain', which would later throw itself at the air passages and lungs.[150] Or they complained of an 'acrimony' of the mouth or on the tongue.[151] Or they felt how an 'acrid viscous mucus' was continually flowing down from the brain into the throat, piercing it like 'a bundle of needles'.[152] In the case of the elderly Monsieur Gouët, the 'head-flux' ('fluxion de tête') was so acrid and fervid that, according to his daughter's account, his tongue swelled up and blisters developed; he was hardly able to eat any more.[153] When acrimonies exited the body through the eyes, the urethra, or the anus, they could also in patients' experience make themselves felt as burning pains. One man suffering from bladder stones even learned to assess the changing 'acrimoniousness' of his urine based on the varying intensity of his pains.[154] For women, acrimonies often manifested themselves as an itching, burning vaginal discharge. Insofar as such discharge freed the body of harmful morbid matter, it was health-promoting. But women also suffered from it and, if it persisted, it was difficult to cure. Nature, as physicians explained, grew accustomed to emptying all 'the impurities' that developed daily in the blood in this way'.[155] One Frau Hofrätin from Coburg, for instance, was so severely incommoded by 'the whites' and their 'acrimonies' that her concerned husband consulted various medical handbooks.[156] Menstrual blood could also, in the experience of women, take on a burning quality due to the acrid, impure substances that menstruation, according to the dominant conviction of laypeople, washed out of the body. The Comtesse de Mouroux, for instance, would immediately get a strong burning sensation in her genital area if ever she did not wash herself thoroughly during her periods.[157] When menstruation was suppressed, such 'acrimonies' would almost inevitably make themselves felt in other parts of the body. Thus when Mme Rostaing developed painful, pus-filled ulcers in her mouth and throat, she resolutely disagreed with her attending physician who assumed a local disease process in the head: 'I believe that he is wrong and that it [the evil] lies in the acrimoniousness of the humor'. For the past five months, she explained, she had not had her period, 'and all of it repairs to the throat'.[158]

Perspiration also served to evacuate 'acrimonies'. After repeated washing of her clothes, one of Haller's scorbutic patients was still seriously troubled by the effects of the remaining traces of sweat. Even a year later, when she put on a dress she had worn the previous summer, she immediately felt 'a horror' all over her body and especially under her arms, where she usually perspired most, and needed to vomit'.[159] One of Friedrich Hoffmann's patients, after a febrile rash illness, reported: 'my sweat during the heat rash was sometimes so acrid that it not only eroded the skin of my chest and temples and made it sore, but it also made my shirt so brittle that when you touched it, it ripped right away, as if vitriolic acid had been smeared on it.'[160] Some patients also believed they could taste the 'acrimonies' in their mouths, and complained of sharply burning, salty or bitter saliva or sputum. Small ulcers or aphthae confirmed the suspicion.[161] And 'acrimonies' could even be noticed in the blood from a bloodletting.[162]

Much more rarely did patients report sensations due to 'acrimonies' inside the body: a burning, for instance, that rose up the back.[163] Usually such 'acrimonies' made themselves felt more indirectly, through their irritating effect on the afflicted body part or organ. As far as his lungs were concerned, related Matthias Kühne in 1575, he was 'very unwell', especially on the left side, where a pulsing, twitching, and wheezing indicated that 'the lung was already being eaten away and was afflicted by a viscous and acrimonious coagulated flux'.[164] 'Acrimonious humors' in the bowels could, together with bile, give rise to severe bouts of colic.[165] In the 18th century, it was thought that the 'sensitive' fibers of the stomach could be affected, a suspicion that was corroborated by acerbic, acrid, burning eructation. Following this reasoning, one sick man believed, for instance, that his repeated vomiting and the constant sensation of pressure in his upper abdomen were caused by 'the influence of the acrid humor on the fibers of the stomach'.[166] Physicians also attributed cramps, 'convulsive rheumatisms', and the like to an acrimonious irritation of the nerve fibers.[167]

When patients or physicians discussed the possible origins of 'acrimonies', they usually connected 'acrimonies' with images of over-heating. The consumption of hot and spicy food or alcoholic beverages was frequently inculpated. At other times excessive heat in body was taken to be responsible.[168] When food or bodily humors 'burned', they acquired a certain sharpness which in itself became the source of 'burning' pains. Outward signs of excessive heat in the body or the blood – for example, a red face – could, in turn, confirm the suspicion that an 'acrimony' was at the root of the patient's skin alterations or other complaints.[169]

Of the primary humors, yellow bile – already hot and dry by nature – was most apt to be 'burnt'. As physicians and educated laypeople conceived it, it had a 'dissolving' effect, similar to acid. It could attack the very substance of the stomach and took away the appetite when it rose upward.[170] In the rural medical culture of the 19th century, bile virtually epitomized the notion of acrimony. As a clergyman from Franconia remarked derisively, the village healers in his area would already see that 'his bile was sitting in his skin and flesh' as soon as a farmer entered the room'. 'This is indeed a lucky time for bloodletting', they would comment; 'add to it a small pot of buttermilk and this will cleanse your body of its acrimoniousness'.[171]

Humors like lymph or serum that were considered as cold and watery by nature could also be 'burnt' or 'acrid', however. According to his physician's account, Ludwig Albricht, for instance, had not only heated up his stomach with 'fervid' wine but had also 'corrupted' his brain with 'acrid phlegm' in so doing.[172] The notion of burning, acrid lymph or serum may have reflected the common experience that 'serous' secretions from wounds, for instance, or a simple catarrh sometimes seemed to irritate or even erode the skin. One of Tissot's female patients described how her measles had been accompanied by such severe 'fluxes from the brain' ('fluxions de cerveau') and copious nasal discharge that her upper lip became inflamed and burned like fire.[173]

Like fluxes and other kinds of morbid matter, acrimonies were highly mobile. They could move from one place to another within the body, giving rise to different symptoms as they did so. Basically every body part could be afflicted. The manifold consequences of a reduced or suppressed menstrual period illustrated this well. Because her 'menstrua' had not come at the right time and only 'in a very reduced way', one young neighbor of Michael Folckhamer began to suffer from an 'acrid salty flux from the head'.[174] In the experience of laypeople and physicians alike, however, 'acrimonies' thrust their way most frequently to the skin. 'Acrimonies' were regarded as the major cause of numerous pathological skin alterations, and 'acrimonies' and the skin were so closely linked that often a rash alone was enough to warrant a suspicion of an 'acrimonious humor'.[175] Accordingly, the apothecary Viton considered his wife's recurring 'lichen-like' skin alteration on her thigh to be sufficient grounds for believing she had an 'acrimony' in her blood.[176] In view of her poor complexion and pimples – she also had pain in her joints – Mme de Constable was led to believe that she was suffering from an 'acrimony in the blood' from her mother's side of the family.[177] One physician even saw a simple reddening of the face and particularly the nose as 'proof' of acrid blood.[178] Conversely, the absence of pathological skin alterations was an important indication of 'clean blood', without any 'acrimonies'.[179]

Skin and Rashes

In modern medicine, the human skin is seen as a highly complex, heterogeneous structure. It is comprised of several tissue layers in which numerous specialized organs and structures have a place – tactile corpuscles, for instance, or sebaceous and sweat glands. The skin, in phenomenological terms, also plays an important role as the body's boundary and skin changes can be extremely traumatizing also in psychological terms. Thanks to the skin more than anything else, we perceive ourselves as both taking up space and as spatially delimited. Only occasionally – for instance in intense physical contact with others or when the skin is injured or pierced – does the sensation of a partial dissolution of the body's boundary arise, accompanied sometimes by powerful feelings of desire, fear, or pain.[180]

In many cultures, including that of the West, the skin is also something which can be 'read', which carries signs or messages.[181] This is true not only for intentional changes to the skin, including everything from moisturizing cream and make-up to tattoos and branding. On just seeing people's skin in its natural state – doughy, bloated, pimply, taut, or tanned, for example – we tend to draw far-reaching conclusions about their personality and character. From another perspective, psychosomatic theory sees the skin as a medium which can give bodily expression to psychological tensions and conflicts and thus makes them visible from the outside.

The meanings conveyed by the skin in the early modern period correspond only to a limited extent to such modern conceptions.[182] There are some notable differences. First, skin in texts and images of the 17th and 18th centuries appears, much more than today, as a separate layer, as a kind of removable sheath or cover, rather than as the outer margin or boundary of a compact body mass. Contemporary anatomical drawings of flayed cadavers – the so-called 'écorchés' and pictorial representations of, for instance, the martyrdom of Saint Bartholomew, the flaying of Marsyas, or the corrupt judge Sisamnes – illustrate clearly how the skin was thought of more like clothing which could be quite easily pulled off.[183] Human skin became in this way more closely related to the hide of animals.[184]

The degree to which the skin was perceived as connected only loosely to the flesh underneath emerges clearly in the notion of a distinct, extensible space 'between flesh and skin'. Though evidence of this conception already exists for the Middle Ages and though the idea seems to have been quite widespread in the early modern period, historians of the body, to my knowledge, have totally ignored it.[185] In this space 'between flesh and skin', it was believed, harmful humors and excremental matter could accumulate before they were excreted through the skin as sweat or a rash, or found their way back inside of the body. Perhaps the notion was inspired by the experience of slaughtering animals, whose skins or pelts could be separated from the flesh relatively easily. Or the image was brought forth by the fact that morbid matter seemed to accumulate visibly under the skin in the form of abscesses or pustules or similar skin alterations, before making its way out.

Among patients, this space 'between skin and flesh', or in French 'entre chair et cuir', was in any case mentioned quite frequently.[186] Wulf von Closter, for instance, complained that the 'poison' was running 'between skin and flesh' ("zwischen Felle und Fleisch"), and in all of his blood as well.[187] And one 28-year-old patient of Tissot reported having pimples below her chest 'entre chair et cuir'.[188] Some patients even located concrete physical sensations in this space, for example, an itching, a pricking, or a burning,[189] or even a sensation of movement. Every day, Caspar von Hobergk sensed 'great disorder' of the blood that was running to and fro 'between the skin and flesh' and he feared that a fever might develop from it in the future'.[190] Another patient felt an 'undulation' or 'shuddering' ('frissonnement') in his shoulders, 'as if my skin were peeling off'. It felt 'as if someone was blowing air between skin and flesh' and he supposed that his sweat was making itself felt in this way as it pushed to the outside.[191] Some patients located a similar space within the layers of the skin itself rather than between flesh and skin. One of Tissot's female patients, for instance, suffered from a 'state of death' ('état de mort') 'between the skins of the face' ('entre les paux de visage') every time she imbibed warm drinks such as coffee.[192] Similar descriptions of a defined space underneath the skin where morbid matter accumulated are to be found in 19th-century German ethnographies.[193]

Early modern interpretations of rashes fit into this explanatory framework. Rashes reflected first of all the body's effort to free itself of dangerous morbid matter. The morbid matter was driven away from the vital parts to the body's periphery, underneath and in the skin, so as to flow from there across the body's boundaries when blisters, pustules, or boils opened up. Thus a perception of 'skin diseases' in the modern sense existed only to a limited extent. In cases of prolonged, painful, or disfiguring skin alterations, the changes acquired a pathological value of their own and had to be treated. But rashes, pustules, and ulcers usually only reflected a pathological change inside the body. They were the mere result of the effort made by the body or the healing power of Nature to get rid of this morbid matter. This also meant that as long as the skin eruptions persisted, one had to assume that there was still morbid matter in the body, which was seeking its way out in this manner. Hence it was often considered wrong, and indeed dangerous, to suppress rashes, pustules, and the like. In many cases it was, on the contrary, imperative to promote them, for instance by applying irritants to the skin. Thus Claudia Benthien's claim that in the 19th century, skin ailments were taken not as arising from inside the body but as affecting only the skin and therefore had to be treated with local, external remedies, stands in complete contradiction to what we find in patient letters and medical literature.[194] In fact, the opposite is true: rashes and skin lesions were usually seen as part of a pathological phenomenon that affected the entire body and attending physicians and laypeople alike treated them accordingly.

From this perspective, rashes were a cause of concern because they indicated the presence of morbid matter in the body but at the same time were welcomed as a means to evacuate that matter. Patients and, especially in the case of children, relatives felt relieved when a rash appeared. After all, it indicated that morbid matter was successfully being driven out.[195] This also accorded with physicians' views.[196] It is telling in this context that the Greek and Latin etymological roots of the modern technical terms for 'rash' have decidedly positive connotations: both 'exanthema' and 'efflorescence' refer literally to a 'flowering' or 'blossoming'. Frequently, patients and relatives were therefore worried that a rash might be too weak or disappear too quickly, leaving part of the morbid matter inside the body. Years and decades of pain could result if the morbid humor then threw itself at a different body part. Particularly feared in this respect was the measles – a disease which modern medicine incidentally also associates with a substantial risk of serious complications. Experience showed that the remaining morbid humor could easily change its site and 'throw' itself at another body part, the eyes or lungs for instance.[197] The mother of one five-year-old girl, for example, related that following an attack of febrile catarrh, the girl had developed a rash similar to cradle cap. But the 'expulsion' had been too weak, she thought, because the girl suffered from bouts of severe pain in her belly and was tormented by pruritus, affecting mostly her head and abdomen. In the end, the mother reported, a whooping cough, accompanied by heavy vomiting, freed the child at least in part from 'this acrid and thick humor' that 'in my opinion is the cause of all her ill'. When the girl got the measles, this awakened new fears. But this time, only three hours after a moderate bloodletting, 'the measles fortunately came out'.[198] Many letter-writers put their hopes in the healing effects of a rash and they often saw their expectations fulfilled: the condition began to improve markedly when the rash appeared. Conversely, others dated the beginning of a prolonged illness – a chest ailment, for instance – to the premature disappearance of a rash.[199] Experience showed, for example, that patients with smallpox risked turning blind when the acrid morbid humor was not sufficiently evacuated through the skin and threw itself to the eyes instead.[200]

Only sometimes were skin alterations perceived as diseases in their own right, especially when they were massive or disfiguring or when they were attributed to morbid matter that by its very nature tended to accumulate above all in the skin.[201] In such instances, people in those days would speak of skin 'diseases'. Even then skin complaints were considered only as the most obvious result of an accumulation of impure, generally 'acrid', morbid matter inside the body, however, and external, topical treatment was usually complemented by efforts to free the body of its impurities. One female English patient, for instance, who, in her own words, was suffering from a 'skin disease', tried her luck with bloodlettings, warm baths, cupping glasses, and blistering plasters.[202] Another patient was so fed up with her rashes that she turned to a renowned Father at the local Hôtel Dieu who was said to know a proven, fail-safe remedy for her disease. It was composed of an acrid fluid that drove the lichen-like humor ('humeur dartreuse') out from the space 'between the two skins'.[203]

Chronic skin lesions, in particular, became a major problem in their own right when they were painful or itchy or when they resulted in a repulsive appearance. This was the painful experience of young Marie Thérèse Dumoulin, who developed inflamed, purulent swellings on her cheeks, from which blood and secretions of changing colors flowed. She reported that this had 'caused great disgust' for visitors. Many of them did not dare to get too close or did so only 'with the greatest reluctance'.[204] In another case, the attending physician was quite satisfied with the rash, which covered the patient's entire body. He took it as a sign that the blood was being successfully cleaned of 'acrimonies'. The patient, however, was so disgusted by it that she began rubbing certain 'remedies' into her skin. Her friends and relatives were not surprised when before long the rash 'turned back on the body' and she died.[205] In the worst cases, a nauseating smell accompanied the disease. The 11-year-old Marie Bourquin, for example, whose mouth had been eaten away by scorbutic abscesses, reeked so badly that she had to be removed from other patients.[206]

Closely associated with notions of impurity, skin disorders took on a further dimension in social life in such cases, one we are no longer familiar with today: by their mere presence the patients could inflict lasting harm on others.[207] Like the sight of epileptic seizures, the repulsive, frightening appearance of disfiguring skin lesions and oozing ulcers could provoke strong, negative emotions, a powerful cause of disease. Even worse, their impact on the maternal imagination was believed to be so strong that the baby risked being born with similar disfigurements. If they did not want to heap guilt upon themselves, the patients therefore had to keep out of sight of pregnant women. Since pregnant women were virtually everywhere this ultimately implied a life in seclusion.

Red Murrain (Erysipelas)

Somewhere between symptomatic rashes and specific skin disorders was red murrain or erysipelas. Today it is understood as an acute bacterial infection of the skin and the clinical picture is clearly circumscribed. But in the early modern period, the terms 'erysipelas' and 'red murrain' were used in a much wider and more varied manner. In general, they comprised complaints characterized by more or less extensive reddish and painful swellings, often accompanied by heat and prostration. Even in scholarly medical literature, quite different complaints fell into this category, including even ergotism or 'St Anthony's fire'.[208]

Terms like 'erysipelas', 'erysipèle' (in French), and 'Rotlauf' (in German) were used only occasionally in patient letters and other personal testimonies. Rétif de la Brétonne attributed the 'erysipelas' on his leg to an injury he had sustained when carrying his daughter, in which he was quite close to modern medical understanding.[209] But like other pathological alterations of the skin, red murrain was seen to point above all to a general disease process, to some acrid morbid matter which could change its location in the body. When it appeared on the skin, complaints elsewhere were likely to disappear. Conversely, when it receded from the skin, other, possibly more severe, symptoms might emerge. Thus Marie-Jeanne Orget's red murrain, which was accompanied by fever, 'relocated' itself from her right leg to her genitals.[210] Another female patient with recurring 'red murrains' in the face had first been ill with the smallpox and then suffered a painful, purulent ailment of the ears before the humor – as is implied – threw itself at her face.[211] Similarly, one of Helvetius's female patients was greatly concerned after two episodes of facial red murrain that the morbid matter might fall upon her chest. She was convinced 'that it is the same humor that caused my minor nose complaint, my flux from the chest and my erysipelas'.[212] For similar reasons, Mme de Menou even had an artificial ulcer made with a cauter: for years she had been suffering from swellings, an itching 'between the skin and flesh', and rashes, especially before her period. And when the symptoms shifted to her scalp, she feared the fluid 'might play an evil trick' on her. In fact, she reported, the fluid subsequently found its way to the artificial ulcer. She did, however, retain a periodic effervescence of her blood, and whenever she did not have a rash on her face, she suffered from an unbearable heat in the blood, which nothing could cool. Hence she requested a remedy 'to dispel this humor or this acrimony that has taken root'.[213]

Scurvy

Another much-discussed illness, one that was often connected to 'acrimonies' and similar morbid matter such as burned black bile, was scurvy (French: 'scorbut', German: 'Scharbock', Dutch: 'schoorbuick' or 'schuerbuyck').[214] Physicians regarded scurvy as a widespread illness, particularly around the North Sea and Baltic Sea, as well as in Westphalia. Some authors deemed it a new disease, and Johann Baptist van Helmont dated its beginning precisely to the year 1556.[215] Others believed the typical symptoms could be found in earlier accounts.[216] In any case, at the end of the 16th century, scurvy became a kind of 'fashionable illness'. In letters to Thurneisser – many of his patients came from the 'endemic' areas of northern Germany – it is mentioned frequently. But still in the 18th century, many a physician was said to diagnose 'scurvy' every time he was 'confronted with a disease whose symptoms are unusual or unknown'.[217] Even unauthorized healers resorted to this diagnosis at times. One such healer, in 1647, explained that a patient who had died in his care 'was full of scurvy, had consumption too and his liver and lungs were eaten up'. Asked by a physician how he had recognized this, he said: 'from his completely stiff arms and limbs'.[218]

The symptoms of scurvy at the time coincide only to a limited degree with what we today attribute to scurvy as an illness caused by vitamin C deficiency.[219] A multitude of complaints were linked to a diagnosis of 'scurvy', some of which might also more generally be attributed to an 'acrimony'. One of Friedrich Hoffmann's patients, for instance, suffered from serious headaches, itching and burning skin, tightness in the chest, and a stiff neck. He had a bad, sour taste in his mouth and his urine looked as if it contained red grains of sand. Thus the patient thought 'that scurvy is the main cause, and [it] is accompanied by all kinds of pulling and tensing in the outer parts of the skin, and in the shoulders and back it burns and pricks'.[220] Another patient complained that, particularly in the morning and in cold air, he was overcome by 'a scorbutic burning and blistering in the face, hands and feet, alongside a biting in the stomach and hunger'.[221] In individual cases, 'scorbutic' symptoms could also include pains in the heart and calves, dizziness, cramps, insomnia and anxiety, fever, dropsy, paralysis, arthritis, and red murrain.[222] The surgeon Peckatel was not able to put any weight on his feet for nine weeks and had to stay in bed when attacked by 'Schorbuck'.[223]

Two sets of symptoms did, however, stand out. At least in the 18th century, bleeding from the gums and the loss of teeth were regarded as particularly typical.[224] People spoke of 'Schurmundt' or 'Scormunt' ('scurvy mouth' or 'scorbutic mouth'). The gums appeared to disintegrate. According to Zedler's famous encyclopedia, it was therefore 'no wonder that such patients reek from the mouth, making it impossible for any person to remain close or around them due to the disgusting smell'.[225] Consequently, when the aforementioned Marie Bourquin got painful, smelly cankers in the mouth at the Hôtel Dieu of Chalons, those around her considered her 'scorbutic'.[226] One of Tissot's female patients complained that her gums were 'so livid and cankerous that one might think I have much scurvy'.[227] Another patient concluded on the basis of her diseased gums that she had scurvy and thought she had inherited this from her mother.[228] And one of scorbutic Samuel Downing's leading symptoms was gum bleeding upon the slightest touch.[229]

A second symptom which was then considered highly characteristic of scurvy but is far less familiar to us today was the appearance of blotchy skin alterations and itchiness. The aforementioned Samuel Downing had previously suffered from hard, red, flaky blotches on his hands, elbows, and calves, which had been interpreted as 'scorbutic'.[230] In 1574, Ludolf von Closter related that he had had brownish black patches on his legs for the past five days, which were interpreted by some 'good people' as possible signs of scurvy.[231] In the French patient letters of the 18th century, a diagnosis of 'scorbut' was only rarely mentioned by laypeople and, when it was, it was usually in reiteration of a physician's diagnosis. This may have been because scurvy was regarded as a disease of the North and Baltic Sea regions. But it is also possible that French patients rejected a diagnosis of scurvy because they held it, for reasons that are still unclear, to be stigmatizing. It is striking how vehemently individual patients resisted the diagnosis. The Comtesse de Champagne refuted the authority of the famous Bordeu when he interpreted the small red pimples ('boutons') on her legs as a sign of scurvy. In her opinion, they were merely the result of a minor effervescence, brought about by great heat.[232] Another patient was convinced that she was suffering from a 'lichen-like humor' ('humeur dartreuse') and not, as her physician believed, from a scorbutic or bad hemorrhoidal fluid.[233] The Comtesse de Vougy resisted the idea that her husband, who was suffering from catarrh and a bloody cough, had lost teeth due to a 'scorbutic humor'. She felt that it was possible to lose teeth for other reasons as well.[234]

Patients and their relatives rarely explained what they assumed to be the immediate cause of scurvy. Their ideas become somewhat clearer, however, when we look at the reasons why they favored certain types of treatment. Hieronymus Birckholz, for instance, thought it was important in treating his incipient 'Schörbock' 'to bring the spoiled melancholic humors into a better state'.[235] Most physicians, too, assumed that the cause was a pathological contamination or thickening of the blood due, in particular, to 'corrupted' or 'burnt' black bile.[236] It was via black bile that 'scurvy' was linked to the semantic network surrounding the concept of 'acrimonies'. Black bile was traditionally thought to tend toward 'acrimoniousness', particularly when it was heated excessively. Thus leg ulcers ranked among the major effects ascribed to morbid or burnt black bile in melancholics, and especially in patients suffering from a particular type of melancholy which was called 'lycanthropy'.[237] Furthermore, medical treatises on scurvy almost routinely made reference to the viscosity of scorbutic blood; similarly black bile was consistently regarded as especially dry and viscous. In the 18th century, some authors also suspected scorbutic 'acrimonies' in the phlegm or in the serum. For Lorenz Heister, for example, the scurvy of one female patient originated 'in an acidic, salty, quite acrimonious and viscous serum or watery liquid'.[238] 'Acrimonies' also made lymph and blood form clumps. Geoffroy, for instance, explained the cramps of one of his patients as coming from her 'coarse, viscous and very acrid blood that for this reason flows only with effort through the vessels and whose serum is like a brine, which clings to the nervous membranes, throws them into disarray and pulls them into folds, irritates them and puts them in a convulsive condition.'[239] And in a similar fashion he interpreted a 'gouty rheumatism' as having been caused by acrid, thick, and coarse blood that moved through the vessels only laboriously and caused stagnation in various places.[240]

In this way, the concept of 'scurvy' linked images of a peculiar, usually acrid, morbid matter to those of a 'physical' change in the consistency of the blood, of an obstruction or of a local deposit. Along these lines, the blood of patients with scurvy was said to be of a particularly evil consistency in that it was 'partly too thick and coarse and partly too acrid and runny in some places'.[241] Geoffroy thus interpreted the complaints of one sick woman as 'the symptoms of a slight scorbutic affection, caused by thick, clumpy blood that circulates slowly, whose poorly connected elements give off this acrid serosity, which sometimes causes rheumatic pains when it throws itself at the muscles, and sometimes makes for a slight cough, either without sputum or with an expectoration of liquid mucus when it throws itself more or less abundantly at the chest.'[242] Of another patient it is said that his blood was generally 'mottled with scurvy and very dry'.[243]

For physicians, the principal external causes of scurvy were the climate and a diet that promoted 'acrimonies'. This assumption was supported by the fact that certain regions were affected more than others. The author of the entry on 'scurvy' in Zedler's encyclopedia blamed above all too salty food, such as smoked or roasted sausages, ham, smoked fish, salt meat, dried and salted fish such as salt herring, old cheese, hard and impure water, spoiled, yeasty beer, wine that had turned sour, and the like; therefore the Lower Saxons, who enjoyed such foods with the greatest appetite, generally had scurvy.[244]

The Therapy of Acrimonies

The medical treatment of diseases which were attributed to 'acrimonies' reflected their similarity to other fluid, mobile morbid matter and impurities in the body. The ultimate goal was to expel the morbid matter. This was often achieved through bloodletting, enemas, mineral waters, and laxatives. The peculiar affinity of 'acrimonies' to the skin, however, made their evacuation through the skin a particularly appealing option, and many patients experienced at least a temporary improvement following such treatment. For this purpose, diaphoretic medications and warm baths were used in the first place, which were meant to both open the pores and assist perspiration. Subsequently, at least in persistent cases, a permanent passageway for the acrid humor could be opened, by applying a glowing cautering iron to create a fontanelle, a chronically festering skin ulcer. For patients and their relatives, there was apparently no doubt as to the necessity and efficacy of such treatments. The mother of a young man recounted how the rash her son had acquired from a maidservant at the age of nine had been carefully treated with baths and herbal infusions until all the 'morbid humor' had been eliminated.[245] When rashes appeared subsequent to warm baths and when secretions ran off from a fontanelle, irritating the surrounding skin, she felt confirmed in her belief that 'acrimonies' had been mobilized and evacuated.[246] In the case of Anna Eisenberger, according to the chronicler, certain potions succeeded in 'driving' the morbid matter from the arms into the thighs. When this threatened to render a leg amputation necessary, the barbers cut into the flesh and burned out the wound to create drainage. After the woman had returned to her cloister, the wound was not well cared for. It closed up and no longer permitted the evacuation of morbid matter. The patient got sick again and died.[247] Sometimes patients themselves or their relatives explicitly requested the painful application of a cautering iron. The sanious or even malodorous secretion that flowed from a fontanelle for months and years to come might be unpleasant but it was palpable proof to them that such an artificial 'issue' was necessary. Monsieur Muroz de la Borde, for instance, who had been suffering from a severe, disfiguring rash around his mouth since his 17th year, had been living with a sanious fontanelle on his arm for no less than seven or eight years. Every morning and evening he had to renew the dressing. Yet on no account did he ask Tissot to close the wound; he merely wanted some additional remedies to soften ('adoucir') and diminish this 'acrid humor'.[248] Quite similar was the case of the elderly Monsieur Gouët: when the weeping rash ('dartre vive') on the back of his head, with its accompanying symptoms, did not recede even after his drinking 25 blood-cleansing bouillons, his daughter saw no other solution but to use the cautering iron and create a fontanelle. Nothing else, she said, would make sufficient amounts of morbid matter flow out and prevent it from further mixing with the blood.[249] Geoffroy confirmed that the man had very acrid blood containing ample amounts of bilious particles.[250]

There was similar agreement between physicians and laypeople – at least, educated laypeople – when it came to prophylaxis. Dietetic measures above all were to be used to prevent further 'acrimonies' from developing. Patients had to avoid any food or drink that could increase the inner heat and cause excessive 'concoction' and a burning-up of body substance. In particular, salty, overly spicy, and smoked foods, as well as 'hot', spirituous drinks, were to be avoided and the blood was to be kept as 'mild' as possible.[251] The necessity of dietary restraint in such cases appears to have made some sense to the patients. The Chevalier de Belfontaine, for example, initially attributed his rash to frequent, even excessive, indulgence in coffee during his sojourn in Turkey.[252] The degree to which people followed dietetic advice when sick – let alone in times of health – is, of course, a different matter. The late 18th-century writer Marmontel, for instance, said of Mme Filleul, who was ill with fever, that the cause of her suffering was an 'acrid humor'. The water cure prescribed by a physician had had a positive effect as it had caused a rash to appear all over her body – the visible expression of dispelling the 'acrimony' successfully. The patient, however, according to the writer, brought ever new 'acrimonies' into her blood by eating spicy foods and ragouts and ultimately succumbed to her disease.[253]

Miasms and Contagia: Plague, French Disease and English Sweat

'Miasms' and 'contagia' figured strongly in early modern learned medicine.[254] In contrast to other types of morbid matter they produced largely the same disease whenever they entered a person's body. Epidemics had been explained in this way ever since antiquity. The word 'contagion' here primarily referred to the idea of a material transmission of a disease from person to person or via objects to which the morbid matter, the 'contagium' or seed of the disease, adhered. This also explains why the notion of 'contagion' was much closer than today to that of 'hereditary' disease. Both surmised a morbid principle of some sort which was passed from person to person. In fact, in German the term 'erblich' ('hereditary') was often used to describe diseases caught from others, through physical contact or by wearing used (or 'inherited') clothing. When a servant traveling in Samuel Kiechel's entourage came down with a suspicious disease, his master prohibited anyone from visiting him because he feared that the servant had 'inherited something of that evil illness'.[255]

'Miasms' by contrast were associated principally with impure or vitiated air. Their main place of origin was believed to be swamps, stagnant water, rotting vegetables, human and animal excrement, cadavers, and the like.[256] Arising from processes of organic decomposition, they became inseparably mixed with the air and 'infected' it in a very literal sense, namely changing its nature in the same way that tiny amounts of dye could color a large volume of water.[257] People and animals that breathed or absorbed such 'infected' air were thus almost inevitably 'infected' themselves. Perhaps the 'infected air from Zeeland' was to blame for his illness, thought one of Jan Heurne's patients.[258] To no avail, related Vincentz from Breslau, had people left their houses to flee the devastating plague of 1567/68; they were unable to escape the epidemic because 'the air was infecting them'.[259] Justus Eberhard Passer was impressed by the results of an experiment during a devastating epidemic in Vienna: physicians tethered a dog above a pit filled with more than a thousand corpses. Four hours later, the dog was dead.[260]

In contrast to 'contagia', 'miasms' could thus arise anew under appropriate conditions and they could usually be suspected from their stench.[261] Nevertheless, the boundary between 'contagium' and 'miasm' was fluid.[262] Patients could infect those around them with the vitiated breath and perspiration caused by the processes of decomposition and putrefaction within their bodies. Hieronymus Wolf's grandfather, for instance, was said to have pulled the young Hieronymus over to him 'with imprudent caresses', thus infecting him with his diseased breath, and in the end the boy became seriously ill as well.[263] If educated laypeople used the terms 'contagium' and 'miasm' at all – which they did rarely enough – it seems they used them almost synonymously. While traveling in Italy, Monsieur Marcard, for example, feared that a 'scabies miasm' had crept into his body, although he had always slept under his own wool blanket on white sheets and thus did not know how he could have acquired 'this poison'. Perhaps, he said, it came from the wool of his newly purchased coat.[264] In the same context, laypeople and physicians occasionally also talked about a 'virus'. Of course 'virus' must not be understood in its modern microbiological sense here. In Latin, the term generally referred to plant and animal fluids and in particular to those that were poisonous or stank. The aforementioned scabietic Monsieur Marcard even used the terms 'miasm' and 'virus' interchangeably.[265] And some physicians, too, thought that venereal diseases, scrophula, and cancer might be caused by 'viruses'.[266]

In the 16th and 17th centuries, the plague was dreaded above all else.[267] It was the 'evil disease' par excellence. In patient letters it is mentioned only occasionally, presumably because its rapid, dramatic course rendered seeking advice by letter pointless. But in autobiographies and similar sources, the plague is ubiquitous. Time and again, cases of infection by the plague in the family or among friends find mention, along with descriptions of its impact on social and economic life.[268] There was a great fear of contagion: 'friends are avoiding each other', wrote Weinsberg about the Cologne plague of 1553.[269] The frequent observation that several inhabitants of one house got sick was enough to suggest the danger of transmission. Measures taken by the authorities such as quarantine and isolating patients fostered such fears. In the Breslau of 1523, for example, nobody was allowed to leave the house, not even to go to mass or to the market.[270] Even children might be isolated from their parents if they showed suspected symptoms of the plague: when Glückel von Hameln's daughter developed a swelling under her arm, the mother argued to no avail that it was merely a flux which had moved there from the head. The girl was separated from her mother and given into the care of two elderly people and a maidservant.[271]

In retrospect, the sense of panic which the plague aroused is entirely comprehensible. Mortality rates were high. Sometimes whole families or communities – of monks, for example – were wiped out. Those who were able to flee therefore did so. And those who had to stay behind tried to protect themselves with bloodletting, amulets, or plague medicines such as theriac, garlic, vinegar, or 'plague pills', and they smoked out their houses to purify the air of 'miasms'.[272] All too often, however, all efforts failed. 'Nothing, neither bloodletting nor medication, helped', recounted Vincentz of Breslau, 'because even those who most diligently looked after their bodies every hour for three days straight were still taken by the poison [and] fell into great heat and ravings'.[273] The reputation of learned physicians was harmed considerably. It was plain to see that they were powerless, and some of them – when speaking privately – admitted that the treatments of even the uneducated 'empiricists' were possibly superior to their own.[274]

The most prominent symptoms of the plague were thought to be headaches, fever, massive prostration, clouding or even loss of consciousness, ravings, and, especially characteristic, boils, which are today explained as swollen lymph nodes. When Weinsberg's sister developed bumps on her body, she thus had good reason to fear 'she had the evil already on her body'; she became severely ill the following day but ultimately recovered.[275] That Christian von Weinsberg had the plague was doubtful, on the other hand, as long as he suffered only from 'great heat', shrieked, and did strange things. Only when a boil appeared on his leg, 'it was certain that it was the pestilence'.[276] Accounts of illness by those who survived the plague underline what a dramatic course the disease took. The 11-year-old Alexander Bösch was attacked so suddenly that his father had to carry him home from the field. In his own words, he became 'so sick and weak that nobody thought I would get up again. Many weeks I had no awareness of myself; my mother and six of my siblings died in the meantime without my knowing it.' Subsequently, he had to learn to walk again.[277] With similar vividness, young Andreas Ryff described his illness during the Basel plague epidemic of 1563/64. From one day to the next he was overcome with the chills. His father arranged for a bloodletting and Ryff went to bed. Later, he was hardly able to remember what had happened in the four weeks that followed: 'the heat and headaches came over me quickly so that I knew nothing of myself and instead fantasized and raged'. He did, however, remember very well the pain when the barber treated his boils or plague spots, since he had thought 'nothing else but that someone had torn my heart out'. From early November until Christmas, he lay 'in great illness, with serious fighting, ranting and raging, and many times was I given up for dead in this time'.[278]

Boils and plague spots on the skin indirectly pointed to the cause of the deadly disease, namely to the plague poison, which threatened to overpower the heart as the center of all the vital faculties. Through the boils and spots, Nature tried to take the morbid matter away from the heart and to free the body of it. Consequently, it was essential to assist Nature in her efforts. If possible, the blood was to be freed preventively of the morbid matter and from any impurities in general through bloodletting. For this reason, Weinsberg would frequently have a vein opened and his blood 'freshened'.[279] In the case of the plague epidemic in Breslau as well, Vincentz recounted how 'everyone wanted to have his blood renewed right away so that death might spare him this time'.[280] Additionally, a large number of remedies were available that were ingested to 'drive the poison away from the heart', as Weinsberg put it.[281] The poison's effect could also be fought using one of the many antidotes that were thought to be 'proven'. Once the typical boils had formed, it was imperative to open them in order to drain the morbid matter from the body.[282]

Another disease that spread fear and terror was the 'French disease'. Sometimes, in the wake of Girolamo Fracastoro's work, also called 'syphilis', its symptoms were, compared with the disease we know by that name today, often much more dramatic.[283] Those affected seemed to rot and molder alive. They stank horribly and died, as was said about the Landgrave Wilhelm, a 'miserable, wretched and horrifying death'.[284] It was common knowledge that there was a particular danger of contracting the disease if one associated with whores or other 'loose' womenfolk. Hieronymus Fröschel in the 16th century, for example, said that, heeding the warnings of his mother, he had carefully avoided 'evil company and careless women' and was therefore spared the disease that so many brought back from Italy.[285] Considering that he had been drunk more than once as a young man and visited a whorehouse four or five times, Hermann von Weinsberg was grateful that he did not get the 'French pox' or 'Spanish disease', which was 'still widespread and many people got miserably sick from it'.[286] There was also a concern about other possible ways of transmission, however. The 'French disease' was an important cause of the generally increasing fear of the sweat, saliva, and excrement of others. Sufferers' skin lesions and obtrusive smell gave rise to fear and disgust. Graf von Zimmern described vividly the visit of Erasmus von Schenk to Archbishop Berchtold. The archbishop had 'had the disease that is called French' several years previously and was left with a 'mark in his face next to the mouth'. Even though he was otherwise considered polite and wise, he still 'without much discretion or deference ate and drank with everyone' and did not take the 'dread or revulsion' of his guests into consideration, who could not express their distaste 'because of his high standing'. When Erasmus had to drink from the same cup with him, he had 'such a dread and aversion [...] that it seemed to him a pain went through his whole body as soon as he had drunk'. Upon returning from this trip, he was bed-ridden.[287] Because morbid matter was also thought to be excreted through the skin, similar fears were stoked when a bed or bedding was shared. As late as the early 18th century, Bövingh, despite his 'clear conscience', was gravely concerned when a barber diagnosed his rash as syphilis, because he had slept in an 'unclean' bed while traveling some time earlier.[288]

A third great epidemic disease, the 'English sweat', must be mentioned. It was at the time a 'new' epidemic, the nature of which has sparked numerous speculations but has so far not been conclusively identified in today's medical terms. Swarms of people became its victims. It was a 'horrifying plague', recounted Lucas Rem, who became ill himself, as did his wife. According to Rem, many took flight from it, and most of those who stayed behind became ill, many of them dying.[289] The typical symptoms were severe flu-like complaints, unbearable thirst and an intense, stinking flow of sweat – palpable evidence of large quantities of morbid matter draining from the body. People tried to keep patients warm, as Oldecop from Hildesheim recounted, so that the 'Swet nicht insloge', that is, to prevent the morbid matter from turning back toward the inside and affecting the heart and other vital organs.[290] It was also believed that patients should not be given anything to eat or drink and that, come what may, they were to be kept awake and sweating for 24 consecutive hours. It was an ordeal. According to Vincentz's retrospective account, one could hear 'shouting and pleading for a drink of wine or water in every house because the sick person's heart was about to burn up'.[291] Weinsberg described how an acquaintance of his by the name of Gymnich became sick with the 'sweat'. He was put to bed, buried under a thick comforter, and forced to stay there. He begged the women and children for something to drink, but, for fear of killing him, they did not give him anything, no matter how hard he pleaded. Finally, he implored his neighbor for some air and above all a small glass of wine, because 'my heart will otherwise burn away'. When the neighbor angrily berated the man's relatives, saying they were about to kill the sick man and proceeded to give him some wine, everybody yelled 'murder, murder'. But Gymnich soon felt better and recovered.[292] Several others were said to have hidden for fear of this treatment until 24 hours had passed and they felt better again.[293]

A similar 'new unheard-of epidemic' that went by the names 'Böhmischer *Ziep*', that is, literally, 'Bohemian twinge', and 'grüner Schafhusten'('green sheep's cough') raged around 1600 in Silesia. The illness began with headaches and a temperature, followed by a cough, hoarseness, a sore throat, and attacks of sweating which sometimes soaked the whole bed, as well as 'anguish of heart'. Survivors felt weak and tired for a long time afterward, 'as if a mild poison had got into the blood'.[294]

The 18th century was largely spared the big waves of epidemics of the previous centuries. The last major plague epidemic was in Marseille in 1720. The reactions to it – like the reactions to the cholera which swept through Europe for the first time in the 19th century in several devastating waves – show that the belief in and fears of transmissible diseases remained very much alive among the general public. Against the opinion of physicians from Paris, people in Marseille, along with the local physicians, insisted that this epidemic was of a contagious nature. According to a contemporary account, patients were abandoned to their fate without hesitation, because everyone saw that often whole families got ill, and surgeons, nurses and gravediggers died in droves.[295]

During the cholera epidemics of the 19th century, wealthy people in many places fled to the countryside or to towns that were allegedly free of the epidemic. Physicians and magistrates sometimes did the same, causing, in some places, the near collapse of medical care, the economy, and the municipality. Again the general public usually remained more convinced of the contagious character of the epidemic than did some physicians. In the physicians' eyes, the empirical evidence was contradictory. The epidemic broke out in many places where there had been no known contact with affected places. On the other hand, many physicians and nurses who had been in close contact with cholera patients did not become ill.[296]

Aside from such large epidemic waves, people were also confronted with frequent local or regional epidemic outbreaks of diseases such as smallpox, 'spotted fever', and 'red dysentery' with its bloody stools. Transmission was feared here as well. Bövingh, for instance, talking about a dysentery epidemic in 1719, felt it was worth underlining that he visited patients 'without dread'. It was 'due to God's shield' that he stayed fresh and healthy, although he 'was at times surrounded by many who were already afflicted with the contagion'.[297]

The extent to which laypeople reverted to the concept of contagion or 'miasm' to explain individual disease when no epidemic was taking place cannot be gleaned from patient letters or other personal testimonies with sufficient clarity. Patients rarely referred explicitly to contagion as the purported cause of their illness, as did a patient of Verdeil who thought he had possibly contracted his 'pneumonia' through a 'contagium'.[298] In some cases, the transmissible character of a disease must have seemed so apparent or was taken so much for granted that no explicit mention was called for. This would have been true, for instance, of smallpox, measles, and whooping cough, or rabies, which one of Tissot's patients feared contracting merely by touching with her lips an object that might have come into contact with the saliva of a rabid animal.[299] Similarly, men attributed their venereal diseases to intercourse with a 'suspect', 'impure' woman, though they did not explicitly call it a 'contagion'. The medical literature of the time also frequently alludes to widespread fears among the general public of becoming infected with consumption or falling sickness. And it was apparently common knowledge that overcoming a contagious disease gave a certain protection against it for the future. Hieronymus von Holsten, while on a military expedition, was alarmed when he encountered people who were sick with smallpox in a monastery, 'because I had not had it yet'. Soon afterward, he was sick himself.[300] For a similar reason, Duchess Sophia, ill with smallpox in 1593, advised her godmother against visiting her, because the godmother's young daughter 'may not yet have been burdened with that same disease'.[301]

A further element which formed part of the dense 'semantic network' surrounding such terms as 'contagium', 'miasm', and 'virus', was the notion of pathogenic animalcules of varying sizes. Frequently, diseases were attributed in particular to worm infestation. The symptoms could be dramatic. A patient of Geoffroy, for example, believed that she was suffering from a worm in her stomach. For four months, time and again, the worm moved up into her chest and throat and took her breath away, and when she ate, it went down into her stomach to feed.[302] Similarly, a 54-year-old man attributed the tension and heaviness in his stomach to a worm because he felt a movement inside himself and suffered from ravenous hunger whenever he had not eaten anything in a few hours.[303]

Cramps as well were attributed to 'worms' and, therefore, treated with vermifuge.[304] The intense dizziness, fainting spells, pains in his right side, and melancholy moods of a 63-year-old businessman led his physicians to diagnose a nervous disorder. He, however, reckoned it was a 'flat worm' and in the end the physicians respected his wish to be treated for it. They prescribed him the most varied vermifuges and, finally, he did 'give birth' ('accouché') to a worm and felt considerable relief in his right abdomen. He was, however, disappointed in his hope that all his complaints would end.[305]

In the lay medical culture of the German countryside various kinds of worm were known, too, though in retrospect their significance for the everyday experience and interpretation of illnesses is difficult to gauge precisely. The 'tooth-worm' as a major cause of toothache sometimes seemed to become visible at the tip of the root when a tooth was pulled. The 'heart-worm', which caused complaints in the upper abdomen, and the 'consumptive' or 'hectic' worm ('Zehrwurm') were sometimes deemed responsible for emaciation in children.[306]

Indigestion, Winds and Slime

Nutrition and digestion have always been paramount in Western medicine. Old Galenic medicine grouped foods – as it did medicinal substances – above all according to their primary qualities, their degree of heat, coldness, moisture, and dryness. In the early modern period, however, the focus was more on the degree to which various types of food burdened the stomach and the body in general, on how easily they could be digested or concocted. After all, food and drink entered the body as foreign matter. The body had to assimilate them in a literal sense. It had to adapt their quality and consistency to its own substance and to integrate it. The more the food – or medication – was similar to the body, the easier it could be digested and assimilated. 'It seemed as if from my own body', a sick nun said to emphasize how agreeable a medication was.[307]

Usually, however, the assimilation remained incomplete even in the case of agreeable nutrition and healthy digestion. The stench of feces and urine provided tangible evidence of how much impure, even rotten, matter the body ingested along with food. In this sense, even milk could be said to create 'much dirt and debris in the body' and therefore to make frequent 'purging' necessary.[308]

Physicians held somewhat contradictory notions about the exact processes of digestion. As we have seen, according to the old Galenic doctrine, digestion was a process that went through several steps of heating or 'concoction' of the food, achieved by 'inner heat'. In a first step, food became chyle in the stomach. In a second step, the warm liver concocted the chyle to good, pure blood. In a third step, the individual body parts or organs assimilated those parts of the blood which were appropriate to their respective natures.[309]

The various steps of this Galenic model of concoction were only sporadically mentioned by laypeople. The 'second digestion', thought a patient of Samuel Hahnemann, was in his case particularly laborious.[310] The 'second digestion, whenever the first one happens to go well', was also at the root of the severe stomach aches from which a 37-year-old female patient of Tissot suffered after eating.[311] For most members of the general public, however, it seems that it was sufficient to know that raw food needed to be heated in the body so as to lose its 'raw', impure, and harmful nature. For this reason a 'cold' stomach was a matter of particular concern.[312]

In the 17th and 18th centuries, chemiatric and mechanistic physicians developed new interpretations of digestion such as fermentative, chemical breakdown or mechanical crushing. In the patient letters, these newer models left hardly any traces, however. The interpretation of digestive processes continued to be bound up in the traditional Galenic kitchen imagery in which food was 'cooked' in the body like vegetables and meat in a cooking pot. Only occasionally patients or their relatives mentioned a 'ferment' in the stomach or intestines, usually only to account for flatulence or other presumable effects of a preternatural 'effervescence' or 'fermentation' in their belly. 'It must be assumed', a patient suffering from burping who felt pressure in his upper abdomen wrote, 'that a yeast in the stomach makes digestion difficult or ruins it entirely'.[313]

If the digestive heat was too 'weak' or the stomach too 'cold' or the body's ability to digest was overtaxed by too much and/or indigestible food, various complaints loomed. They fall roughly into three categories: cachexia, winds, and the accumulation of mucus or slime.

If the body was unable to assimilate all the food ingested, physical decay followed in the form of cachexia. Cachexia was an important subject of medical debate, but it played almost no role in the patient letters. Massive weight loss was sometimes mentioned in the letters but it was usually associated with the effects of a 'consuming heat' or of the loss of vital substance due to excessive expulsion.

A second, more typical, result of an overtaxed digestive process came in the form of flatulence, wind, and acid reflux, and this was indeed frequently experienced by patients. The patients complained about pressure and pains, particularly in the upper abdomen, and were noticeably relieved when the air had found a passage out. To them the constant release of wind seemed in fact indispensable to good health. Wind was produced even when digestion was good and, like humors, had the capacity to roam throughout the body and do mischief. A French infantry officer, for example, complained about the terribly strong wind that rose from his upper abdomen to his chest, at times forcing him to open his collar in the evening.[314] With other patients, the volatile vapors rose even higher, as in the case of the medical councilor Greiffenclau in 1665, who commented on his headaches by saying: 'To my mind, the present state can be ascribed to nothing else but weakness of the stomach ['imbecillitas ventriculi'], from which an abundance of vapors rise into the head, bringing about the cruelest pains as well as ceaseless buzzing and droning.'[315] Similarly, Johann Lindt wrote that his 'ailing in the head' came from evil vapors for which his weak stomach was to no small degree responsible.[316] One patient even had to wrap his head tightly in a scarf, because, he wrote, 'I believe the wind, if I didn't wrap something around my head, would split it apart by attacking it impetuously'.[317] Similar notions of morbid wind straying through the body were reported as current among the rural population as late as the 19th century. Country folk, it was said, regarded 'deviated' wind, alongside fluxes and worms, as the most important cause of disease and had it 'roaming in the most diverse, remotest corners of the body, causing all kinds of trouble'.[318]

It thus becomes clear how passing wind and burping could be praised for centuries as beneficial to health. Wind was not to be retained, said Johannes Wittich, because 'many diseases can result from this'.[319] Even a leading humanist like Erasmus of Rotterdam recommended that it was better to hide the noise politely with a cough than to retain the wind inside the body.[320] Accordingly, the lack of self-restraint with which Dutch travelers in the 17th century let go of their wind in the presence of others is reported to have struck the native population of West Africa as odd indeed.[321] Only gradually, as the more 'animalistic' aspects of human life were increasingly banned from the public sphere,[322] breaking wind and belching in the presence of others came to be considered indecent. Only sighing and yawning – at that time also interpreted as a means to expel vaporous waste – remained admissible, provided the mouth was covered with a hand.

The third and most frequently mentioned result of 'weak digestion' or a 'weak stomach' was the accumulation of undigested, liquid food remains and the development of 'raw', 'cold' phlegm, mucus, or 'slime'. This accumulated at first in the stomach and surrounding area. Accordingly, a 30-year-old patient of Hoffmann wrote that the medication had made his 'stomach and both hypochondria completely slimy and as if glued together'.[323] Because his stomach 'was piled up high with a lot of old slime', Hans Ulrich Krafft was 'not [able] to eat or drink much'.[324] And a third patient complained that his meals all turned into 'acid and slime'.[325] In the end, the mucus could spread throughout the body. Brother Placidus in St Gallen, for instance, suffered from stomach problems, lack of appetite and insufficient 'concoction' of his food as well as from 'fluxes from the head' and dizziness.[326] And Anna Post wondered if the 'phlegm' that had for years been descending from her head into her throat and chest did not perhaps originate in the stomach.[327] Hieronymus Birckholtz thought it likely that the 'superfluous watery moisture in the head and stomach' resulted from 'a lack of natural warmth'.[328]

When stools or other excretions were covered with mucus, this was thought time and again to prove such causal links.[329] Treasurer Monsieur de Soulas, according to his own estimation, had lost more than a hundred pounds of mucus thanks to the administration of numerous emetics and clysters he had been prescribed for his eructation and complaints in the upper abdomen.[330] A 78-year-old Benedictine monk was convinced that not only his mucous sputum but also his urinary pain was due to 'bad digestion', and he believed he could actually see the mucus in his urine.[331] Similarly, an aristocratic patient of Hoffmann reported that either in his urine swam 'a great quantity of viscous slime, which you can picture as sludge in standing water, or there are many red particles in it'.[332] Others saw similar mucous admixtures in menstrual blood or when their veins were cut in bloodletting. 'The blood ordinarily looks very black and knotty', it was said, for example, 'with a blue, slimy skin on top'.[333] 'Direly blue blood' was found in the seriously ill Frau Bösch.[334] Depending on the case, mucus could also be malodorous: according to Philip von Farnrode's account, several people thought that his intensely bad breath came 'from loose slime and moisture that supposedly stuck to the inside of the pharynx and stomach'.[335]

Even if, in German texts, the terms 'Schleim' and 'Phlegma' are often used interchangeably to refer to mucus in the sense I have just sketched out, they had very little to do with the idea of phlegm as one of the four natural bodily humors. French patients and their relatives often made the difference clear by using the term 'glaires' for this kind of mucous discharge instead of 'pituite' or 'phlegme'. In the German language, mucus was often also qualified as 'old' or 'viscous'.[336] And insofar as 'viscous' mucus could easily build up,[337] the transition was made to a further concept central to the experience of disease and the body: obstruction.

Obstruction and Disrupted Excretion

As we have seen, an essential attribute of early modern ideas about the body – one that stands out particularly in comparison with today's conceptions – was the belief in the body's permeability. Fluids, wind, and vapors seemed to move freely within the body, virtually uninhibited by any anatomical boundaries. In individual cases, patients even believed that solid matter – pieces of their inner organs – were able to wander throughout the body and would ultimately be excreted. One of Tissot's patients, for instance, found that his stool contained pieces the consistency and taste of which were strikingly similar to that of cooked liver; worried about his liver, he had apparently not even shied away from putting some of it into his mouth.[338] The body's boundaries were similarly permeable. Krafft, whose slime-filled stomach would no longer tolerate warm food, recounted: 'Whenever I used to eat even a little minestrone or warm soup, a visible vapor would rise from the back of my neck, surprising those who saw it.'[339]

The permeable body also offered a convincing explanation for the many cases where symptoms changed during the course of an illness or even an entirely different picture emerged. The morbid matter had simply changed its location within the body but it still was essentially one and the same disease. Since the beginning of her uterine dropsy, recounted a patient of Tissot accordingly, she had hardly suffered any more from her cerebral catarrh ('rhume de cerveau') and chest ailments; apparently the liquid had gone to the uterus, she thought.[340] One of Geoffroy's patients suspected that she had a whole warehouse of humors ('un magazin d'humeurs') in her head. First, they had given her severe headaches, then they had 'thrown' themselves upon her neck and between her shoulders and, finally, they had 'fallen' into her loins, thighs, and legs. She hoped that an artificial ulcer, a fontanelle, might be able to stop the flow of humors.[341] And a third patient, Mlle Darmenon, retained a swelling above the ankle subsequent to a serious fever disease. The swelling 'sometimes rose up into the stomach and then into the throat and finally to behind her ear, where it could be seen'.[342]

The permeability of the body was also a prerequisite for good health. The orderly performance of the bodily and mental functions and even life itself were founded on the continuous movement of fluids, spirits, and vapors within the body, as well as across the body's boundaries. This movement was, however, always considered to be in peril and all the more so as, in the course of the early modern period, it came to be increasingly located in discrete anatomical channels, above all the blood vessels. The flow could be slowed or even blocked altogether by any thickening or agglutination of the fluids but also by anything that caused a narrowing of the channels or ducts. The result was stagnation or obstruction (French: 'obstruction'; German: 'Stockung', or 'Verstopfung', Italian: 'ostruzione' or , 'oppilazione'), which had far-reaching consequences on one's health and life.[343]

Roughly speaking, disruptions in the flow of fluids came in two varieties. The life-preserving stream or circulation of blood and other humors in the body itself could be affected, or the expulsion of a superfluous or harmful humor to the outside of the body could be impeded. I will first take a look at the latter case, which took on far more importance in lay experience and interpretation of illness.

If the substance was corrupted and dangerous, grave consequences threatened to come sooner or later if its excretion was prevented. The substance accumulated in the body and, in the worst cases, putrefied even further. The excretion of stool – whose smell and consistency left no doubt about its impure nature – was rarely entirely blocked, though if ever this did happen, the dramatic result was fecal vomiting, still known today as 'miserere'. But a certain sluggishness of the bowels was a widespread complaint and gave some patients great cause for concern. One of Tissot's patients allegedly stayed at times three to four hours on the toilet due to his obstinate constipation.[344] When an excretion of feces was late and the rotting vestiges of food remained in the belly too long, danger seemed imminent. Keeping the body 'open', as was a common phrase at the time, was therefore an important concern. The 69-year-old local magistrate Stambke, who had gouty or 'podagrical' 'debris' stuck in his blood, feared all the more for his health as his nature, in his own words, 'tended very much toward obstruction' and he sometimes did not have 'an open body' for two days in a row.[345] Laxatives and clysters were accordingly widely used for the prevention of disease, and some patients even applied them on a daily basis. In France, they were reported to have become part of people's regular personal hygiene.[346] While the somewhat milder clysters were popular among the upper classes, common folks apparently preferred drastic laxatives.[347]

Stone diseases were considered the main reason behind a severely reduced or, in the worst case, interrupted flow of urine. Patients described the small and large stones they excreted with their urine,[348] and some ultimately agreed to have the stones removed in a painful and dangerous operation. Conversely, other patients experienced the beneficial influence of a plentiful flow of urine on their health. His frequent flow of urine must have been what kept him alive, it was said about 63-year-old Monsieur Laval, who had 'great cacochymy', that is, vitiated humors and 'degenerate blood'.[349]

The skin as well, as described above, was an indispensable organ of excretion in the early modern understanding of the body and its diseases. Skin was literally 'porous', riddled with minute pores or ducts that connected the inside of the body with the outside world. Physicians and laypeople alike believed that, through the skin, the body was able to rid itself of liquid and volatile waste, impurities, 'acrimonies', and morbid matter. The belief in the purifying, liberating effect of sweat and perspiration had a deep impact on the contemporary body experience. Frequently, people believed they could feel how an increased flow of sweat or breaking into a sweat improved their condition significantly. 'I experience relief only when I am able to transpire or sense that I am producing white discharge', recounted one Dutch patient, adding that she unfortunately sweated only very little.[350] She had been suffering greatly in her head, chest, and limbs, wrote another patient, but 'a plentiful sweat appeased my pains'.[351] At times, the harmful substance that left the body with the sweat revealed itself through a change in the smell, taste, or consistency of sweat. In this sense it was not only victims of the 'English sweat' who complained about their 'acrid', 'sour', or 'fatty, sticky' sweat.[352]

If, on the other hand, one's sweat did not flow sufficiently, the harmful, impure substance remained inside the body. With this in mind, numerous letters discuss the 'interruption', 'suppression', or 'halt' of their perspiration as well as the grave consequences of this. One patient, for example, used to have very sweaty armpits as a child. She recounted that this had stopped four years ago, and she assumed that the humors whose flow was 'obstructed' in this way had much to do with her illness. Monsieur Laval wrote of something similar: he used to sweat so much in the past that he soaked several shirts each day. Since the previous fall, however, the sweat had stopped and he had been feeling worse ever since. He was suffering from headaches and backaches, dizziness and a slight fever, and intermittent swelling of his feet and legs. In the end, plagued by shortness of breath, an irregular heartbeat, and an overwhelming weakness, he saw death approaching.[353]

Particular attention was given to sweating from the feet. Maybe the patients and their relatives considered the feet an especially auspicious exit point because they were furthest from the vital organs. Achatius Trotzberg, for example, plagued by 'abdominal pain' and 'shooting pains in the joints', thought that if only his feet 'could be made to sweat again as they did in youth', he would 'be well on his way to recovery'.[354] Around 200 years later, the 33-year-old bailiff Bruckner described the harmful consequences of preventing the excretion of sweat from the feet. His feet had been very sweaty and quite a nuisance to him, and even though he knew about the dangers of 'suppression', he had unfortunately put fat on his aching feet during a long walk. Promptly, the perspiration had decreased and it had remained reduced ever since. Since that time, his eyes had ached and grown ever weaker; images of spiders and flies were buzzing before his eyes.[355]

Similarly, people were disappointed when their expected recovery failed to materialize in spite of plentiful sweating. He had sweated in the night, complained Monsieur Marcard, but nonetheless 'suffered all morning from the most horrible melancholy'.[356] In the case of Philipp Jacob Spener as well, his breaking into a profuse sweat fostered the deceiving hope of recovery from his serious illness after months of a reduced flow of sweat; he died soon afterward nonetheless.[357] Another patient feared that nothing could help her with her throbbing head and ear ailments because she had been sweating constantly for a long time and still experienced no relief.[358]

The main cause of a 'suppressed' flow of sweat was a cooling of the skin, especially when it happened suddenly. This notion was based on the idea that skin and fibers contracted in the cold, causing the excretory ducts to become narrow or even close up entirely. The excruciating pains of 37-year-old lieutenant Roussany, for example, had begun after a night out at a ball. That night 20 years ago, he thought, his sweat had been suppressed by the dancing or the cool of the night. Since then he had been suffering from intense pain in the legs, had at times been confined to his bed, and had been living a life of 'pain and sadness'.[359] Monsieur Lavergue dated the beginning of his complaints to public ceremonies in Lyon, which he had watched for hours from his open window out of 'stupid curiosity', exposing himself to the cool evening dew. Even prior to this, he had noticed how working long hours in his cool office had suppressed his perspiration. The consequences had been coughing and bringing up phlegm, sometimes with blood in it.[360] Similarly, a baron from Bamberg complained that sleeping in a cold room had made his usual morning sweat run dry, promptly causing his old, intense pains to return. For this reason, he now asked for a remedy to free himself from the 'bad matter' that had spread throughout his body.[361]

Apart from external cold, it could be dirt or fat that clogged the pores. This was one of the main reasons why health guides of the 18th and early 19th centuries advised their readers to wash regularly. Washing, it was thought, not only prevented those bodily excretions that had accumulated on the skin's surface from reentering the body;[362] it also kept the pores themselves open and permeable, thereby guaranteeing sufficient perspiration.[363]

In view of the beneficial effects of sweating on health, the obvious thing to do in order to prevent and treat diseases was to promote perspiration. Some patients stayed in bed all day to 'entertain the sweat'.[364] The salubrious effects of a warm bath were also considered by patients (as they were by physicians) to lie principally in assisting excretion via the skin; the effect was further aided by subsequently wrapping oneself in warm blankets.[365] Mme Neider was therefore disappointed when warm baths, despite their heat, had produced no perspiration and had given her no relief; quite the opposite, she was even less capable of walking than before.[366] So-called 'sweat baths' were also part of the common repertoire of treatments.

The belief in the beneficial effects of sweating permeated all social classes. To promote sweating, recounted a physician from Franconia in 1828, people subjected themselves to the 'greatest external heat'.[367] The use of sudorifics, of sweat-promoting remedies like tea, spiced wine, and pepper was also firmly established in rural areas.[368] Even with women in childbed, complained one physician from the Upper Palatinate, the belief that they had to be kept warm was 'carried out in the most exorbitant way', by keeping them sweating all the time with hot tea and blankets.[369]

In learned medical writing from around 1600, 'insensible' perspiration, in contrast to visible, palpable sweat, attracted considerable attention. Santorio Santorio had demonstrated with his famous weighing experiments that human beings constantly gave off matter via the pores, in an amount that far exceeded all other excretions.[370] In academic medicine, such notions soon met with wide recognition and were also mentioned by physicians in letter consultations.[371] The patient letters, however, only vaguely suggest, at most, that educated laypeople were aware of the difference between visible sweat and imperceptible perspiration. Occasionally patients used the term, writing, for example, 'it seems that my imperceptible perspiration is suppressed'; but they may well only have been repeating a physician's words.[372] Ultimately, the new concept of imperceptible perspiration apparently remained largely foreign to lay medical culture. This may in part have been for linguistic reasons. In French and English the terms 'perspiration' and 'transpiration' were etymologically much more closely related to 'respiration' than the German 'Schweiß' to the corresponding German term 'Atmung'. They were already fraught with images of volatility which left little need for a new word. But the new concept hardly appears in German patient letters either. Maybe the notion of an 'imperceptible' perspiration was too disconnected from people's everyday bodily experience. The palpable and visible sweat, which in many cases of illness increased or changed, offered an entirely sufficient explanation that was easier to grasp.

With girls and women of child-bearing age, menstrual bleeding constituted an important additional route of elimination. As mentioned earlier in the context of plethora, physicians' notions about the nature and the value of menstruation came to differ markedly from those of women. This is one of the few areas in which we encounter a chasm between the interpretations of physicians and those of educated laypeople. From the late 16th century, the large majority of physicians understood the menstruation of a healthy woman as simply a relief from superfluous but otherwise good, nutritious blood, resembling the blood of freshly slaughtered animals.[373] This view went against the previously prevailing cathartic interpretation, which saw menstruation as a purification from harmful poisonous matter via the uterus, the 'cesspit' of the female body, as it was sometimes drastically characterized. The harmful, poisonous nature of menstrual blood was not only clear from the manifold diseases that were observed to befall women who accumulated it in their bodies when they stopped menstruating due to sickness or age. According to the writings of Pliny and other ancient authorities a few drops were enough to make plants wither and to kill insects and worms. Dogs that licked it became rabid. Secretly mixed into a love potion, it made men go mad. The sheer presence of a menstruating woman made wine and beer turn sour, spoiled preserved meat, and caused bread-baking to go awry. Her gaze made mirrors lackluster and swords dull.[374]

In the letters of sick women, and to some degree in those of their relatives as well, menstruation played a central role.[375] In numerous cases, the period was mentioned, if only briefly, for example, to assure a physician that it was regular and sufficiently copious, or that the woman in question was experiencing none of the 'indispositions' 'that usually go along with it'[376] and hence the complaints had to stem from a different source or the diagnosis of a uterine cancer must be wrong. Indeed, because menstruation was considered so important for women's health, it seems in some respects to have been less a taboo topic than today. Fathers, husbands, and even male neighbors wrote in fair detail about the amount and consistency of a sick daughter's, wife's, or neighbor's menstrual blood.

Occasionally excessive menstrual bleeding was the principal complaint. It was, above all, associated with the danger of weakening the body and with dropsy. Most of the time, women and their relatives were worried, however, because the amount of menstrual blood had diminished or the periods had shortened. Even in the absence of any other symptoms, this was sometimes reason enough for a letter consultation. Indeed the sheer number of cases in which a disrupted menstrual period was the main or even sole reason for consulting a physician leaves little doubt that most women still adhered to the old interpretation – by then dismissed by most physicians – that menstruation was a catharsis, a purging of bad, impure blood. This conviction is also evident in the widespread fear of menopause, the natural cessation of menstruation with age. In the middle of the 19th century women were, according to physicians, [377] still extremely worried about this change, because they believed that menstruation normally served to rid the body of foul, corrupt morbid matter at regular intervals. When periods stopped, this almost inevitably put a woman's health and life at risk.[378]

By continuing to insist on to the purifying, cathartic function of menstruation, women held on to a view which, in many respects, implied a far more negative image of women than that held by physicians. It was an image that associated women and their sex with dirt, corruption, and sin. However, women's insistence on the old cathartic view cannot simply be explained by outright resistance to medical innovation. The new doctrine of the sensibility of the nerves and of the particularly nervous excitability of women, for example, became established – as we will see later on – among educated women of the 18th century in a matter of a few decades and, to some women, it even opened welcome new possibilities for self-fashioning. Women's insistence on a cathartic understanding of menstruation is probably better explained by deep-seated, long-lasting cultural influences. It appears that the old interpretation of menstruation as purging or cleansing simply suited women's subjective perception and experience of their bodies better than the new medical theories. In retrospect, this female experience of a beneficial cleansing was the result of a long-standing historical process. It had created a particular 'habitus', to use Pierre Bourdieu's expression, which shaped women's deeply embodied experience and made it appear as a simple fact that their bodies needed regular cleansing, a fact proven countless times by their feelings of relief, of purification when the monthlies set in.[379] Even today, long after the decline of humoralism, many women experience menstruation as liberating and purifying. And even the beliefs in the harmful influence of menstruating women and in the poisonous nature of menstrual blood survived until very recently. Far into the 20th century, menstruating women were in some areas of Europe kept away from wine cellars or bakeries for fear they would make the wine turn sour or keep the dough from rising.[380] Apart from that, common sense seemed to contradict the physicians. Women knew from experience that menstrual blood did not actually look and smell exactly like 'pure' blood, such as that from a bloodletting.

In addition to its natural cleansing effect in healthy women, the body could use menstruation in times of sickness to eliminate the morbid matter of the disease in question. As in the case of skin lesions this evacuation as such was highly welcome but also raised great concern. After all clots, mucus, pus and similar changes in the color and consistency of the menstrual blood were only the visible manifestation of a serious disease inside and might indicate the most horrendous of all female diseases, cancer of the uterus. For eight weeks, wrote Johannes Hancke in 1581, his wife had not had her 'usual red flux'; then it came back but looked 'like pus from a bloody ulcer' and 'smelled bad'.[381] For the Comtesse de Mouroux, a thick, ill-colored menstruum and a period that lasted only two days were her greatest chagrin.[382]

Besides feces, urine, sweat, and menstrual blood, other routes of elimination also came to the fore in some cases. We would regard these routes today as pathological, but at the time they were primarily interpreted as part of the body's or Nature's efforts to dispel harmful, poisonous, impure matter from the body. We have already looked at some of them: nasal discharge, vomiting, bleeding from hemorrhoids, the nose, the stomach or the lungs, and rashes. These excretions, too, could cease as a result of 'obstruction' and, above all where habitual excretions were concerned, similar consequences were feared as with the 'suppression' of sweating or the cessation of menstruation. The body, it was assumed, had become accustomed to excreting harmful substances via this route. If it was now blocked, these substances would necessarily turn back toward the inside of the body. Hence 'suppressed' hemorrhoids or catarrhs could, in the experience of those concerned, have an extremely negative effect. His father died at the age of 58 from the 'consequences of an obstruction of the hemorrhoids', wrote Monsieur Bruckner, for example.[383] Snuff was therefore considered conducive to health as it aided nasal excretion.[384] And even an unhindered discharge of ear wax could seem helpful for maintaining good health. For 30 years, Alexander Bösch cleaned his ears every morning with the help of a silver instrument and was convinced that if he did not continue to do so, he 'would have to suffer particular fluxes, headaches and earaches'.[385]

Stagnation and Deposits

Patients' and physicians' fears about the consequences of a disrupted or obstructed flow of blood and humors centered on excretions, on the flow across the body's boundaries. The movement of fluids within the body could also be disturbed, however. The consequences could be harmful, above all for two reasons.

It was possible in the first place that the stagnating blood or humor would lose its natural healthy condition. It would go bad or rot and be transformed from a natural substance into harmful morbid matter. If this substance did not soon find a way out of the body, it threatened to spread through and 'infect' the entire body, or it could form a local 'deposit', an 'abscess', or a so-called 'metastasis'. In this sense, obstruction and stagnation were part of the same semantic sub-network as the 'fluxes' and 'catarrh'.

Second, obstructions and stagnation could have quite dramatic, harmful local effects, at the place of the congestion. Especially in lay accounts, obstructions appear almost like a mechanical, hydraulic problem. The stagnating matter accumulated, extending and overstretching the vessel or the space in which it was contained. Some patients concluded that they had an 'obstruction' just because they felt tightness or pain in a certain area. In many cases 'obstructions' were also palpable from the outside, especially in those areas where they seemed to occur most frequently, namely in the belly and its organs. Physicians and laypeople alike feared in particular an 'obstruction' of the liver and spleen.[386] Accordingly, the main reason for asking a physician or surgeon for a physical examination was a suspected 'obstruction'.

Harm could also arise from the backlog of fluids behind the obstructed duct or organ. A healer explained to a relative of Vincentz that her complaints came from, among other things, the 'stagnating blood in her liver'. He therefore recommended physical exercise and 'medication for the pain, swelling, impurity, obstruction of the liver' as well as to 'drive out the coarse slime and bile that is in her liver'. Even the buzzing in her ears, thought the healer, came from the bile that had risen to her head.[387]

Cancer

The most terrifying consequence of an 'obstruction' and the ensuing local accumulation or deposit of morbid matter was the development of hardened, scirrhous knots or tumors and cancers. Cancer thus serves as a particularly vivid illustration of the central importance of the undisturbed flow and purity of humors in the early modern experience of the body. This notion of a hardening deposit was complemented by other explanatory elements which helped people understand the symptoms and effects of cancerous growths. Some of these we have already encountered. 'Acrimonies', in particular, were often evoked in order to account for the ulcerated degeneration typical of the advanced stage of the disease.

In modern Western societies, cancer is an illness fraught with emotional, metaphorical, and political meaning. Its causes are frequently seen in factors which lie outside of the individual's control: in hereditary predisposition and harmful environmental influences ranging from nuclear power and 'electro smog' to carcinogenic substances like asbestos and benzene. Like few other diseases of the body, cancer also tends to be attributed to psychological traits. With cancer, according to this idea, 'bottled up' emotions like anger or grief attack one's own body and literally consume it.

Cancer is so much seen as emblematic of our time that there is little awareness that what was then also called 'cancer' was quite common already in premodern society.[388] In the early modern period, cancer ranked among the diseases which aroused the greatest fear. It was considered 'the disease that is the most miserable among all the diseases to which the human body is prone', especially because of 'the cruel pain and the unbearable putrefaction which consumes the body bit by bit while it is still alive', and because the course of the disease often stretched over several years.[389] Personal testimonies of cancer patients such as the horrendous accounts by Anna von Österreich and Margarethe Milow communicate very well why there was such fear.[390] Few other diseases were characterized by such drastic signs of decay, for everyone to see and smell, as cancer. And few other diseases caused such unbearable and persistent pain.

The learned medical understanding of the nature and genesis of cancer differed fundamentally from that of today, however. Medical authors did not always agree on the details, but generally they assumed that cancer originated from an accumulation of morbid matter, due to a local obstruction of humoral flow or due to morbid matter which moved there from elsewhere in the body. Traditional Galenic medicine associated cancer closely with black bile.[391] In the 18th century, more attention was given to pathological lymph and to 'acrimonies' which, as we have seen, shared with black bile an acrid, biting nature.[392] According to physicians' experience, cancer usually developed from a hardening, a scirrhus, which was sometimes palpable, for example in the breast. Once it developed into cancer, it caused pain and increasingly affected the surrounding area until it finally broke through the body's surface, releasing putrid, purulent morbid matter.[393] Cancerous ulcers provided the best evidence of the role of 'sharp' black bile or 'acrimonious' matter in the genesis of cancer. After all, ulcers on the skin or mucous membranes could best be explained in terms of an acrid, corrosive substance. What else could eat flesh and skin away so dramatically? Also, the secretions from festering cancerous ulcers seemed to irritate the surrounding skin. Thus the major reason why a husband was not prepared to believe that his wife had uterine cancer was that, following intercourse with her, he had experienced no adverse consequences. Tissot, he thought, would surely agree that 'uterine cancer as the cause of all these discharges would certainly have given the whites enough poison and acrimony to infect me, if it really were that'.[394]

Today cancer appears above all as an inscrutable, uncanny, hidden disease process inside the body and common in both men and women. In order to grasp the early modern meanings of cancer, however, it is important to keep in mind that the disease was at the time associated much more closely with images of a decaying surface. Pathological changes inside the body were, in most cases, inaccessible to patients and physicians. They could be recognized only indirectly, from the pain and other subjective sensations which patients reported or from the effects on the general well-being of the patient. Only those forms of cancer which were close to the body's surface and, in many cases, sooner or later destroyed it, would usually be diagnosed and experienced as cancer.

This also explains another striking difference in comparison with today: cancer was experienced as a highly gender-specific disease. It is no coincidence that early modern letter consultations for (presumed) cancer patients almost exclusively related to women. Aside from skin cancer, which is comparatively rare, it is largely breast and uterine (cervical) cancers that manifest themselves through palpable knots, ulcerous decay, and ugly, smelly discharge from the body. Diagnosing cancer of the intestine, kidneys, stomach, lungs or brain, on the other hand, which may as frequently have occurred in men, was possible to only a very limited extent while the patient was alive, and these forms of cancer could easily be taken for other diseases with similar symptoms. Probably many patients who were said to have died, for example, from 'consumption' in the early modern period would be diagnosed today as cancer patients.

The time following menopause, when the impure humors were denied their natural route of elimination, was feared as a period of particular risk of cancer in women. Galen had already underlined the increased danger of breast cancer in women who were 'no longer cleansed by the natural purgation'.[395] This concern is reflected in the letters of elderly women. The 46-year-old Mme de Chambery, for instance, was highly alarmed when she sometimes passed menstrual blood that contained lumps and smelled bad precisely at the time of her life when her periods seemed to be coming to an end.[396]

The perception that women were diagnosed much more often with cancer than men fitted in perfectly with the widespread lay assumptions about the differences between the female and male bodies, assumptions that similarly formed the basis of the cathartic interpretation of menstruation. Far more than men, women were subject to disruptions in the flow of humors, more precisely of impure fluids, which they had to eliminate from their bodies every month to stay healthy. The fact that mainly the uterus and breasts were in danger of developing cancer acted as striking confirmation of this assumption. Both organs were closely connected to menstrual blood. The uterus served directly to expel the blood, and when the menstrual flow was disrupted, impure humors almost inevitably accumulated in the uterus. A vicious circle could begin: the scirrhous hardening of the uterus further hampered the discharge of menstrual blood. But the breasts as well were closely related to menstruation since – according to traditional medical thinking – the blood that nurtured the child in the belly during pregnancy flowed to the breasts following birth and became breast milk. Ultimately cancer's predilection for women, like their need to menstruate, supported prevailing cultural beliefs about women's natural tendency to inner impurity. When the cancer was accompanied by putrid secretions, this impurity became more than obvious to the senses, too. Ultimately, far more than men, women were in danger of a disruption to their inner bodily order, just as social order was, from the male perspective, continually endangered by women.[397]

The diagnosis of cancer remained often difficult and controversial until the disease reached an advanced stage and began to ulcerate. Yet a correct diagnosis was crucial for the appropriate treatment. Medical writers were increasingly convinced that especially with breast cancer only a complete removal of the cancer – and of the breast with it – could bring a definitive cure.[398] For according to Herman Boerhaave, the disease had 'never been healed to date whenever it was not possible to take out with the disease the part that was afflicted'.[399] What is more, the notion began to take hold that only a radical removal of the surrounding tissue and of the lymph nodes could prevent a renewed growth of the cancer, because 'when after the extirpation of the cancer only the merest hint of it remains behind, it sprouts again before long and brings on evil that is just as horrible'.[400]

Some surgeons designed large instruments, looking more like hedge shears, specifically for the purpose of taking the breast off quickly with little blood loss. Nevertheless an operation was mutilating, painful, and very risky due to the bleeding. Moreover, the prospects of healing remained doubtful even in the case of a radical removal. Experience taught 'that if in one part of the body the glands become sick, others, often in distant locations, are affected as well'.[401]

Understandably, patients and their families were reluctant to consent to radical surgery. The 25-year-old Elisabeth de la Loë decided that she wanted 'rather to die peacefully than suffer the pain of the amputation', all the more so as she thought the chances of success were slim. At the grave of Monsieur de Paris, she was later 'miraculously' healed.[402] Other women had the necessary courage and optimism to undergo the dangerous and painful operation. Mme d'Arblay, better known as Fanny Burney, is a well known case, though in retrospect the operation may well have been unnecessary because she was not really suffering from cancer after all.[403] At the age of 45, Margarethe Milow as well was confronted by her physicians with the horrible choice between living another year and then dying the 'most horrid, most painful death', or undergoing surgery. A sympathetic cure and the treatment of an obscure healer had already failed and she agreed to the operation. Fully conscious, she experienced her breast being cut off and placed on a board next to her, and how the knife scraped along her rib bones. She survived the intervention but died barely two years later.[404]

Alternatively and much more appealingly, especially with tumors that were less advanced or which could not be safely removed due to their location, various medications could be tried. He was convinced, the husband of a 30-year-old woman diagnosed with breast cancer declared, that what had to be done now was to 'find the remedy to strengthen the stomach so that my diseased wife may regain her strength, to assuage her pain and to prevent the evil from spreading while waiting to destroy it'.[405] The goal in this case was to soften the morbid matter that had hardened into a tumor, to liquefy and mobilize it in order to drive it out. 'Melting' or 'dissolving' medications were therefore of primary importance in cancer therapy, at times combined with 'specifics'. Most notably hemlock and mercury compounds, when dosed accurately, where credited with a curative effect. Other remedies were applied to keep the blood and humors liquid. Dietary recommendations aimed at reducing the production of sharp, acrimonious humors.

In breast cancer, a topical, local treatment with blistering plasters was also sometimes used to create and maintain an exit for the carcinomatous matter through the skin. As Pieter van Foreest recounted in various case histories, patients and unauthorized healers seem to have put great faith in this method and resorted to it even against the explicit advice of the learned physician, who feared that this would only further irritate the cancer.[406]

Pathological Heat

Among the most common complaints of patients in the 18th century was an unusual and unpleasant sensation of heat. Such sensations were described much more frequently and in a much more differentiated manner than Western physicians today are familiar with in their daily medical practice. We encounter here another nodal point within the complex 'semantic network' formed by early modern interpretations of illness and the body. To a certain extent the outlines of something similar to illnesses we know by terms such as 'El calor' from other cultures begin to emerge.

Anything that increased one's inner heat, in the patients' experience, also had the potential to bring on 'hot', 'fervid' complaints. Too much exterior warmth, as resulting from an extended hot bath, could heat the body excessively, while a walk in the cool air or a cold bath promised relief. Physical or mental labor also heated the body. The councilor of commerce Cherot des Marois, for instance, found that his daily business activities heated him in such a way that he felt a constant fire burning inside.[407] But above all, patients blamed foods and medicines for creating heat. The antimonium, related a cavalry captain who had a 'dry' and 'bilious' disposition, had flared up 'truly hot embers' ('brasier') in his stomach.[408] A 'horrible heat' took hold of Mlle de Maltzan after the family physician had pressed upon the reluctant woman supposedly fail-safe pills, which, as it turned out, contained a large quantity of mercury. Only a cooling herbal infusion and cold footbaths brought relief.[409] Some patients attributed their intense sensations of heat to drugs like cinchona and stopped treatment because of that.[410] In fact, some patients experienced excessive heat even from chamomile tea that a physician had recommended or from the medicinal water from Spa.[411] At times, patients hardly knew any longer which remedies and food to consume that would not heat their stomachs because they could not even tolerate a hot drink, let alone a glass of wine.[412]

Of the bodily humors, the hot and dry yellow bile was particularly closely associated with the physical sensation of heat, and this is one of the rare instances where we hear certain echoes of the old idea of balance. Some patients explicitly associated their 'bilious' temperament with a certain tendency to become heated. Her blood caught fire very easily, complained a female patient of Tissot.[413] And Schwitzer de Buonas knew from his own experience that 'heat-causing' medication did not suit him due to his bilious disposition.[414] It often remains unclear, however, whether those patients really attributed their easily heated disposition to a mere excess of natural rather than pathological bile. Some pointed explicitly to a previous excessive heating or burning of their bile or to another qualitative change in it. The bile in question was no longer one of the four natural humors but a more or less specific morbid matter, its nature approaching that of the 'acrimonies'. According to a nun in Geoffroy's care, her horrible pains, acute vomiting, and a 'fire' that vehemently rose up into her face and took away her 'pulse' came from a 'great heated up bile' and from wind. During her colic attacks, there was 'a fire throughout the body' and all hot medicines brought her to the brink of death.[415]

There were different kinds of heat sensation. Patients and their relatives connected localized sensations of warmth or heat to an assumed passage or a local accumulation of hot, acrid or irritating morbid matter. Any body part could be affected but most often patients complained about such sensations in the stomach and in their skin and mucous membranes. While such cases were characterized by a localized 'burning' sensation, other patients described the heat in distinctly dynamic terms. Similar to humors, heat – or hot substances – roamed throughout the body but they often moved much faster and, following the imagery of fire and flame, were apt to rise upwards. Heat originated predominantly in the abdominal area. This was where food was 'concocted', and this 'cooking process' naturally required an inner fire, the 'innate heat' of the medical tradition. Fires could, however, burn more or less intensely. They could flare up uncontrollably, to such a degree in the worst cases that patients felt as if they had glowing embers in their bellies.[416] The humors, for their part, were transformed by the excessive heat. They boiled, thickened, or evaporated.

Numerous patients described heat sensations in these ways, using images of a sudden boiling or flaring up. The intensity of the heat increased or decreased rapidly, or it changed location, at which point it became 'flying' heat.[417] People wrote of 'flushes', or of 'bubbling up', as if their blood had been put on a stove.[418] The Comtesse de Lucinge, for example, complained about 'great fires' inside her and literally felt the 'ebullition' of her blood.[419] A 'horrible heat' and sudden 'boiling up' is what Mme le Pin experienced in her head alongside a burning in her stomach. She could no longer tolerate salt or hot foods except soup; even the slightest warmth caused her intense pain.[420]

Women in the 18th century described complaints of this kind strikingly often around the time of their menopause.[421] Mme Viard d'Arnay, for example, suffered from such flushes several times a day. The heat rose into her face, her face turned red and she began to sweat. Over time, she feared, the heat might dry out her blood altogether.[422] Similar descriptions can be found in letters of other menopausal women. Day and night, complained the Marquise de Brichanteau, she was gripped by 'heats' ('chaleurs') that rose to her head and made her flushed. This lasted only a short time and afterward she was sweaty. Geoffroy, when commenting on her state, wrote that it was in fact all too common for ladies of her age to complain about 'heats and fires [...] that rise into their faces and are followed by sweating which disappears again right away'. He explained that this was due to vapors which 'accompany this state and are very burdensome but not dangerous at all'.[423]

Hot flushes followed by sweating count among the characteristic symptoms of so-called peri- or post-menopausal syndrome in today's Western societies. This reaction of the female body to menopause is, however, not universal nor can it be explained on purely biological grounds. In other cultures, the 'typical' complaints of peri- and post-menopausal women take on different forms. In Japan, for instance, menopause is associated much more often with joint pains than with hot flushes.[424] This may be due to some degree to different nutritional habits: foods rich in soy especially are said to counteract the decrease in the body's estrogen production and ease in this way (post-)menopausal complaints. The prominent position given to this day by Western menopausal women to hot flushes, however, probably also reflects the traditional importance of sensations of heat and images of rising vapors in early modern somatic culture which has to some degree remained part of our Western cultural heritage. Growing up in this culture and taking its conceptual framework for granted, Western women learn to focus their attention on sensations of rising heat and hot flushes that in other cultures might well go unnoticed.

Vapors

As we have just seen, images of an inner heat or fire were closely linked to the notion of 'vapors' (Latin: vapores, French: vapeurs, German: Dämpfe) or trails of smoke (French: fumées). For early modern physicians and laypeople, the concept of 'vapors' or 'vapores' was a key to understanding various physical and affective changes and complaints before it was gradually replaced in the 18th century by the new model of the 'nervous diseases'.

Talk of 'vapors' or 'vapores' was by no means merely metaphorical. To patients and physicians it was literally vapors or trails of smoke which rose up from the abdomen into the chest, throat, and head. In a way, the human body in this model resembled a chimney in which hot air ascends but, as opposed to a chimney, the body had no opening at the top. Vapors and smoke could not escape through the skull and so collected underneath. This explained why 'vapores' affected the brain above all. In the brain, they blended with the subtle, ether-like spirits of the soul. These 'animal spirits' were, according to traditional Galenic understanding, the immaterial, rational soul's immediate tools and responsible for the intellectual faculties. Vapors could critically disturb the movement of the animal spirits, causing them, for example, to revolve – which was one of the accepted early modern explanations of the origin of dizziness.[425] Above all, vapors and smoke, due to their cloudy or even sooty consistency, compromised the natural purity and translucency of the spirits. This necessarily had an effect on operations of the mind. Condensed to specific forms and figures like clouds of smoke, vapors and fumes could even trick the brain into taking such internal images as external reality. Melancholic people were most at risk. From their burnt black or yellow bile rose blackish, sooty 'fumes' that could lead to the typical melancholy hallucinations. The Devil knew how to harness this potential. By skillfully gathering and shaping such sooty fumes, he made melancholics believe that they had been transformed into werewolves or had been riding to a witches' Sabbath on a broom.[426]

The notion of hot vapors moving or wafting through the body can be found in numerous patient letters. It is indeed only in the light of this notion that some of the sensations which patients reported become comprehensible to us today. Insofar as people knew from experience that warm air moved upward and expanded, rising vapors offered an evocative explanation for feelings of oppression and shortness of breath in the belly, chest, and throat. Everything came together when one female patient complained simultaneously about 'vapors', feelings of suffocation, and 'heats' that rose up from her stomach,[427] or when it was said about another sick person with a 'hot liver' that smokes ('effumigationes') rose into his chest and took his breath away.[428] The head and the brain were even more susceptible. Numerous patients described disturbances to their well-being and mood, convulsions and other complaints which they connected to vapors or to their condensation to a liquid in the cold brain. Anna von Bradowa, for example, complained that 'the rising of unnatural fumes' and the 'liquid in my head' did not abate,[429] while Vincentz related that sloe wine strengthened and cooled the stomach and, if consumed after eating, prevented 'the fervid fumes from rising into the head'.[430]

The notion of 'vapors' represents one of the most important examples of a medical concept that underwent a marked change in the course of the early modern period – with profound repercussions on the lay experience of the body and its diseases. The concrete image of hot vapors which rose upwards was gradually abandoned in favor of the new concept of nervous complaints in which the notion of vapors retained a purely metaphorical value. I will be looking into this in detail later.

Fever

In a society still dominated by agriculture, everybody knew that heat was released when plant or animal matter rotted or fermented. On cold mornings, steam could be seen rising from dung heaps. It was only reasonable to presume that very similar processes of putrefaction or fermentation took place in the abdomen, where foul organic matter collected (and was periodically evacuated) even in healthy people. In this way, the frequent sensation of increased heat in patients with fever was connected to notions of corruption and putrefaction: fever came from rotting, decaying matter inside the body and this released heat. Putrefaction and decay, which caused heat and fever, could arise wherever humors accumulated in the body and turned foul. Sometimes the stomach or the vital heat was too weak to assimilate food. One dyspeptic patient believed that 'insufficient cooking caused the fever movements'.[431] At other times, the flow of blood within the body or the elimination of waste matter was disrupted.[432] Several patients attributed their fever to a chill they had caught, for instance, when wading through a river.[433] Cold water, as we have seen, made the body's fibers, pores, and ducts contract. The perishable waste could no longer be eliminated through sweating and deteriorated into morbid matter.

Though both fever and heat were often attributed to foul, putrefying matter, the two concepts were not nearly as close as they are today. Our modern understanding of fever as defined by a pathologically raised body temperature essentially goes back to the late 19th century only, when the clinical thermometer found its way into medical practice and soon spread to the general population.[434] 'Fever', in the early modern understanding, was a disease not a symptom and a range of different 'fevers' could be distinguished. Sensations of heat were seen as quite typical for some of these 'fevers', but they were not in themselves sufficient for the diagnosis and there were 'fevers' that were characterized by cold shivers rather than by sensations of heat.[435] In 1604, Frater Placidus, for example, recounted the case of a fellow brother whose disease had beleaguered him with such coldness ('frigore') that the sufferer thought it was a 'fever'.[436] The coldness had not been particularly intense with her 'nerve fever', related a patient of Tissot, but had shown all the symptoms of a 'fever', such as headache and exhaustion.[437] Other patients complained about an unpleasant coldness of their hands and feet at the beginning of a 'fever'. And because Monsieur Gouchuat felt no more than the hint of a cold sensation in his back, he was uncertain if he really had a 'fever'.[438]

'Fevers' could thus be accompanied by sensations of cold or heat. Educated laypeople and physicians seem to have largely agreed, however, on one characteristic which all 'fevers' had in common: a changed, quickened heart or pulse. Apart from the patient's general prostration, the key to diagnosing a 'fever' was not the use of a thermometer but feeling the patient's heart beat or pulse, and sometimes the patient reported the subjective sensation of an inner movement or agitation. Hence patients and their relatives often referred to a 'feverish pulse' or described its qualities in more detail to either substantiate the suspicion of a 'fever' or, depending on the case, call it into question. Her pulse did not feel so good this morning, said the rheumatic Mme du Bourgneuf.[439] Her pulse remained 'feverish', wrote Mme de Verdun.[440] A 'small', 'thick', 'fast', or 'high' pulse, to the layperson, was the most characteristic sign of a 'fever'. A healthy pulse by contrast was slower and 'broader'. 'Those who feel his pulse', wrote the wife of the sick Monsieur Decheppe de Morville, 'consider his state better by the day and assure us that no fever whatsoever is present'.[441] It was 'never observed that she had a fever' during her illness, it was said about another patient. 'Rather, her pulse is usually quite slow and her extremities are always ice-cold'.[442]

The close connection between 'fever' and a quickened pulse found expression in people's readiness to diagnose even extremely brief spells of 'fever'. A 'fever' could sometimes denote a quickly passing physical sensation, as when a patient complained about the 'fever attacks' that would befall him daily after lunch.[443] Another patient was not sure that he had a real 'fever' when his pulse was irregular and quickened over a period of several hours and he felt 'emotions' (understood here as a movement of the blood) after going for a walk or when it was warm – but he clearly thought it was a possibility.[444]

Another important symptom of 'fever' was a general feeling of illness. Such a 'sensation of fever',[445] however, was less significant for the specific diagnosis of a 'fever' than the pulse. It was only as an exception that the diagnosis was based entirely on it: Monsieur Darmenon, for instance, had 'no fever at the pulse' but felt 'the heat and the weakness that fever commonly produces'.[446] Conversely, one sick man did not at first want to believe his physician that his pulse was 'heightened', because he felt fine – until he too finally recognized the 'fever' attacks.[447] Incidentally, even the more general 'sensations of fever' tended to be described in dynamic terms, that is, not so much as a prolonged heat and exhaustion but as 'fever movements'.[448]

The intense heat that often accompanied 'fevers' or even characterized them was not necessarily harmful in itself. It could in principle also be understood in a positive way, as a sign that the body was trying its best to concoct the morbid matter, thereby assimilating it or preparing it for elimination. In this case, the heat of the 'fever' would simply reflect an increased development or deployment of vital heat needed to 'cook' and transform the foul morbid matter. For it was crucial for recovery that the decaying or fermenting morbid matter was rendered harmless and/or excreted – otherwise the matter accumulated and putrefied further or led to local obstructions and tumors.[449] With 'fevers', according to a long-standing medical doctrine, this 'maturation' and evacuation of morbid matter took the shape of a 'crisis' or a 'critical excretion'; the term 'crisis' is to be understood here in its original Greek meaning as a decision that leads to change for the better or for the worse.[450] People knew from experience that a 'fever' often got better with intense sweating. Other excretions could also fulfill this task,[451] and it was up to the skillful physician to facilitate them at the right moment. 'After the body was cleansed well, the fever went away swiftly', recounted a patient of van Geuns.[452] When morbid matter was only incompletely concocted and remained inside the body, on the other hand, a 'critical deposit' could be formed.[453]

'Fevers' were also among the few illnesses for which there was a generally recognized and (supposedly) specific remedy: cinchona bark or Peruvian or Jesuits' bark, as it was sometimes also called. Coming from South America in the 17th century, it became widespread as the most important 'fever' remedy apart from bloodletting. From today's point of view, cinchona has no antipyretic effect but acts against malaria parasites. In the majority of 'fever' cases, we would therefore not expect any beneficial effect. To patients and physicians then, however, it was a tried and tested remedy and many experienced a considerable improvement.[454]

As we have seen, early modern medical writing often referred to 'fevers' in the plural. Like 'cancer', 'scurvy', and 'smallpox', 'fevers' were a distinct disease entities which could come in different types, each characterized by certain symptoms and a typical natural course or periodic recurrence. In this sense, laypeople, like physicians, wrote about 'slow' or 'continuous fevers', for instance, or about 'bile fevers' and 'putrid fevers'. Cramps, pains, and mental clouding, it seems, tended to be interpreted as signs of a 'nerve fever', while yellow skin suggested a 'bile fever'. Tertian (three-day) and quartan (four-day) 'fevers' were characterized by the typical rhythm of recurrent attacks.

As with 'Gichter', laypeople in particular tended to understand fevers as independent agents or to even personify or demonize them. In much detail, Johannes Butzbach recounted how a maiden finally freed him from a long fever illness. She walked with him to a meadow before sunrise. Once she had evoked and cursed the fever, she peeled bark from a tree and tied it around his naked belly. He stayed wrapped up like this for three days. Then she threw the bark, 'together with the fever so to speak', into the fire and 'this is how I became healed'. Only when his confessors had dissuaded him from 'such superstitious things' did the disease attack him again.[455] Up into the 19th century, the 'fevers' – and here especially the 'cold fever' and the related 'ague' – counted among the illnesses for which pious and sympathetic healing practices such as 'praying away' were often preferred. As late as 1860, a physician from Donauwörth reported that the incantation of fever was 'known and practiced by one individual in almost every village'.[456] A physician in rural Franconia even thought that a physician had to avoid calling something a 'fever' because 'the latter, in the opinion of the country folk, cannot be healed by any physician'.[457]

Consumption and Consumptive Fever

In contrast to the dynamic sensations of flushes, agitation and quickening of the pulse, which were characteristic of most 'fevers', some patients complained about something like a consuming fire inside their bodies which often came without any noticeable change of the pulse. This was a typical symptom of the so-called 'hectic fever' (Greek: 'helko' = to draw, tug), a complaint that was closely associated with 'consumption'. Along with cancer, consumption or phthisis (French: phthisie, German: Schwindsucht, Italian: tisi) counted among the most dreaded diseases. The term was used in medical literature for a number of different chronic and consuming illnesses, including 'spinal consumption'. It referred above of all, however, to the widespread form of pulmonary consumption or, as it was usually more vaguely called in France, the 'chest disease' ('mal de poitrine'). The fear of this disease is not surprising. There are no exact numbers but contemporary censuses and church registers, as well as statistics recorded in the early 19th century, indicate that a very high portion of the total number of deaths were due to consumption. On top of that, consumption often affected people in their prime and, unlike almost every other disease, often took a treacherous course, starting inconspicuously with a cough and a sensation of moderate heat. Many victims looked healthy for a long time due to a lively complexion. But once the disease had taken root, all therapeutic efforts were in vain.

The typical symptoms of consumption were apparently well known among the general public. Slightly red sputum was already enough to stir grave concern in anxious patients.[458] And when the complaints were more serious, patients and their relatives were well aware that the outcome might be fatal. Considering his mother's meagerness, her cough, and her glowing red cheeks, Jean-François Marmontel, for example, was almost certain she was suffering from 'this deadly pneumonia', from which his father had died earlier.[459] Other patients mentioned nocturnal sweats, bloody sputum, and rib pain as indications of a presumed chest or lung ailment.[460] These are largely the typical symptoms that physicians of the time also associated with consumption. 'Consumption', in Buchoz's succinct definition, 'is a chronic lung disease that is accompanied by a 'slow fever' with bouts in the evening and after dining, by nocturnal sweat primarily on the chest, slight breathing difficulties and a cough that becomes worse in the evening and toward dawn, and in which one renders sputum which is initially mixed with blood and then with pus. This disease is always followed by weight loss or emaciation of the whole body.'[461]

In the late 17th century, medical writers began to describe pulmonary consumption as resulting from a pathological alteration of the lung's substance: post-mortem knots or 'tubercles' of varying sizes could be seen in the lung and sometimes in other body parts as well.[462] Occasionally, physicians also used the new term ('tubercle') in their exchanges with patients. Geoffroy, for instance, in a letter consultation, ascribed a patient's cough and sputum to 'several ulcerated tubercles in the chest'.[463] The talk of lung 'tubercles' (German: 'Tuberkel', French: 'tubercules') did not take hold in lay medical language. Even in patient letters from the late 18th century the term can be found only sporadically.[464]

The typical symptoms of consumption as they were experienced and perceived by patients and their relatives pointed, rather, in a different direction. Like many other diseases, consumption was generally considered by laypeople – along with many physicians[465] – as an illness of the entire body and not just the lungs. Its typical symptoms indicated two parallel or subsequent pathological processes. The mucous, purulent, or bloody sputum and the copious sweat fitted into the familiar model of a beneficial evacuation of harmful or superfluous substance. And the feverish heat, red face, and physical deterioration pointed to the inner 'fire' that Marmontel, for instance, thought he could see in the eyes of his consumptive mother, the 'fire' that 'consumed her blood'.[466]

The typical feverish heat and agitation of many patients linked consumption closely to 'hectic fever', which was also called 'consumptive fever'. 'Hectic fever' was one the three basic forms of 'fever' in the classic Galenic teachings. The preternatural heat in the body made blood surge and rise to the head and increased the movement of the spirits in the heart and vessels. 'Consumptive fever', however, did not merely cause turmoil of the humors and spirits; with time it literally devoured them. Just as an oil-lamp needs oil, according to the oft-quoted Galenic simile, vital heat needed a special fuel, the 'radical moisture'. Even in times of health, it gradually consumed the 'radical moisture' and the vital spirits, which could only be insufficiently replaced from food. In the end, the 'flame of life' necessarily had to go out because it had consumed its own fuel. This was dying a natural death. Consumption, or the 'hectic fever' so typical of it, seemed to accelerate this process enormously. The lamp of life burned stronger than usual and made the cheeks red. But this intensification was ultimately synonymous with a premature death.[467]

Patients only alluded to this theoretical model. Joachim Breckow in 1574 described his subjective sensations of an 'interior burning', an unnatural, excessive heat inside the body and expressed his fear that his 'spirits are running out', because he kept 'losing weight, becoming more and more meager by the day'.[468] Mme Faugeroux also felt a great fire in her chest and intestines, accompanied by a 'slow fever'. She was so hot that, even in the winter, she did not want to light a fire.[469] 'I was always on fire', wrote another patient after months of coughing and several episodes of hemoptysis. She also complained about 'the fire' that 'consumed' her.[470] Mme de Launay likewise complained about the 'terrible heat' of her blood. Following one of several miscarriages, she had bloody sputum for a year and a half, caused, she thought, by a morbid milky humor. A fontanelle brought only temporary relief. After an extremely traumatizing birth, a 'slow fever' 'consumed' her, causing her to lose a terrible amount of weight and leading to emaciation.[471]

Expenditure and Exhaustion

Pivotal for the maintenance of both health and life, it was believed – and this has been a leitmotiv in this study – was the body's capacity to purify or cleanse itself of harmful substances. Even repeated, intense evacuations could thus appear as a favorable sign that the body's natural healing powers were active, and if the disease and the evacuations persisted, the obvious explanation was that the body had not yet managed to free itself completely from the harmful matter. For this reason, many physicians, as late as the 19th century, treated the typical copious watery diarrhea resulting from cholera with emetics and laxatives and advised bloodletting to consumptive patients with bloody expectoration.

When unusual evacuations persisted over an extended period, however, it was feared that they could eventually threaten life and limb in their own right. After all, the body inevitably lost not only harmful morbid matter through evacuations but also some of its vital matter. Thus, the fear of an accumulation and putrefaction of excessive or harmful humors in the body went hand in hand, at times, with the fear of excessive excretion. The extent of the danger varied with the amount and the kind of matter which was eliminated. Different bodily substances were believed to be unequally rich in vital, balsamic matter. While urine and stool were basically just waste matter and contained, if any at all, only traces of valuable, balsamic substances, even a moderate loss of fluids such as blood or even worse semen could be dangerous.

The fatal consequences of an excessive evacuation or depletion of vital matter could be witnessed in cases of sudden, intense blood loss, as from a severe injury, a traumatic birth, or massive pulmonary, intestinal, or uterine hemorrhage. It could quickly bring a patient to the brink of death. But patients could also feel weakened by repeated losses of smaller amounts of vital fluid over longer periods of time. Her menstrual bleedings were so abundant, wrote one English patient of Boerhaave, that she was no longer able to stand.[472]

Excessive, unhealthy loss of valuable, vital matter was also a dangerous consequence of evacuative treatments. For her intense menstrual pains, Mme Millière, only 19 years old, had had 60 bloodlettings in only three years.[473] Often patients themselves, sometimes against all warnings, demanded a drastic, necessarily debilitating, treatment or took certain known purgatives on their own accord.[474] Among the rural southern German population up into the 19th century, the number of substantial evacuations produced by a laxative or emetic was said to be the principal criterion by which its quality and effectiveness should be judged. But a treatment which was too intensely evacuating could, in the patient's experience, be debilitating.

Hummel had to throw up 17 times after taking the emetic his physician had given him. Two days later, a similarly drastic purgative made him produce 15 stools, and later in the same week a bloodletting was done and he was given a sudorific medicine. In his own words, all of this left him 'very worn out'.[475] Sometimes there was no choice because of the severity of the illness. After she had been miraculously healed from her 'erysipelas', 57-year-old dressmaker Marie-Jeanne Orget died because a 'flux from the chest' had ultimately turned into dropsy brought about by the many bloodlettings she had to have.[476] An acute, inflammatory chest disease forced the physicians of another patient to take 60 ounces of blood from her in only four days. According to the patient, she did not recover from this loss for a long time; yet she does not seem to have questioned the necessity of the bloodlettings as such.[477]

In some cases, patients and their relatives were convinced that physicians did go too far with their bloodlettings and purgatives or did not take the delicate constitution of the patient sufficiently into account. Bitter reproaches were occasionally heard. 'This is too dangerous a means', wrote a woman suffering from migraines about the possibility of a second bloodletting a mere two months after the last and in spite of the fact that the first bloodletting had not helped stop the blood from permanently rising to her head.[478] The attending physician of a 55-year-old female patient prescribed more than 300 enemas in only five months for the intense pains in her lower abdomen. In the end, her stool was full of mucus and had finger-long fibrous coverings. The patient believed that retained sweat was largely responsible for her disease, which in principle argued in favor of the necessity of an alternative evacuation. The massive discharge of mucus, however, did alarm her. She had become very weak and emaciated and feared she might become consumptive.[479] Mme de Konauw likewise thought that, considering the numerous purgatives she had been given, 'the balsamic substance went out of my blood from all these remedies, and that my nerves are weakened'.[480] And the Marquise de Louvois asked Tissot to weigh in with all his authority against the local surgeon's adamant intention. Although her husband's body and nervous system were already extremely weakened, she said, the surgeon absolutely wanted to create a fontanelle, a man-made running sore, and he also insisted on warm baths – which were at the time commonly considered weakening and evacuating because they opened the pores. He entertained the opinion that the Marquis had a venereal disease, even though there was not a single pimple or other symptom that indicated such a disease.[481]

Dropsy

Weakness and weight loss were frequently mentioned as consequences of excessive evacuations. Another greatly feared effect of a continued loss of valuable fluids, in particular of blood, was dropsy, though the disease, according to physicians, could also derive from an insufficient production of blood due to poor digestion. This was one of the few illnesses for which bloodletting was generally considered dangerous because it only thinned and watered down the blood even more. Some German patients, in particular, also held excessive drinking of water responsible for it; the German term for dropsy, 'Wassersucht' translates literally as 'water disease'. Franz Wolf von Zollern, for example, was unwilling to accept the physician's diagnosis for a long time, asking 'how he could possibly have caught dropsy, since he had not had a drink of water in many years'.[482]

Typical signs of dropsy were bilateral swelling of the legs, arms, face, or belly. 'I am very afraid of dropsy', wrote the apothecary Vogel from Windsheim, 'because my thighs are swollen almost every night'.[483] If only a single limb was swollen, this was not considered a sign of dropsy: 'if it were to be dropsy – may God protect me from it – it would be on both sides and not on only one', thought the same Vogel.[484] Some patients recognized further symptoms such as a changed, sickly complexion and reduced urinary excretion. One man said he was able to hear a noise in his chest as if water were shaken in a bottle when he threw himself on a couch.[485] Others observed that their legs became swollen in the evening and were thinner again in the morning.[486] In dire cases, the skin even broke open and water flowed out.[487]

We can barely surmise which diseases recognized today by modern medicine were formerly diagnosed as 'dropsy'. Such swellings, according to today's understanding, can occur with many different diseases, ranging from heart failure caused, for example, by a valvular defect, to liver cirrhosis and kidney failure. Localized swellings on the arms and legs can also point to impaired lymph and blood drainage. Nevertheless, seeing the historical 'dropsy' as simply the name for a symptom would be a mistake. 'Dropsy' was considered a discrete entity – and together with cancer and consumption it topped the list of incurable, fatal diseases. Even a minor swelling of the hands and feet was enough to raise intense fears. 'We used all kinds of remedies, including acidulous mineral water, but to no avail', recounted Hummel about his stricken daughter. She became bedridden.[488] In another case, the patient's concerned husband consulted a learned physician only to hear that there was nothing to be done 'about the swelling and dropsy'.[489] And Franz Wolf von Zollern was told by his physicians that 'he should turn to God because he had dropsy, an incurable disease'.[490] Hermann von Weinsberg described the suffering and fears of his dropsical wife in particular detail: in March 1573, her legs began to swell monstrously and soon her belly too became bigger. 'Oh, I was worried long before March; I'm terrified, I'm so terrified,' moaned the sick woman. 'And because she had had a horrible cough for ten years, winter or summer, it was said that it was an affliction of the lungs and that dropsy would follow.' She dragged herself about with a cane in the house and sat at the table for all meals or on a lower special resting chair. She ate and drank what the physician recommended and what the apothecary prescribed her. The physician was not allowed 'to tell her anything about dropsy or melancholy' but 'it was considered a certainty that she knew very well she had the water inside her, since she had seen many sick people'. She died after two months of suffering in May 1573.[491]

Seminal Economy

The most valuable substance in the body was, according to a long-standing tradition, not blood but semen. Hence excessive loss of seminal fluid could arouse particularly intense fears. In the 18th century, the balance between concern about the effects of retained semen and fears of excessive seminal loss shifted towards the latter. It was widely accepted in pre-modern medicine that semen would spoil when it remained in the body too long. Traditionally sexual intercourse thus was therefore seen as important for health in that it removed superfluous semen from the body. A patient of the 17th-century physician Theodor Konerding in Hanover, who was frequently away from his home and wife on duty, believed his large, hard testicular tumor came from 'retaining the semen'.[492] Such notions were still widespread after 1700. Semen first accumulated in the seminal vessels, as Baron de Beaucourse's painful experience at the ripe age of 67 told him. He was at an age where doing without the 'use of women is natural and does not cost anything', but when he abstained for an extended period, his blood became thick in such a way that he got swellings in the scrotum.[493] If the accumulated semen spoilt, dire consequences could be expected for the entire body. One of Tissot's patients saw in it the cause of his convulsions. His sexual desire had awakened very early, he said, but he had always resisted it. He had had his first seizure after his frequent nocturnal emissions had suddenly stopped. A 'too great amount of semen', he assumed, had entered the 'blood mass' and had caused a 'fermentation' there and a 'collapse'.[494] In this light, coitus (practiced in moderation) was an indispensable part of a healthy lifestyle, while voluntary or forced abstinence was dangerous,[495] if Nature herself did not manage to free the body of superfluous semen via nocturnal emissions.

Learned physicians, however, considered semen to be particularly rich in valuable, vital matter, in innate heat or 'spirit', or in 'quintessence'. Excessive seminal loss was accordingly at least as dangerous. The disastrous consequences had already been described by the ancients, and early modern authors repeatedly painted them for the reader in lurid colors: growing weakness, headaches, poor eyesight and memory, pale complexion, loss of appetite, impotence, infertility, and spinal consumption, followed by ever more severe pains and complete bodily degeneration.[496] Beginning in the late 17th century, such fears gained momentum and ghastly case histories illustrated the agony that awaited the victims. Shaken by convulsions and tortured by horrible pains – this is how Diderot's and Alembert's famous 'Encyclopédie' described the consequences in the 18th century – the patient would be confined to bed for months. He would have constant erections as well as distressing ejaculations that caused him to cry out in pain and that were followed by a state of utter weakness. With sunken eyes and a disturbed expression, haggard and completely exhausted, such a man would finally die a terrible death which, compared with the wretched life he had in the end been condemned to, appeared as a salvation.[497]

There were different reasons for a chronic, excessive loss of semen. If sexual excess was the cause, patients themselves were held responsible. Some of Tissot's patients wondered if debauchery in their early years had contributed to their present affliction. It might have been better, one of them wrote, if he had fulfilled marital duties 'with more moderation'. For years, he had also preferred the time after meals for this purpose. But he had had to realize that this time, which was dedicated to digestion, was particularly inappropriate because, especially in the heat of summer, he often felt tired afterwards instead of refreshed.[498] Another man got even thinner than he already was and his complexion began to look unhealthier when, at the age of 14, he overindulged in the pleasures of the flesh with a maidservant.[499] Giving him various aphrodisiacs, the last mistress of the 44-year-old Marquis de Louvois had fired up his already abundant desire. Now, according to the descriptions of his wife, he suffered from a marked loss of memory, spells of dizziness, failing eyesight, and other signs that his nerves and brain had been massively weakened, even though he had only slept with his wife seven or eight times in the previous 18 months because of her pregnancy and subsequent nursing.[500] Because of his affected nerves, the melancholic Monsieur Chillaud would feel discomfort whenever he was with his 'beloved wife'. Ejaculating weakened him severely, and after the act he would often cry bitter tears.[501]

Even greater dangers loomed with sexual self-gratification, which I will discuss in more detail further along. A partner was not necessary and it easily became a habit, detached from the body's natural need for evacuation.

Besides these 'self-inflicted' ways of losing semen, there were pathological forms of involuntary excessive semen loss, or what were regarded as such. According to physicians, it was already dangerous when nocturnal emissions became more frequent.[502] Some men shared this worry as early as the 16th century. In 1593, law student J. B. von Zweybruck sought medical advice most of all because of the emissions that had haunted him, along with sexual dreams, since his 14th year. He knew that nocturnal emissions happened to healthy people, especially in youth, but in his case, he wrote, it happened only when he slept on his left side, which was already weakened, and afterwards he suffered from head complaints and sensations of heat, as well as a pulsing pain in his left side.[503] By the 18th century, this concern was apparently fairly widespread. 'It is true', wrote one of Tissot's patients, plagued by convulsive fits, 'that it is necessary for Nature to relieve herself from time to time at a certain age and in the case of a man who, like me, has no intercourse with women and does not even think about it; but is it natural that the emissions are so frequent and followed each time by attacks of dizziness?' [504] One of Haller's patients complained about exhaustion, extreme weakness, and impaired vision as the 'saddest consequences' of his emissions.[505] 'What burdens me most', wrote another patient of Tissot who was consumptive, coughed blood, and suffered from bloody stools, 'are the nocturnal emissions'.[506] Others recounted that their complaints worsened as a result of nocturnal emissions and that they experienced dizzy spells, shortness of breath, an irregular heartbeat, headaches, and other 'outrageous afflictions'.[507] In the case of Gauteron, an 18-year-old law student, all his problems revolved around his nocturnal emissions. They had been going on for four years, he wrote. He was always crestfallen the morning after, felt lightheaded, and developed strange red spots on his knees. At present, he was still well and felt strong, but he feared the ill might take hold and lead to an increasing, life-threatening exhaustion and debilitation.[508]

Alongside nocturnal emissions, 'gonorrhea' stoked fears of the severe consequences of a chronic semen loss. Today, 'gonorrhea' denotes a sexually transmitted, bacterial disease. In the past, however, the term was used more generally for various forms of genital discharge. As the Greek etymology of the term suggests ('gone' = seed, 'rrheo' = 'to flow'), there was no clear distinction between genital discharge in general and semen loss. Insofar as, following Galen, women were assumed to have their own semen – a position still taken by some physicians in the 18th century[509] – it was possible to also see female discharge as semen loss or at least as a loss of valuable vital matter. Gonorrhea and genital discharge were distinct from nocturnal emissions primarily in that they also occurred during the day and without being accompanied by sensations or images of sexual pleasure.[510] As for the causes of such an unnatural loss of semen, physicians named in particular semen that was watery or too thin, a decreased 'ability to retain' the semen and, closely linked to this notion, a slackening of the genital fibers.[511]

This little sketch of diseases from semen loss concludes my exploration of the most important conceptions of, and terms for, diseases that were current in early modern lay culture. It would certainly be possible to add further explanatory elements which can be found now and again in letters or personal testimonies. At times, such complaints combined to form a disease pattern that we are no longer familiar with in modern Western culture; 'chlorosis' or 'virgin's disease' would be an example.[512] In the vast majority of cases, however, the 'semantic network' sketched out above, with its various nodal points and conceptual agglomerations, was a sufficient basis on which to make sense of a given illness, to explain its causes and to choose the right treatment.

Part III

Dominant Discourse and the Experience of Disease

It is one of the basic premises of this book that the experience of illness and the sick body is always and inextricably framed and shaped by culture and society, by prevailing images and notions about the body and its diseases, as well as by the linguistic conventions and practices through which these notions come to be expressed. These notions not only point the way to appropriate diagnosis, prophylaxis, and treatment, offer an orientation, and instill confidence in the face of the threatening changes that are taking place in the body; they also have a decisive impact on how physical phenomena and pathological symptoms are experienced – indeed, on whether they are perceived at all.

The predominant medical terms, images, explanatory models, and practices of a society or social group are to a large degree the product of the cultural, social, economic, and political circumstances at a given time and place. They reflect a specific understanding of man and the world around him, a given set of hopes and fears, values and norms, economic, political, and military interests, and preferred forms of linguistic, visual, and ritual symbolization.

In the relatively homogeneous cultures which cultural anthropologists have traditionally studied, dominant medical concepts and practices can often be closely linked to an ethnic group's particular, relatively homogeneous, world view and can be regarded as an integral part of this view.[1] Post-medieval Europe, however, was remarkably pluralistic. Different fundamental religious and philosophical tenets, and different notions about man and his relationship to his natural and social environment existed side by side and sometimes in opposition to each other. In respect to the body and its diseases, academically trained physicians achieved for themselves, from the Middle Ages, a growing degree of interpretive influence, first among the upper classes[2] and then, in the long run, among the less educated classes as well. In Western societies, 'orthodox' medicine, though in itself quite pluralistic, has thus for centuries reflected, above all, the norms, values and interests of learned physicians and of those classes or social groups to which the physicians belonged and felt their loyalty, or whose support they sought. These groups varied considerably, however, from case to case and depending on the historical period.

Because medicine not only shapes perceptions, creates meanings, and provides practical guidance but also implicitly or explicitly imparts a specific world view rooted in the power structure of the day, medical conceptions and practices have attained a cultural and political significance that far surpasses the realm of medical theory

and practice. By turning an ever growing range of aspects of individual and social human life into the privileged, if not exclusive, object
of medical expertise, medical discourse has become a powerful and
efficient means to 'naturalize' the values, norms, and interests of the
dominant elites (or of groups aspiring to this status) and to thus keep
them, to some degree, beyond criticism. In recent decades, historians
of medicine and science have increasingly studied the 'social construction' of medical discourses with a view to the values, norms and interests that are, either implicitly or explicitly, transported via 'official'
medical discourse.[3] This work has been instructive and enlightening
but, as emphasized in the introduction to this book, the predominant
approach in this area, that is, the historical analysis of the 'dominant'
discourse of a time on the basis of the products of elite culture, also
has serious methodological limitations. It relies on the supposition
that the 'dominant' discourse in question, including its implicit moral
and political messages, was recognized and accepted in more or less
large parts of society, that is, that the discourse was truly a dominant discourse and not one limited to a small ruling minority whose
notions and practices might have exerted virtually no effect on the
everyday life of the majority of contemporaries. Discourse analysis is
likewise hardly able to determine the degree to which medical innovations were themselves owed to the changing everyday experience
of the body and diseases. Only if we supplement the analysis of elite
discourse with sources that are closer to the everyday life experience
will we gain a more encompassing view and be able to assess the actual
impact of 'dominant' discourses.

Taking this tack, this third and final part of the book will focus on
two case studies to analyze in greater depth the complex relationship
between the 'dominant' medical discourse of learned physicians and
the experience and interpretation of the body and its diseases among
ordinary people. This relationship can be analyzed best by looking at
medical theories and practices which were new at the time and can
thus be linked more specifically to changing values, norms, and interests which dominated in medicine and society at large at the time they
emerged. At the same time, their actual impact on society and everyday
life, on the experience of the body and its diseases and on the subjective
sense of an embodied self can be assessed more precisely. My examples
are taken from what were probably the most significant and socially
influential developments in 18th-century medicine: the new theory of
nervous complaints and nervous excitability, and the massive campaign
against masturbation.

The Sensible Body

Throughout this study, I have repeatedly spoken against the fashionable postmodern notion of 'the body as text' and insisted on the necessity of taking the body seriously also as a discrete entity which is ruled by laws of its own. Of course, the body and our embodied experience are framed and influenced by culture but this influence clearly is limited and shaped, in turn, by the body's natural, biological properties, even if we can grasp these properties only through the lens of our own culture. The perceptions and sensations human beings from different cultures have when they are, for example, exposed to a poison, to electric shock, to smallpox, or to cancer cannot be meaningfully understood as mere cultural or social constructs. These sensations and reactions also reflect – albeit not exclusively – the given biological condition of the body, which accounts for the similarity in the ways people from very different cultures experience it.

If, against the claims of a radical constructivism, we take the body seriously as an agent in its own right, it does not follow, however, that the role of 'biology' is always the same. On the contrary, the relative influence which the body's natural condition on the one hand and the socio-cultural setting on the other exert on the genesis, manifestation, and experience of diseases may vary quite dramatically. Not all diseases and ailments are shaped by culture to the same degree. Phenomena which we designate in modern medicine with terms such as 'smallpox', 'syphilis', or 'bronchial carcinoma' can be observed by the Western physician in numerous other cultures in a quite consistent, almost identical way, even if those afflicted may perceive, experience, and interpret their symptoms quite differently or, at times, may not even perceive them as abnormal or pathological at all.[4] In the same way, early modern descriptions of diseases like 'pulmonary consumption' permit a tentative, if also often questionable, translation into today's medical terms, even though different perceptive and interpretive frameworks tinged the accounts of patients and healers. Admittedly, for individual cases, a retrospective diagnosis can rarely be arrived at with any certainty. Yet there is still good reason to assume that many of the 20- to 30-year-olds in early modern times who coughed up blood and phlegm, had night sweats and fever and lost weight and were diagnosed with 'consumption' or 'chest disease' suffered, according to today's criteria, from pulmonary tuberculosis rather than, say, malaria or stomach ulcers. There is also reason to believe – again according to today's understanding – that older women with heavy genital bleeding, a malodorous discharge, and a palpable hard lump in the abdomen, women who grew increasingly weak within only a few months and finally died, likely succumbed to uterine cancer rather than, say, to chronic heart failure or palsy.[5]

Certain illnesses, in contrast, manifest themselves in very different forms in different cultures and historical times, at times in such a way that we can come up with no plausible modern Western equivalent for them.[6] Among the conditions and diseases over which cultural factors seem to exert a particularly strong influence are those that modern Western medicine describes in terms such as 'functional complaints', 'masked depression', or 'psychosomatic illness'. Conditions of this type are known, it seems, in most cultures, even though they are named differently and exhibit different forms. They have in common – from the viewpoint of modern Western medicine – that they cannot be explained by structural, organic change and that vague sensations of malaise or mood swings are prominent. Common symptoms, to name only the most significant, are dizziness, feeling weak, tired, and heavy-limbed, restlessness and agitation, trembling limbs, cramps, insomnia and bad dreams, pains without apparent cause, disturbed appetite, an irregular heartbeat, tightness in the chest, anxiety, despondence, and sadness.

There are also specific patterns of complaints or 'pathological' behavior which are found to exist only in very narrowly defined geographical and cultural contexts. In medical ethnology and transcultural psychiatry, the term 'culture-bound syndrome' has become widely accepted to describe such illnesses.[7] Koro, amok, latah, and susto are some of the best-known examples. Characteristic of the Southeast Asian koro, for example, is an all-encompassing fear that the genitals are shrinking and threaten to completely retract into the abdomen, resulting in death.[8] The Malay amok, to cite another example, is noted for its characteristic sequence of deviant behavioral patterns: the typical amok fit begins, often provoked by a slight or insult, with a sudden mood swing following a normal mental state. A burst of more or less random violence against by-standers is followed by a deep sleep and no memory of what happened.[9]

In certain respects, the term 'culture-bound syndrome' is a misnomer. All diseases are to some degree 'culture-bound', even if they are attended by 'organic' alteration. Culture always has its share in the way diseases are perceived and described, let alone explained.[10] Moreover, even ostensibly culture-specific illnesses such as koro have been shown to occur in other places in a similar form.[11] Conditions such as koro, amok, latah, susto, and brain-fag[12] provide a particularly striking illustration, however, of the extraordinary influence that culture can have on the genesis, manifestation, perception, and interpretation of illness. In this way, one could also speak of 'culture-bound syndromes' within our own culture and history. Among the most prominent and intriguing examples are 'acedia'[13] and 'melancholy', 'pica', 'chlorosis', 'hysteria', 'hypochondria', and 'nostalgia' as well as more recent diagnoses like 'chronic fatigue syndrome' and 'anorexia nervosa'.[14] Even when the names of diseases like 'melancholy' or 'hysteria' remained in use over long periods, not only their interpretation but also their typical manifestations changed. The dramatic fits, cramps, and palsies of 19th-century 'hysterical women', for example, have all but vanished from modern Western society.

In what follows, I will focus on the 'vapors' and 'nervous complaints' of the 18th century. They were and remained among the most common complaints of upper-class patients, and the influence of culture on these diseases can be shown to have been particularly strong. But their interpretation and even the way they presented themselves also underwent a fundamental change in the course of the 18th century. I will start by describing this process of change in detail. Then I will explore the degree to which this change reflected the changing social and cultural context of the time and will outline the impact on lay medical culture of the new discourse on 'nerves'.

My analysis will focus in large part on France and francophone Switzerland. The large body of extant French patient letters, which span the entire 18th and the early 19th centuries, provides a solid basis for an analysis of the influence of a defined cultural setting on the ways in which patients presented and explained characteristic complaints, made them part of their identity and used them as a means of public self-fashioning. For the potency of cultural factors in influencing and overlaying these complaints is expressed in the varying forms they seem to have assumed in different countries. While in France they tended to be associated more with sensations that evoked images of movement and agitation (with cramps, excitability, fits of rage, and the like), letters and case histories from Germany and England revolve around the more subdued mood changes and physical complaints traditionally associated with hypochondria and melancholy and reminiscent more of depression in a modern sense. Such national differences were already underlined by physicians at the time and a comparison between French and German patient letters points in the same direction. Melancholy or spleen (from the Latin 'splen', referring to the spleen as the primary locus of melancholy), for example, was regarded as a predominantly English illness. In fact, it was the proverbial

English malady.[15] In France such illnesses, according to contemporary observers, slowly came 'in vogue' only in the late 18th century.[16] In Germany, Heinrich Matthias Marcard described 'the exceedingly widespread weary constitution of our day, especially among the better class of people'.[17]

Making a clear distinction between a more hysterical, overexcited French nervousness and a more neurasthenic or depressive German enervation would be venturing too much. In pathophysiological terms, the conditions could be seen as two sides of the same coin. An excessive stimulation of the nerves, it was believed, resulted in weariness and weakness. But weakness could result in nervous irritation, in turn, as a patient of Friedrich Hoffmann's found: 'all the nervous parts in my entire body have been weakened and become so sensitive that nothing happens but a great deal of disorderly and cramping movement, tension and pulling in all my body parts.' [18] Nevertheless it might be worthwhile to pursue such national differences over longer periods. The implications could be far-reaching. Joachim Radkau has shown how a culture of irritability, of nervousness, developed in Germany around 1880. This culture, he believes, found its fatal expression in the bellicose enthusiasm leading up to the First World War.[19] In this light it is noteworthy that the late 18th-century 'era of nervousness' in France culminated – or found release – in the turmoil of the French Revolution and the Napoleonic Wars.

A New Disease: The Vapors

Early 18th-century French physicians faced a new disease of epidemic proportions: the vapors. While until the mid-17th century the complaint had been virtually unheard of,[20] there was now hardly a household, they found, without at least one member suffering from the illness; one could even say that the entire human race was to some degree afflicted.[21] Patient letters and other 'ego documents'[22] give ample testimony to the remarkable career of the 'vapors' in ancien régime France. Even among the less educated, the term had taken hold. For instance, for the master glazier Jacques-Louis Ménétra it was the most normal thing in the world to refer to his 'cousine à vapeurs', who often vented her bad mood on him.[23]

Characteristically, the vapors began with symptoms such as convulsions and cramps. Initially, in the first treatises on the disease[24] as in ordinary layman language, the term 'vapors' seems to have been linked to these symptoms almost exclusively, in fact.[25] 'It is said that a woman suffers from her nerves – or has the vapors, as they call it – when every now and again [...] she suffers complete convulsions or at least has cramp-like symptoms', writes Lorry.[26] In early 18th-century letter consultations, the term 'vapors' is sometimes used similarly to refer to convulsive movements and twitches.[27] They were at times accompanied by muscle twitches in the arms, legs, or face that moved from body part to body part at an astonishing speed.[28] Depending on the case, a plethora of further complaints could be present as well. A sensation of tightening, for example, was described, along with pressure in the chest area, shortness of breath, or even serious choking and coughing fits, intense palpitations, an irregular heartbeat, pulsing arteries, strange sensations of hot and cold, an excessive urge to empty one's bowels, pains, the sensation of a bloated and heavy head, headaches and dizziness, ringing ears, and weakness and pains in several body parts, as well as complaints of a more general nature such as exhaustion, tiredness, insomnia, and lack of appetite.

The moods and emotional life of people afflicted with the vapors often changed as well. 'Without any apparent reason', patients 'break into excessive fits of laughing or crying', Dumoulin reported.[29] When Mme Graffigny once erupted in laughter, her nurse came running because she thought her mistress was suffering a fit.[30] In other cases, patients or relatives thought it necessary to emphasize that the fits were not accompanied by crying and yelling.[31] Moods could change abruptly and for no apparent reason. 'Anything makes them afraid', said Dumoulin; 'at the slightest suspicion or even for no reason whatsoever, they are gripped by wrath, jealousy and the most intense passions; hope, joy and other pleasant feelings are only fleeting with them; they love to the extreme and a moment later detest the very same person for no apparent reason; they are fickle in their intentions and stop doing something after they have only just begun.' Disconcerting dreams, especially about death and burials, haunted such people and turned their sleep into a source of terror.[32]

Historical Roots: 'Vapores', Hypochondria and Hysteria

In the early 18th century, elements of the medical tradition's various explanatory models and theories on illness found their way into the imagery that was linked to the discussion of the vapors at the time. These elements had a multifarious effect on how the vapors were perceived, experienced, and interpreted, and at the same time were themselves repeatedly reinterpreted and transformed.

The term 'vapors' or, in French, 'vapeurs' originally referred to the fumes and vapors which, according to the Galenic tradition, could quite literally rise from the belly towards the chest and head. In effect, the new pathological term 'vapors' was applied to several symptoms which, in traditional physiology, were explained as the consequences of vapors in the old sense, of fumes, plumes, or winds that rose or wafted through the body: flatulence, for example, or tightness in the chest, shortness of breath, and headaches could be understood as the direct consequence of an abnormal accumulation of such volatile matter in a confined, narrow space. Convulsions and mood changes, too, could be explained as resulting from vapors or fumes in this literal sense. The 'fumes', for example, 'which spread everywhere' and rose up into her throat, were held responsible for the convulsions and twitches from which the sister of a Strasbourg dignitary was suffering.[33] According to some physicians, trembling, twitching, and convulsions, as typical symptoms of the vapors, even pointed directly at the underlying patho-physiological process: they were ascribed to an intense fermentation of the 'animal spirits'. The resulting rapid expansion of volume inevitably led to a disorderly action of these 'instruments of the soul' and hence to uncontrolled muscular movement.[34]

The new conception of the vapors touched upon and overlapped in multiple ways with the traditional concept of 'hypochondria', a concept which remained significant alongside the vapors, though apparently more in German-language areas than in France.[35] The term 'hypochon-dria' originally pointed merely to the presumable seat of the disease: since Antiquity, the area of the abdomen below the 'chondral', carti-laginous parts of the costal arch was called the 'hypochondrium'.[36] The typical hypochondriacal complaints concentrated on this area. They included colic-like pains, feelings of bloating and tenseness, and 'winds' or flatulence.[37] Like the vapors, hypochondria then was closely related to the notion of wind and fumes that developed in the abdominal area and it was also typically accompanied by changes in the mind and the emotions.[38] Because the spleen was located in the right upper abdomen, the traditional location of black bile, the vapors and hypochondria were both connected to melancholy. Julie de l'Espinasse, for example, com-plained that she had become 'hypochondriacal and black-biled' ('atra-biliaire') and was tormented by 'melancholy vapors'.[39]

In patients' accounts, 'hypochondriacal' complaints are frequently located primarily in the upper abdomen. A 30-year-old patient of Friedrich Hoffmann, for example, reported that he had never expe-rienced the intense constipation that lasts for days 'which other hypochondriacs usually complain of particularly'. But he did suffer from strong flatulence and 'sour belches', felt languid, and found himself yawning before noon; yawning and sighing were in those days thought to serve the expulsion of volatile waste products. Also, he related, something 'pulls and tickles' in the right side 'above the stomach, where the bile might be, in such a way that it makes my lips move with it [...] and it is as if something were still closed in this spot and as if something wanted to work itself loose; at the same time, it also pulls my right foot up and down.'[40] Another 'hypochondriacal' patient of Hoffmann vividly described the attacks of 'alarm', fainting, and nausea he used to suffer from. The attacks always began 'with strong, violent throat clearing because my windpipe was squeezed together; then the wrist artery began to contract and became as fine as a pin: then came 'Angst' – in this case, as in many other instances, used in the original sense of 'tightness' (in Latin: 'angustia') – and pressure in the stomach as if a peg were stuck in it. The blood moved around in my extremities and sometimes paralyzed my left hand. I had a prickling sensation in my arms and feet and in my torso too.' More and more often, it seemed to him as if 'all my blood were retreating to my inner parts and I felt as if I had an apoplexy'.[41] With another attack, he felt as if 'something terrible were happening in my body, whereby my head was puffed up, or so it seemed, and my eyes became dim.'[42] Even against the diagnosis of his attending physicians, another of Hoffmann's patients assumed that 'hypochondria' was at the root of his manifold ailments. He suffered from cardialgia (pain in the stomach pit or heart region), nausea, irregular appetite, colic, and serious constipation. The last problem provided the diagnostic key, since, when his body was 'open', he wrote, 'I feel best and don't suffer'. The upper abdominal ailment was accompanied by restless sleep 'with all kinds of useless and unpleasant dreams', and a considerable, as he supposed, 'enlargement and limpness of the seminal vesicles', so that, when he conversed with a women, semen would drip out imperceptibly, a problem that had rendered him 'almost impotent, noticeably frail and quite weak of memory'. Not least, his emotions suffered, 'so that, for no reason whatsoever, I'm sad and disagreeable, but at times extraordinarily cheerful or find myself thinking about unnecessary and farfetched things; I also like to be alone.'[43]

Hypochondria played a less important role in the French letters. If it was mentioned at all, it came closer in nature to the new concept of the nervous complaints. One of Tissot's patients, for example, believed that, following years of solitude, excessive reading, and psychological torment, his 'nervous system' had been damaged to the extent that he had fallen into the onerous state called 'hypochondriacal vapors'. As proof, he listed his physical complaints, shedding light, at the same time, on his understanding of the disease. His main symptoms – flatulence, constipation, copious urine – were linked to the abdominal region; with his other complaints, however, the imagery of 'hypochondria' became entangled with that of the vapors and the new concept of nervous disorders. He complained of a rising facial rash, frequent sighing and yawning, headaches and a buzzing sound, anxiousness, and an 'extreme sensitivity' to the slightest annoyances.[44]

In the worst cases, hypochondria was accompanied by serious mental aberrations and even hallucinations traditionally deemed typical of melancholy, to which hypochondria also had close historical ties, last but not least because both were linked to the spleen as the principal site of the black bile.[45] In 18th-century medical language, however, the term 'hypochondria' became increasingly connected to a particular form of delusion, namely a disproportionate, exaggerated fear of disease and death with which it has become almost synonymous since.[46] Such fears were mentioned as a symptom of hypochondria as early as the beginning of the 17th century.[47] Physicians in the 18th century encountered such fears on a daily basis and, it seems, their patience was sometimes put to a hard test. For instance, a female patient of Geoffroy, after a bad nightmare and suffering from a sensation of heat in her chest, was unable to free herself from the conviction that she was soon to die of consumption. On one occasion, she even called for both a physician and a priest. However, as far as the physician could tell, her only problem was that her legs were shaking.[48] Some physicians were sympathetic. They felt that their patients' need to conceal their suffering in order 'to avoid exposing themselves to ridicule, which is even harder to bear' might contribute significantly to their suffering.[49]

In medical discourse, then, hypochondria increasingly became an illness of the imagination.[50] Yet this new understanding of hypochondria hardly played a role in the narratives of patients even in the late 18th century,[51] except when patients were confronted with this view from the outside. Thus, a Parisian patient who suffered from indigestion, headaches, and vapors wrote that all of his complaints only led to the physicians 'taking him for a hypochondriac' and him 'finding hardly any relief'.[52]

Like hypochondria, hysteria was an important historical precursor of the vapors and shared with them characteristic symptoms. The history of hysteria goes back a long way.[53] It was originally closely associated with the female genitals: the term 'hysteria' derives from the Greek word 'hystera' for 'uterus'. According to traditional ideas, the uterus could wander freely around inside the abdomen, making patients feel as if a sphere or even an animal were pushing upward against the diaphragm. If it rose into the chest and throat, it caused the characteristic complaints of so-called 'uterine suffocation'. These included tightness in the chest, serious shortness of breath, an irregular heartbeat, and fainting. Such images or sensations of a wandering uterus, or at least of a moving (foreign) body inside the abdomen, occasionally also appeared in 18th-century patient letters. A 47-year-old female patient of Geoffroy, for example, not only suffered from 'frequent vapors and a tapping in all my body parts', and from belching and headaches, but also felt a kind of ball inside her abdomen which rose into her stomach.[54] Mme Disse described feeling as if a 'foreign body' were rising from her stomach into her throat, causing her heart and breathing to stop for an instant,[55] and the Marquise de St Innocent experienced the sensation of something rising into her throat, 'as if to strangle me'.[56]

The most significant alternative to the conception of hysteria as resulting from a wandering womb, one that eventually came to prevail in early modern medicine, explained the disease as the result of a corruption or fermentation of female semen or menstrual blood.[57] Lorenz Fries, for instance, attributed the 'suffocatio matricis' to 'poisonous vapors from polluted matter in the uterus' and to retained and spoiled (female) semen – especially in the case of nuns, widows, and virgins, who had 'never taken an interest in men'.[58] Going into still more detail, Daniel Sennert elaborated how 'vicious vapors' rose up from the female genitalia via body cavities and blood vessels, stifling the breath and voice, causing an irregular pulse, or even resulting in a loss of consciousness. Like Fries, Sennert located the source of these vapors in the menstrual blood, or in the (female) semen. He also described similar, if less pronounced, symptoms as the result of retained semen in men. The great diversity of hysterical complaints was explained by the variable nature of the said vapors and their affinities to individual organs.[59] In this way, medicine around 1700 still connected numerous symptoms to hysteria. To some, like Sydenham, hysteria was the chameleon among the diseases. Giorgio Baglivi, for example, listed shortness of breath and the sensation of suffocating alongside heart palpitations, convulsions, vomiting, diarrhea, cold extremities, urinary retention, stroke, fainting, and pains as the typical symptoms of hysteria. This, he noted, could all too easily lead to confusing the disease with others of an entirely different nature.[60] However, certain symptoms were seen as particularly typical of hysteria. Alongside the aforementioned sensation of a rising foreign body, the so-called 'clavus hystericus' or 'clou hystérique' must be mentioned here: an acute, localized headache and a peculiar sensation of coldness at the cranial vertex.[61]

In this second model, hysteria or 'uterine suffocation' was a special case among the many kinds of disease that originated from harmful fumes or vapors. Many 18th-century physicians thus related the vapors to hysteria or used the two terms almost synonymously[62] or understood 'hysterical vapors' as a particular (especially serious) type of vapors[63] which could culminate in a temporary loss of consciousness or cause patients to emit harrowing screams or pull out their hair.[64] Long-lasting affective change, by contrast, was less typical of hysteria.[65]

In the eyes of some German physicians, hysteria counted among the most widespread diseases. J. H. Jungken's *Wohlunterrichtender Medicus* of 1725, for example, claimed that there were 'few women who can consider themselves free from the fortuities of this illness'.[66] For England, Thomas Sydenham made a similar observation: only the 'fevers' were more common.[67] Patients and their relatives seem to have used the term only rarely, however, even in Germany. Probably this was due to the negative connotations of hysteria, particularly its close association with female genitalia and ungratified sexual desire.[68] In contrast, as we will see shortly, the new nervous complaints – of which the vapors, in a reconfigured sense, soon became the most significant representative – were thought to evidence a particularly refined intellectual and moral sensibility far removed from the realm of the lower, animalistic drives.

The Rise of the Nerves

In 18th-century medicine, the nerves became a central explanatory model for disease. Everything became a nervous complaint.[69] A new type of imagery emerged, along with a new system of meaning, which affected for a long time to come the manner in which patients perceived, experienced and described both their bodies and a wide range of physical and, in modern terms, psychological complaints. Already since the late 17th century, the lines between the vapors, hypochondria, and hysteria had become increasingly blurred in medical discourse, in France even more so than in Germany. Numerous symptoms and illnesses were now ascribed to the nerves which had previously been associated with other diseases. This was above all the case with complaints that had until then been interpreted as the result of rising vapors and fumes. But hysteria also began to be reconfigured as a nervous disease by some physicians as early as the beginning of the 17th century,[70] and from the late 17th century onwards this notion became more and more established. Hypochondria too was increasingly seen as a nervous or brain disease and put on a par with hysteria. In the process, the traditional interpretation of hysteria as a predominantly female disease and of hypochondria as a predominantly male disease increasingly lost in significance, even if there were still those who insisted on a fundamental difference between the two.

In the long run, medical understanding of the role of the nerves changed as well. Initially, many authors held to the view that the nerves resembled small tubes through which the animal spirits flowed, communicating between the brain and the different body parts.[71] To them, nervous diseases were only the result of a disturbed, pathological movement of these animal spirits. Increasingly, however, physicians focused their attention on the very substance, the solid matter, of the nerves. The nerves and the brain came to be seen as the seat of the disease. A crucial impetus for this development came from new experimental research into the sensibility and irritability of nerve and muscle fibers, which Albrecht von Haller and others began to undertake on animals around the middle of the 18th century. Endowed with a unique sensibility, the nerve fibers gained paradigmatic value in a new 'vitalist' view of the body that emphasized the specific properties and abilities of the living organism and held them up against mechanistic interpretations.

Within this model, the vapors as well as the related concepts of hysteria and hypochondria once again acquired new meanings. Many physiological and pathological phenomena were now explained as resulting from changes in the substance or consistency of the nerves – be it from atrophy, drying-out, or excessive limpness – or, even more importantly, from an increased 'flexibility' or 'excitability' of the nerves. Almost any symptom in any body part could now be attributed to the nerves, since, as Tissot noted, 'the nerves are everywhere'. The lists of ailments that, in spite of their variety, were thought to be typical of nervous complaints were accordingly long. Pierre Pomme, for example, prefixed his much-read book with several pages listing the symptoms from head to foot.[72]

In their writings, physicians spread their new understanding of the vapors and nervous complaints as the result of an excessive irritability or irritation of the nerves among the wider public. For example, the author of the *Encyclopédie* entry on the vapors proclaimed the old notion of 'vapors', which allegedly rose from the abdomen to the brain, to be long disproven. It was not trails of smoke ('fumées') or vapors that caused disease but an irritation of nerve fibers in the abdominal organs, which, sympathetically, affected the remaining nervous system and particularly the brain.[73] Readers of Vandermonde's *Dictionnaire portatif* of 1760 were informed that the vapors were a 'sensitive and irritable disposition of the nerves', causing them to be in constant, spasmodic motion.[74] Physicians disseminated similar views through their written advice to patients who consulted them by letter and above all, we may presume, through their daily work and communication with patients and their families.

From the viewpoint of healers and especially of learned physicians the new concept of 'nervous diseases' was indeed attractive, both professionally and financially. A concept that turned frequent yawning or recurring headaches into a symptom of a pathological nervous irritability created an inexhaustible need for their medical services. And their advice was particularly in demand among the wealthy classes on whose money and support the physicians' status in society as well as their income depended to a considerable degree. Tissot was among those who were particularly successful in promoting this trend and cashing in on it, first with his *Advice to People* and his *Essay on the Disorders of People of Fashion* and even more so later with his work on nervous diseases. It brought him great success: numerous sufferers from nervous afflictions turned to him. Other healers also tried their luck. Mme Deffand, for example, made reference to a healer who attributed all illnesses to a knotting or pathological folding of the nerves.[75] Some physicians even made the treatment of nervous diseases their specialty. The most widely known representative of these new 'nerve physicians' was, in France, Pierre Pomme. He was convinced that any symptoms that had hitherto been ascribed to the vapors, hysteria, and hypochondria, had only one cause: the nerves were dried out and, as a result, excessively tense. The nerves, according to Pomme, behaved not unlike parchment. In their natural, imbued state, they were supple, soft and elastic. When their moisture evaporated, however – as a result of food that heated the body too much, medication or the many irritations of urban life, for example – they dried out and became increasingly tense and contracted. This led to the known variety of complaints. His treatment, accordingly, consisted primarily of prolonged baths, often lasting for hours, which were thought to restore the lost moisture to the nerves.[76]

Among the common populace, the new understanding of vapors, fumes, or 'vapeurs' as nervous complaints seems to have taken hold only tenuously. According to the famous *Encyclopédie*, the vast population and the mob ('vulgaire') remained loyal to the old notion of vapors as volatile matter in the literal sense.[77] For them the shift may have been too great. After all, the new concept of 'nervous diseases' went far beyond a subtle reinterpretation or elaboration of a familiar explanatory framework. It created a fundamentally new paradigm. It located disease in the body's very substance. But most people still associated disease primarily with foreign, impure, corrupted matter that was not actually part of the body and could therefore be expelled, or they attributed diseases – as in the case of convulsions, a typical symptom of 'nervous disease' – to demons or other supernatural forces which similarly had to be driven out of the body.[78] Among the rural population of southern Germany in the 19th century, supernatural influences still played an important part in the explanation of convulsions, 'falling sickness', and other dramatic or bizarre disease patterns. Only in the course of the 19th century does the new medical model seem to have acquired wide and lasting acceptance among the German population as a whole. Nervous diseases also became a familiar phenomenon in hospitals with their predominantly poor and uneducated patients.[79] By the early 20th century, patients from the lower classes were more likely to have 'problems with their nerves' than more educated patients, who increasingly began to interpret – and still do interpret – similar complaints as 'psychological'.

In France, by contrast, the new doctrine of nervous complaints became accepted much more rapidly, within only a few decades. By the late 18th century, a particular sensibility or irritability of the nerves had become the most important and widespread explanation for numerous complaints. This is echoed widely in the patient letters. Vapors, understood as nervous afflictions, as well as excessive nervous irritability were the most frequent complaints in the letters from the late 18th century. This includes cases where a patient explicitly sought medical advice for 'nervous' complaints as well as cases where patients consulted for other diseases but accorded prominent space to symptoms linked to the 'nerves'. Countless people complained about the excessive 'flexibility' of their nerves or described in detail the violent reactions their sensitive body showed to even seemingly negligible adverse influences. Rare is the case that a historian is able to establish such a clear connection between a new medical concept circulating among the general populace and a change in dominant medical thought.

Embodiment

For quite some time now, medical sociology and cultural anthropology have grappled with the striking differences in the frequency with which certain complaints and diseases appear in different societies and social groups. Of course the potential impact that different environments and lifestyles may have on the frequency and presentation of pathological phenomena must always be borne in mind, but they provide, at most, a partial explanation. Even tangible physical diseases are often afforded vastly differing degrees of attention in different societies – sometimes, in fact, none at all: as mentioned above, certain alterations that are considered pathological in Western society may be regarded as normal in other cultures. Even a comparison between today's industrialized nations reveals a remarkable degree of variation. Heart conditions, for example, are diagnosed in Germany far more often than in the United States. In France, in contrast, there seems to be a 'preference' for liver diseases.[80]

Especially in the case of complaints and diseases for which Western doctors find no organic correlate, a crucial pathway through which such cross-cultural differences are mediated seems to be culturally framed selective perception. Everyone experiences numerous unusual and unpleasant sensations in his or her body over the course of a day. In most cases we are not aware of these sensations except if we focus our attention on them. The degree to which people become conscious of such sensations surely also depends on their personalities and personal histories. But the ability and willingness to notice even certain minute changes in one's mood and physical state is also profoundly influenced by culture. In other words, not only how people perceive physical or emotional disturbances and react to them but also whether they become aware of them in the first place varies considerably from culture to culture.

Early modern physicians' accounts suggest that there were such differences in perception thresholds for physical symptoms even within contemporary society, when they compared different social classes. According to Johann H. Jungken, for example, women who were accustomed to regular hard work and thus had 'hardly any opportunity to think about themselves'[81] were more likely to be spared from the symptoms of hysteria (which he interpreted as a nervous disease). Ethnographic accounts written by 19th-century physicians alternated between surprise and indignation about the 'hard-beaten country man' who did not 'pay attention to disease easily'. Through 'months and years', they claimed, he suffered 'the most agonizing ills with an often unbelievable indolence'.[82] Only particularly dramatic symptoms, especially cramps and convulsions, stood a chance of being taken seriously also by those around them.[83]

Culture not only determines which sensations will enter our consciousness in the first place but also profoundly shapes the manner in which seemingly 'natural' bodily sensations and symptoms, once they are perceived, are experienced. Sometimes culture even brings forth physical sensations which are not known in other cultures.[84] Patient letters and other primary sources touching on vapors and nervous complaints illustrate this point extremely well. Among modern Western women the sensation of a ball rising from the abdomen into the chest and throat and taking one's breath away is virtually unheard of. Yet that was a common experience among early modern 'hysterical' women.

The cultural framing of physical bodily sensations by prevailing medical concepts is even more clearly illustrated by the marked change of patients' typical presenting complaints in the transition from the traditional model of material vapors and fumes to the new paradigm of nervous irritability. In the early 18th century, many patients still described how they 'felt' the vapors or fumes move upward within their bodies, eventually filling the inside of their heads. The 60-year-old Mme de la Buretière and her relatives, for example, described in ample detail the various complaints that her vapors caused her. She felt how 'the smoke and the vapors' rose into her head, bloated her eyes and caused them to secrete a yellowish liquid. The phlegm she coughed up was whitish and creamy, which she apparently interpreted as the result of its having mixed with volatile matter; and when the 'vapors' rose up with particular force following a visit to the doctor, she almost 'suffocated' from them. That the fontanelle on her neck, which the doctors had recommended, did not bring improvement was no surprise to her. For, as a relative of hers wrote: 'what relief can a cautering iron bring from fumes that constantly rise up so plentifully from her abdomen to her head [and] that she feels perceptibly spreading to her eyes?'[85] In the case of a 44-year-old pharmacist, the vapors, 'which rose up into his head from below, making him feel chilly', appeared for the first time two days after he had had a fall. Since the incident, he had repeatedly suffered from them and they always rose from his kidneys or his groin or stomach into his head. They occupied his entire head, 'enveloped' it, making him feel as if his head were empty except for the vapors. All this was sometimes accompanied by twitches in his arms or legs or even face, twitches that changed their location at an incredible speed.[86] 'Trails of smoke' which 'spread everywhere' and rose up to her throat were, in her view, also the cause of the cramps and twitching a Strasbourg patient suffered.[87]

Some patients experienced vapors as warmth rising up. In this case, talking about 'vapors' or 'fumes' was especially appropriate and, in all likelihood, reflected the abovementioned images of the human body as a chimney, as a kind of steam engine,[88] or as a distilling flask.[89] Hence Mme de la Buretière's head was 'full and burning', [90] and to another patient, who suffered from serious dizziness and a leaden heaviness in his head, it even seemed as if the hot rising vapors made the inside of his head boil.[91]

People suffering from nervous diseases in the late 18th century, in contrast, experienced their complaints very differently, even though the symptoms appeared, at first sight, quite similar. Numerous patients now described their complaints as the direct, physically perceived result of a change in their nerves. The place that had been occupied by vapors and fumes, formerly experienced by the patients to rise as a material substance into the chest and head, was now taken by sensations that patients ascribed to the nerves. Because of her 'nerve pains', Mme Dollfus was bedridden for several days.[92] Another patient related that at times, he felt 'a kind of constant compression of the nerves' in his head.[93] More often still, patients complained about a certain rigidity or painful 'stiffness' of their nerves[94] or indeed of their entire nervous system.[95] And still others described how their nerves trembled, twitched, or pulsed, or how they experienced a strong, sometimes painful contraction, a tensing of their nerves, or a jerking inside them.[96] According to her particularly dramatic account, Mme de Chastenay felt nothing less than 'violent strokes' in her nerves. Like the strokes of a piston, they sometimes went through her from stomach to brain; a malaise spread throughout her body and she felt as if she were choking.[97] Another patient's 'nerves' contracted so violently during a colic that her fingers went into spasm and turned in toward her wrist.[98]

Some patients reported sensations in their nerves which corresponded to Pierre Pomme's theory of the nerves drying out. The 40-year-old Monsieur de Leune, for example, assumed that the enduring sensation of a weighty pressure ('oppression') and the slightly cramped state of his chest came from the nerves in his chest being 'dried out' and 'tense' like the strings of a violin. When he overexerted himself, he felt a 'fire' in this area and time and again he had to keep almost silent for entire days to rest his chest. It seemed to him as if the chest or lung nerves had lost all the soft or fatty matter which enveloped them and, without a doubt, kept them supple. He wished for the return of his former constitution in which yellow bile had dominated, which, he hoped, would envelop the nerves in his chest and grease them and thus protect them from the fires and put an end to the sensation of tension and pressure. Perhaps, he suggested, it might be possible to bring about a certain relaxation by way of external fatty, oily compresses.[99]

Several physicians also discussed the possibility that an excessive limpness – rather than tension – of the nerves could be the cause of at least some of the nervous diseases. Tissot, for example, pitched the idea against Pomme's theory.[100] Complaints about a weakened nervous system can be found only sporadically in the patient letters, however.[101] In the patients' experience, weakness could follow an overstraining of the nerves or the loss of valuable balsamic body substance, or it was due to the individual's constitution. A patient of Herman Boerhaave's, for example, discontinued her use of strong laxatives and emetics because they weakened her nerves.[102] A patient of Tissot perceived such a pronounced lack of 'nervous substance' in his feet that it seemed to him as if he were walking on naked bone; in this case, images associated with enervation apparently merged with those of weak sinews or muscles.[103] Another patient, who, thanks to Tissot's book, had recognized masturbation as the pivotal cause of his complaints, lamented the 'slackening of my nervous system'.[104] A Dutch patient considered her 'nerves' to have been 'weakened'[105] by an excessive dosage of purging agents, and for another patient, it was due to the considerable blood loss of her increasingly irregular period.[106] In these cases, the physician's task consisted in 'strengthening' the nerves with suitable remedies.[107] In the 19th century, neurasthenia, a closely related disease conception, exerted a widespread impact, drawing more attention to weakness of the nerves and less to their overstimulation.[108]

No matter whether they attributed their complaints to the nerves' drying out or to their tenseness, trembling, or slackening, the imagery to which patients and their relatives resorted was related to the notion of nerves resembling more or less thick, firm strings running through the body. The comparison of the nerves to the strings of an instrument, in particular, conveyed the idea that nerves were like sinews, whose appearance and consistency was familiar from animals and animal meat. The comparison was carried on to the extent that the terms 'nerves' and 'sinews' were sometimes used interchangeably, even in the 18th century – just as they, incidentally, had been in the medical practice of former times. Accordingly, malformations and deformations of extremities, or contractures – the pathological shortening of muscle and tissue – after burns were popularly also attributed to an excessive tension or contraction of the 'nerves'.[109] In the case of one nine-year-old, it was even believed that the 'palpitation' of nerves in the child's neck could be felt with the fingers.[110] In other cases it was described how hard it was for the 'nerves' in the neck to hold up the head.[111] Or it was complained that, after the application of external remedies, 'all the nerves on the left side became as stiff as a board'.[112]

Critique of Civilization

The 'rise of the nerves' owed much to the new findings and ideas of Willis, Sydenham, Haller, and others. Medical concepts always carry with them certain values, norms, and ideologies, however. A model like that of nervous irritability which not only served as a widely accepted explanatory model but also gave shape or indeed rise to what patients experienced as immediate, natural bodily sensations seems especially well equipped to communicate such implicit or indeed explicit messages.

In the eyes of physicians at the time there was, above all, one explanation for the increasing spread of the vapors and nervous complaints: they were diseases of civilization, the result of a departure from a natural way of living. This claim was confirmed by the observation that almost exclusively the better-off, urban sectors of society suffered from these diseases. The lifestyle of these groups, believed the physicians, exposed the body to numerous morbid influences. Lack of exercise, overly spiced food, fashionable stimulants such as coffee, tea, chocolate, and tobacco, a reversal of the natural succession of sleeping and waking through night-time festivities, constant erotic stimulation due to intensive social contact between the sexes, the excitement of gambling, the artificial stimulation of the imagination through music,[113] novels, and drama, and innumerable other unnatural influences put the nervous system in a state of continual tension and excitement and harmed the entire organism.[114]

Conversely, there was widespread agreement among physicians that in rural areas or among exotic peoples who lived closer to nature vapors and nervous complaints were virtually unheard of. According to Pierre Pomme, townswomen were less robust even in youth than hardworking country women in their old age, let alone, say, Scythian women, who were allowed to marry only after they had killed three enemies.[115] Worse still, the acquired susceptibility to nervous diseases was passed on to progeny. Ultimately, an accelerating decline of the human race loomed.[116]

The new medical discourse on nerves was thus part of a broad contemporary medical critique of civilization, a critique with which we are familiar from other contexts, such as the lament about the worsening air quality in cities,[117] or the claim that (pre)menstrual pains was particularly common among the upper classes.[118] Physicians here were partaking in the lively debate about the negative consequences of civilization and promoting the new esteem for 'natural' living, as proclaimed by Jean-Jacques Rousseau and other leading representatives of the Enlightenment. As experts on what was 'natural' and 'appropriate to nature' in terms of the body, physicians lent scientific sanction to the critique. From this perspective, the nervous complaints became above all the well deserved punishment that Nature imposed on a society that had strayed too far from the natural order of life and society.

The critique, however, was aimed not only at new, fashionable forms of urban lifestyle. Particularly in France it was to some degree also a political statement. It aimed not least at the world of the French court, which – seen as decadent and at the same time idealized – was accused of having carried perverse forms of social life and unnatural lifestyles to the extreme. Held up against it was the ideal of the level-headed, proper citizen who did not give himself over to the culinary, social, and erotic excesses typical of courtly life.

The Sensible Woman

Connected, in part, to this general critique of civilization, the medical discourse on the vapors and nervous complaints also acquired strongly gender-specific connotations. Medical authors agreed almost unanimously that women were the preferred victims of nervous complaints.[119] Hunauld even went so far as to claim that hardly a woman was spared them.[120] According to the physicians, this was in part due to the typically unhealthy lifestyle of wealthier women, which was characterized by frequent socializing and lack of physical exercise. But to many physicians, female nerves were also – like female fibers in general – more delicate, flexible, and excitable than those of men. In other words: women were more susceptible to nervous complaints on account of their natural physical condition.

The new discourse on nerves was in this way tied to another major Enlightenment controversy, namely the debate on the 'nature' of women. Medicine had an important contribution to make in this area too. More vigorously than the Protestant reformation 200 years before it, Enlightenment ideals of equality called the principles of a patriarchal society and, more concretely, the exclusion of women from universities and public office, into question. In this situation, the new discourse on nerves, with its gender-specific implications, offered many male authors welcome arguments for defending the status quo. Thomas Laqueur and Londa Schiebinger have argued that contemporary beliefs that the female skeleton and the female genitals were fundamentally different from those of the male should be seen in this context. Laqueur's claim, in particular, that the insistence on anatomical differences between the sexes in medical writing was a new phenomenon in the 18th century is clearly wrong. His widely accepted finding of an alleged shift from a 'one-sex model' to a 'two-sex model' in 18th-century medical writing results above all from a fundamental methodological flaw: he almost totally overlooked the numerous earlier texts on female and male anatomy and 'diseases of women' which were written in Latin – until the 18th century the dominant language of science and medicine. There are literally dozens of Latin works (though some vernacular ones as well) which, sometimes in turn based on even earlier writings, already stressed and illustrated such differences from the 16th century onwards.[121] There is some evidence , however, that the argument for a fundamental natural anatomical difference between the sexes did in fact acquire renewed importance in the context of Enlightenment ideas of equality and universal rights: social discrimination against women could be justified as an inevitable consequence of the biological nature of the 'weaker sex'.[122]

The teachings on nerves fitted perfectly into this discourse. The naturally heightened excitability and sensibility of the female nervous system made women ill-suited to persevere at intellectual activity such as academic studies, let alone the toil of public office. Too erratic and fickle was their nervous system and, with it, their intellect. Moreover, forgoing such tasks and positions was in their own best interest: excessive stimulation or straining of their nerves, it was held, could inflict permanent physical damage on women. With their sensitive nerves and the resulting capacity for empathy, women were much better equipped for acting as loving wives, self-sacrificing mothers, and entertaining companions. In this way, the new concept of nervous irritability assigned women to the house and home as their natural place. 'Above all, raise your children', E. P. Beauchêne urged his female readers, recommending that they grow old with dignity as the mother of the family[123] – and this in a time when some elderly women were making a name for themselves as artists or as hosts of famous literary salons.[124] Only few physicians turned this reasoning on its head, in line with a tradition of medical feminism that can be traced back to the Renaissance.[125] Reverting to the medical critique of civilization, these physicians declared the way-ward and unnatural upbringing of girls to be the ultimate cause of their increased irritability and the higher prevalence of nervous complaints among women.[126] Along the same lines, some physicians had inter-preted menstrual pain, which they saw particularly often among the upper classes and which was quite often accompanied by nervous com-plaints, as the result of an unhealthy upbringing, one that tied girls to the home too early, depriving them – unlike boys – of necessary physi-cal exercise.[127]

The Cult of Sensibility

The predominantly negative assessment of heightened nervous irrita-bility raises some intriguing questions. Why would countless educated women have admitted to a disease that seemed to attest to their physi-cal and intellectual inferiority? And why would men and women have embraced a disease concept that made them appear the victims of their own lifestyles, laying the blame squarely on themselves for their often difficult and sometimes – to use their own words – almost unbearable lives? For laypeople, as I will show in the following, the attractiveness of the new concept of nervous excitability and nervous complaints indeed lay elsewhere. For one thing, the nervous complaints – for all the subjective suffering they entailed – could take on the function of an accolade. They became the physical badge of moral and intellectual superiority. For another, they served as a kind of body language and allowed for a non-verbal, somatic protest against traumatic experiences and life circumstances.

We tend to associate 18th-century high culture, especially that of France, primarily with the Enlightenment. However, 18th-century cul-ture and cultural change were diverse. Of special relevance for the his-tory of the 'nerves' is a new culture of 'sensibility', which began with the pre-romantic writers of the 1720s and 1730s and later had a wide-spread effect, in particular with the works of Jean-Jacques Rousseau. Comparable developments can be seen in England at the same time and a little later with the German 'Sturm und Drang' movement.[128] The ideal of the cool, self-possessed rationalist became juxtaposed with – or even replaced by – that of the emotional person who gave emphatic expression to his feelings: through language, gestures, and tears.[129] The upper classes quickly adopted the new ideal as their own. A notably more intimate tone can be discerned in letter correspondence; even in the comparatively anonymous patient letters, the shift can be observed.

While, in the early 18th century, a plain cover letter with a patient's short description of his medical history was most common, letters addressed to Tissot at the end of the century often begin with elaborate praise of his excellent achievements, his unrivalled knowledge, and his love for suffering humanity.

The medical discourse on nerves, then, lent this new ideal of sensibility a bodily, physiological basis, as it were, and thereby in turn significantly advanced the trend.[130] The new nervous paradigm permitted sensibility to be interpreted as more than a mere moral capacity, a quality of the soul. Sensibility was now located in the substance of the body itself. While the difference between sensibility and irritability became obfuscated, both terms came to refer at once to qualities of the nervous system as well as to a person's character.

Not just anyone was 'sensible' to the same degree, however. From the beginning, sensibility served as a mark of individual and collective distinction. First, sensibility was connected to notions of an individual moral superiority, to an outstanding ethos and to an uncommonly developed capacity for feeling and empathy. The new model of nervous complaints thus offered a way of presenting oneself – by highlighting a special nervous sensibility – as a particularly sensitive personality. Second, the new ideal of sensibility was class specific to a high degree. Certainly, due to their physical condition, all human beings were 'sensitive'. But some social groups were characterized by a much higher degree of sensibility than others. They literally embodied the advanced intellectual and cultural development of the upper classes. The poet or the philosopher who devoted himself all too passionately to his studies was the preferred victim, comparable to the melancholy scholar of earlier centuries.

With the dual nature of sensibility as both a moral quality and a reflection of the individual nature of a person's nervous system, the symptoms of a pathologically heightened nervous sensibility and irritability became a flexible means of self-fashioning,[131] self-stylization, and self-dramatization.[132] Those who put their vapors or nervous complaints on show in the public sphere, or indeed staged them with some degree of drama, could thereby give expression to their individuality, their moral qualities, and, at the same time, their distinguished social position.[133]

The accounts of contemporary observers point precisely in this direction. In the city and at the court, wrote Caraccioli in 1768, it had come to be considered good manners to suffer from an 'affected nervous system', to 'announce publicly that one couldn't sleep and was hardly able to digest, even when in one's prime and otherwise healthy', and, if need be, to faint.[134] He vividly describes the typical high-society woman who started to suffer from exhaustion and sensations of coldness upon learning 'that it makes you original'. She would rather be melancholy than 'be nothing'.[135] Without a doubt, this is a case of ironic overstatement, but the validity of Caraccioli's evaluation is attested to by many patient letters. Innumerable bourgeois and aristocratic patients gave voice to the extraordinary, even unequalled, sensibility, 'mobility', or irritability of their nervous systems and their entire bodies. Ostensibly, pointing this out was meant as a cautionary note to the physician. On the one hand, they wanted to make sure that he would take their nervous sensibility into account when deciding on their treatment. On the other hand, they wished to point to the exquisite, even unique, nature of their physical and mental condition. They were special. Their nervous system might no longer be able to handle even an innocuous joke.[136] In extreme cases, it could take as little as the noise made by silk coming into contact with certain objects, or by the scrape of a fork on a plate to produce 'unbearable sensations' such as gastric spasms or intestinal cramps. A 'nothing' – a trembling leaf, the flicker of a flame, walking on uneven ground – was sometimes enough to bring on a nervous attack. The consulting physician was implored to take the greatest care when prescribing medication; and, of course, the patient's reaction to the treatment attested once more to his or her exquisite nature. Only a few sips of mineral water would throw the nervous system of some patients into complete disorder.

Above and beyond the potential for an impressive staging of one's sensibility and individuality, suffering from 'nerves' also brought with it practical advantages – secondary gains, as we would say nowadays. Patients with irritable nerves required great care and attention, of course. They had to be treated with lenience and sympathy. An upper-class woman who suffered from nerves could be sure to enjoy the regular company of physicians – of men, that is, who cultivated their ability to listen for professional reasons, whose careers depended on a certain aptitude for intelligent conversation. For some, this almost became a lifestyle. Mme de Chastenay, for example, is reported to have constantly surrounded herself with physicians.[137] Taking daily walks or rides, which Tissot, Tronchin, and other advocates of a 'natural' lifestyle prescribed to those with nervous complaints, also justified spending time outside, in public, and – if one were lucky – away from the ever-controlling hand of a mother-in-law or other relatives. Attacks of the vapors or nervous complaints were also a way of legitimately avoiding unpleasant social commitments. According to Caraccioli, diseases constituted the most common excuse, except for business reasons, which women could only rarely appeal to. Because they could befall someone at any time and disappear just as quickly, vapors and nervous complaints were better suited to these purposes than, say, indicating that one had just taken laxatives – which was another accepted excuse – or, for women, the monthly 'indisposition'. Serious vapors or nervous complaints also could entitle the patient to weeks or months in glamorous resorts, as in the case of Mme de Moncharlé, who made no secret of her unhappy marriage and at times saw herself 'forced' to spend her summers at a spa.[138]

The more widespread the vapors and nervous complaints became, however, the more those afflicted by them risked being suspected of feigning their complaints or at least exaggerating them wildly. People suffering from the vapors, Joseph Bressy wrote in 1789, were generally considered as overanxious and were forced to conceal their complaints to avoid becoming a laughing stock.[139] In England, Bernard Mandeville had observed decades earlier that even talking about the 'vapors' had become a joke. They were considered the mere expression of a bad mood, contrived 'by headstrong, extravagant and despotic women', whenever their 'unreasonable wishes' were not granted.[140] In the early 19th century, Mme de Genlis had nothing but derision and contempt for women of an earlier time who, as she put it, would have their nervous fits twice every week, at a certain time, and lasting three to four hours 'as if it were all a play, whose several breaks could have been compared to the entr'actes'. Strangely enough, she said, these fits vanished all by themselves with the events of the French Revolution and the ensuing forced emigration.[141] A French lawyer, Claude Paumerelle, went so far as to write facetious instructions in the form of a letter exchange with a woman suffering from the vapors. Under the author's guidance, the fictitious addressee learned how to play up her nervous symptoms ever more skillfully, particularly to gain the attention and favor of a sweetheart.[142]

As a 'sensitive' or 'nervous' personality was becoming a daily phenomenon also among the middle classes, the vapors and nervous complaints thus began to lose their function as a mark of distinction for women from the upper classes. As women from the lower bourgeoisie – the 'petites maîtresses', as Caraccioli called them – began to suffer from the vapors and could now 'faint on cue',[143] ladies from higher circles who publicly exhibited similar symptoms risked acquiring the faint stench of vulgarity.

Illness as Protest

I have so far discussed the vapors and nervous complaints almost entirely in the context of and from the perspective of their times. At this point, however, it seems useful to view the rise of the nerves from a different angle for a moment. In the following, I will set out to explore the possible causes of such widespread prevalence of the vapors and nervous complaints based on the explanatory models of modern medicine and psychology. An approach such as this certainly comes with its own set of problems. Analyzing other medical cultures using our modern concepts and explanatory models harbors the risk that we will hold them up as an absolute and objective yardstick, disregarding their historical genesis. Radical constructivists will deny the validity of such an approach. Yet I consider a momentary change of perspective of this nature, provided that it is clearly indicated, not only as legitimate but as necessary. In the end, we always view historical events and processes from today's perspective, even if we begin by trying to grasp them using the categories of their day. Diseases are not exempt from this. An analysis of the history of great epidemics such as the plague or cholera does not need to be confined to historical reactions and interpretations – although neglecting them will certainly not lead to a satisfactory understanding of the historical reality of an epidemic either. It can be meaningful to ask, in retrospect, to what extent shipping traffic, the lack of a sanitary infrastructure, or social inequalities favored the spread of epidemics, even though people at the time might not have made such connections.

Based on what we know today, we will find it difficult to share the view held by 18th-century physicians that the spread of the vapors and nervous diseases among the upper classes was due to the unhealthy effects of modern urban life. To us, tea, chocolate, novels, and concerts seem unlikely causes of such suffering. It appears equally questionable when some historians today, attribute the rise of vapors and nervous diseases to overstimulation from accelerated social and economic change. According to G. S. Rousseau, for example, the 'neurological chaos' in the 18th century reflected nothing but the social disorder of the time. Nervous complaints were an attribute of social groups who lived a life 'in society's fast lane' and who were 'brought to their knees by the pressure of society'.[144] This may make sense in the case of late 19th-century entrepreneurs. The social and cultural circumstances in 18th-century France, however, drastically disagree with this thesis. Certainly, the odd citizen would have perceived the dynamic economic development of early industrialization as a time of increasing activity and turbulence. The vast majority of patient letters, however, and other personal testimonies as well, provide no evidence of a wide experience of frantic activity and stress. On the contrary: feelings of emptiness are voiced repeatedly, a boredom that was only imperfectly disguised by widespread gambling, the experience of lacking opportunity for personal development, especially with educated women.

It is much more plausible to understand the vapors and nervous complaints in terms of the modern psychosomatic concept of 'somatization'. The culture-specific forms, images, and concepts in and through which physical and affective changes are experienced, interpreted, and evaluated not only impart a certain color and meaning to these physical and affective changes; they also endow symptoms with expressive power, with a communicative value, which allows patients to convey a message, to use them subconsciously as a kind of symbolic language. The role of symptoms as a means of communication has already become tangible to some extent in the positive self-fashioning of patients with vapors or nervous complaints as endowed with a particular degree of sensibility and morally superior. In a similar way, patients' complaints may give expression to psychological conflicts, anxieties, trauma, or difficult life circumstances.

In recent cultural anthropology, the concept of 'somatization' has been connected in particular with the notion of 'idioms of distress'.[145] While psychosomatic medicine tends to focus on the individual, cultural anthropologists have provided impressive evidence for how certain diseases can give expression to the taxing or even virtually unbearable situation of a whole ethnic group in a specific social and political context. At times, the complaints seems to transform the discomfort or suffering of the members of that group directly into somatic imagery. Let us take, for instance, refugees from El Salvador who suffer from dizziness and fainting and feel that the ground beneath their feet is moving and threatening to topple them. This would be a highly fitting expression of their uprootedness, their disorientation, amidst the threat of losing their employment.[146] In such cases, it indeed seems as if, as Kaja Finkler wrote with reference to Merlau-Ponty, the 'body acted out its life circumstances', replicating 'the contradictions and the confusion of the outside world' inside itself,[147] even though we must keep in mind, of course, that this is the Western outsider's understanding: the concept of 'somatization' arises from and reflects a very specific, Western understanding of the relation between the body, emotions, and thinking. From a cross-cultural perspective the modern Western idea of a 'psychological' sphere which is somehow separate from the body and can thus interact with it is exceptional.[148]

In the case of the 'nervios', a particularly well-known example of such an 'idioms of distress' the disease term as such even suggest a direct historical connection to the pre-modern concept of 'nervous diseases'. 'Nervios' are common especially in Latin American cultures but can also be found in Newfoundland or, as 'nevra', in Greece.[149] Patients report diffuse pains that wander through the body, weariness and listlessness, dizziness, fainting, an irregular appetite, difficult breathing, and heart palpitations, for which no organic cause can be determined. All of these can be accompanied by anxiety and mood changes. To the Western physician this may appear as a farrago of vague and unspecific symptoms. In the experience of the sick, however, their complaints are part of a well defined whole. They are suffering from a distinct disease, which can be given a name: 'nervios' or 'nevra'.

Returning to the world of the 18th century, there is much evidence to suggest that the vapors and nervous complaints can similarly be understood in retrospect as an 'idiom of distress', as a reaction to individual or collective trauma, fears, disappointments, or constraints. In fact, the patients themselves frequently attributed their diseases to what we would call emotional trauma or psychological distress. Some of them seem to have taken such a causal relationship so much for granted that they felt it was sufficient to simply mention their 'grief' or 'fright' or – hardly more informative – their 'domestic worries', without any further explanation. Others provided more detailed elucidations of the traumatizing or distressing experiences which they suspected of being a key cause of their complaints, or which, to their own surprise, had failed to have the negative impact on their health that would normally be expected. In doing so, they also permit a rare glimpse into the experience of emotional trauma and conflict in the early modern period.

Ranking first in the patient letters among such traumatic experiences was the death of loved-ones. Women referred above all else to the great sorrow that the death of one or several children caused them. Their letters show once more that we should not let ourselves be misguided by statistical figures on high infant mortality nor by the often-repeated claim that 'maternal love' was an invention of the modern period or even of the 19th century, which even a brief look at the role of maternal love in entirely different cultures around the globe renders absurd.[150] There is no doubt that early modern parents were faced with the loss of a child much more frequently than today. Hardly a woman was spared the experience and a mother of six could be left with only one child.

Nevertheless, the stories that women told in their letters make it clear that the death of a child, or even a miscarriage, was among the most painful, life-changing experiences a woman could have. And according to their accounts, their strong emotional response could not fail to influence their health. The Comtesse de Non, for example, a mother of seven children living in Turin, dated the beginning of her complaints to the time shortly after the 'deep sorrow' caused by the loss of a child. Her subsequent pregnancy was disturbed and she had been suffering since then from dizziness, yawning, discomfort in the head, bad taste in the mouth, trembling knees, cramps, teeth-grinding, hot flushes, swelling, and a feeling of heaviness and restlessness, as well as other symptoms of extreme nervous irritability that would worsen before and during her menstrual period.[151] According to the attending physician of a different patient, the 'cruel grief' inflicted on her by the death of two of her children had caused the patient's 'morale' to become almost as diseased as her body. She suffered from feelings of pressure, a painful narrowness in the chest, insomnia, and sensations of heat.[152]

The death of adult relatives and friends, particularly of one's spouse, was also often extremely painful. When, for example, Mme Vionnet learned about the death of her brother, whom she loved dearly, her state of health immediately worsened substantially. A sensation of intense oppression gripped her stomach and throat; she shivered violently; her head glowed while her feet were cold as ice; she felt weak and dizzy.[153] Another patient was unable to eat for two days after the death of a close relative. She got a stomach ache and her health deteriorated.[154] The melancholic Mme Vivaux felt, in her own words, as if possessed, haunted by the memory of the death of her adult daughter and she constantly imagined the sad fate awaiting her daughter's daughter, her own grandchild, when she herself died.[155] Grief over the death of her husband literally went to the head of a patient of Geoffroy: she suffered severe headaches, experienced her vessels as extremely tensed up, and temporarily almost lost her mind and her memory.[156]

When a family member died, the trauma of the loss often became combined with the physical and emotional strain of the weeks and months leading up to the death. The disease of a Geneva businessman who suffered from dizziness, heart palpitations, and presentiments of death was preceded by the death of his wife after two years of suffering.[157] In some cases, there was an added fear of contagion. One countess, for instance, suffered a severe relapse after the 'laborious support' she had given her husband during the last three years of his life. When, in the end, her husband succumbed to his consumption (which the physician had said was 'open', that is, accompanied by inner ulceration), she began fearing for her own life and that of her children, whom her husband had wanted to see even in his final days.[158] Widows and orphans, following the death of a husband or father, were frequently faced with economic worries on top of everything else. In some parts of France, for example, the inheritance rights of widows were only insufficiently secured and their social status suffered almost inevitably.[159] The death of a husband could therefore constitute a serious trauma for the widow, even if the marriage had been unhappy. For instance, the two-year marriage of a 24-year-old Strasbourg patient, who was tormented by frequent attacks of the vapors, of anxiety and panic accompanied by cramps, had been anything but happy. It had 'affected' her nerves and made her plunge into a deep melancholy. But when she then lost her husband, she suffered a harrowing 'revolution' in her body, with convulsions that lasted for days, and she could be freed from them only through substantial bloodletting.[160]

Fatalities predominate among the dramatic, life-changing events that patients or their relatives identified as the causes of current diseases. But sometimes other strokes of fate are mentioned in this context. A servant at the French court, for example, saw his marriage plans and indeed all of his prospects ruined by the sudden death of the king, and this made him ill. Political events, and above all the French Revolution, could play a similar role. One patient said that the psychological effect of the 'befouling' of the Revolution by republican forces had contributed significantly to the 'disintegration' of his nerves.[161] Another patient thought that the events of the revolution had aged him, made him more sensitive and restless, and affected his 'nervous fluid'; the people in his village had taken his and his family's side but people from other villages had threatened to loot and burn down his house.[162] Due to the Revolution, 30-year-old Monsieur Dubois lost his position with the Comédie-française and considered this a decisive cause of his declining eyesight.[163] According to the account of a woman whose husband went insane, it was the Geneva revolt of 1782 that had played a key role in the man's medical history. The hostilities of the time had tormented him to such an extent that he suffered from recurring 'compression' of the heart and fainted on a regular basis.[164] As mentioned above, however, some contemporary observers who were not affected by nervous diseases themselves credited the events of the French Revolution, on the contrary, with a significant decrease in the vapors and nervous complaints. According to them, patients could simply not afford to maintain their former sensibilities and caprices when their lives were at stake.[165]

In addition, burdens or insults from long ago – even childhood experiences – were named by some patients as important predisposing factors in their complaints, anticipating, in a certain sense, the insights of modern psychology. 'In a singular manner', wrote a patient of Tissot, 'a word alone or an unfamiliar glance can disturb me, and this is why I consider it the greatest disaster of my life that I lost both father and mother very early and was left to the care of people who in no way responded to my inherently anxious nature.'[166] In a similar fashion, the Comte de las Cases, in the 19th century, traced his illness back to a difficult childhood. His rigorous upbringing had left him in a state of weakness and lassitude early on, he claimed. At the age of 10, he was sent to live with strict relatives, where he was often left alone and had to stay indoors. He became sad and melancholic and physical complaints were soon to follow: his head became heavy, his digestion slowed, his eyesight and appetite became diminished; he had minor epileptic fits. Later, he emigrated to the tropics with his parents and again suffered greatly from strictness, this time that of his father. He had attacks of ravenous appetite and his limbs felt restless. Violent throbbing and cramping around the heart added to his suffering, sometimes forcing him to stop in his stride for fear that his heart would 'burst'.[167]

Men sometimes also pointed out specific professional strains, such as excessive mental labor or a sedentary occupation. Here they were consciously or unconsciously adopting the physicians' conviction that the lifestyle of scholars exerted a morbid influence, one that made itself felt not only in nervous complaints. Others mentioned their tiring office work or their duties as tradesmen as possible causes of their nervous ailments.[168] The complaints of some male patients whose biographies were marked by painful experiences of dependence, insult, or failure in professional life sometimes seem particularly expressive, even allegorical. A patient who suffered from serious headaches, for example, had been forced to resume his hateful and stressful position as a teacher at a boarding school after his failure as an accountant. Although his wife lived in the same town, he was able to leave the boarding school only two days a week.[169]

Several patients, finally, blamed not so much their troublesome experiences as their emotional response to them. Just as the body had to rid itself of impure morbid matter, which might otherwise cause all kinds of diseases, holding back or restraining strong emotions could have disastrous consequences. Mme de Moncharle traced the beginning of the tightness in her chest, her breathing difficulties, wind and cramps to her wedding of 30 years past, when, 'like so many others, I experienced much perturbation', and she thought that 'the force and violence which I used against myself may have contributed to the damage to my health'.[170] Another patient was said to have harmed himself with the effort of 'bringing himself to play with "noblesse", that is, to lose his money with grace'. He suffered from wind, colic, and a heightened sensibility of the nerves.[171]

Ennui and Narcissism

Alongside individual traumatic experiences and burdensome life circumstances which entire social groups were subjected to, cultural comparisons reveal a further factor essential in determining the frequency and intensity of somatization phenomena. Cultures and societies differ markedly in the degree to which they promote and support the tendency toward somatization among their members in the first place. In China, for instance, patients report conditions that we would recognize as depressive moods almost exclusively in terms of physical ailments. Such differences also point to the variation in the extent to which different cultures permit mental illness and, more generally, the open expression of negative or 'abnormal' feelings. To put it somewhat crudely: if someone is allowed to reveal his discomfort, his conflicts, his negative emotions in an unveiled way and can expect to meet with understanding and receive help, he will not depend quite as much on taking the detour via somatic expression.[172]

With regard to the late 18th century, this might seem, at first sight, hardly relevant. In contrast to previous ideals of a dignified demeanor, women and even men, in the age of sensibility, were allowed to cry in public and indulge in passionate expressions of emotion. The 'detour' via a somatic expression of one's feelings, one would think, should have become almost obsolete. But this first impression may be deceiving. If critical contemporary observers are to be believed, the cult of sensibility also imposed rather strict, rigid rules of its own. The more or less open expression of emotion was not only tolerated in many situations; it was expected – hence hypocrisy and falsehood were always just around the corner. The new obligation to emotional truth – to publicly admitting one's feelings – subjected expressions of emotion all the more to the control of the dominating norms and conventions. In retrospect, this jeopardized the possibility of experiencing one's own feelings as authentic rather than as part of a mask. A society that in this way not only impedes the authentic expression of feelings, but at the same time commands the emphatic expression of false, insincere feelings is, from today's point of view, a hotbed for disturbances to self-esteem. Building on the psychoanalytical concept of 'narcissism', Heinz Kohut and others have given a detailed description of the psycho-dynamics of this disorder. They have also pointed to a person's disturbed body experience in this context and to a fragmented, ever-threatened physical self, one characterized by hypochondriacal fears.[173]

The extent to which such modern psychoanalytical concepts can be applied to the past remains, of course, debatable. While the call for a sound psychohistory is nothing new, no explanatory model has found general acceptance. Some support for the interpretation sketched out in the previous paragraph can be found, however, in the judgment of both contemporaries and members of the subsequent generation. They bemoaned a widespread inability to love in those times of sensibility,[174] and even the absence of a capacity for deep feelings in general. Based on their in-depth study of numerous contemporaneous personal testimonies, the brothers Goncourt in the 19th century concluded that an intensified social life had been unable to fill the 'nothingness' in the hearts of women. The world became an empty drama to them. Long before Freud, the Goncourts saw the famed 'ennui' of the French upper classes as the underlying cause of their nervous diseases: 'This boredom of heart and soul affected the woman's body. It brought her suffering, a certain weakness, a decrepitude, a kind of bodily mourning and jadedness as well as that covert discomfort which the time came to baptize with the vague name "the vapors"'.[175] Similar experiences of suffering from an existential emptiness and boredom are in fact frequently related in the patient letters; everyday life appeared determined by mutually obligatory visits, stiffly ritualized manners, and tedious 'chatter'.[176]

Women, according to some contemporary authors, experienced a heightened pressure to veil their feelings and experiences. Jeannet de Longrois, for instance, held that regrettably, they were forced to live behind a perpetual mask. They were raised to show off their affectations, he remarked, and were forced to control their wishes and desires at all times – with disastrous consequences for their health.[177] Appropriate female role behavior was internalized from a very early age. Even young girls had to stay home almost all the time. They were not allowed to run, scuffle, and be noisy like boys but instead learned the subtle language of 'graceful' movement. Its mastery was a prerequisite for later success on the social stage and, not least, on the marriage market.[178] On top of that, grown-up women in the late 18th century were much more subjected to certain moral conventions than men. The literary topos of widespread sexual libertinage among genteel women in Enlightenment France is deceiving. Women's prestige and self-worth remained tied to their reputation, to their moral character – that is, to what was regarded as such in those times.[179] And the constant social interaction between the sexes in polite society made the danger of becoming a target of gossip ever present.

Expectations concerning the appearance of women in the 18th and early 19th centuries seem ambivalent as well. Fashion changed quickly at the time. In the long run, however, over the course of the 18th century, the new ideal of the slender, delicate, and lively woman replaced the older model of the majestic, dignified lady. Liveliness and naturalness of expression became increasingly valued. Inevitably, however, the new pursuit of thinness, as it does today, had the power to damage the self-esteem of the women whose figures did not conform to it. 'Pretty women fall into desperation when they put on weight and there is nothing they wouldn't undertake to lose it', wrote Caraccioli. For fear 'of losing an elegant and slender waistline', they hardly ate or even drank vinegar, he claimed.[180] For similar reasons, the social and emotional consequences of aging became more traumatic for women. Women who were past the 'attractive age' ('l'âge de plaire'), in Sénac de Meilhan's account, no longer knew 'how to fill the void'. Embitterment over their fading beauty became coupled with the 'boredom of an idle soul'.[181]

The new ideal of the love marriage would, at first sight, appear to have improved the situation of married women. Yet even here the effects may well have differed from woman to woman.[182] On the one hand, historical research has shown that a change occurred also in day-to-day matrimonial life during the 18th century. Kissing or embracing in public was no longer taboo for married couples. In France, for the first time, aristocratic husbands and wives addressed each other using the more intimate 'tu' instead of the usual 'vous'. On the other hand, here too, there are indications that emotional expression rapidly became stereotyped.[183] And the high hopes for marriage as a place of romantic love – as embodied by Rousseau's *Héloïse* – may well have rendered the chasm between these hopes and the actual experience of most women all the more painful. Little is also known about the success of sexual relationships in marriage. Accounts like that of Mme de Chastenay who, despite her lively and precocious sexual interest, openly admitted that matrimonial intercourse meant nothing to her, are rare.[184] However, there is no doubt that around that time the dominant views of female sexuality fundamentally changed. The formerly self-evident assumption that female desire was stronger than male desire slowly yielded to the new ideal image of the sexually modest, passive woman, which would ultimately come to prevail in the 19th century. Apart from that, the fear of becoming pregnant would have been enough to spoil the pleasure for many women.

To sum up: women and men from the upper classes adopted the new model of nervous sensibility in more than one way. It offered a coherent explanation for many different ailments that formerly would have been explained as expressions of hypochondria, hysteria, and vapors in the literal, material sense of the word. 'Nervous complaints' also served – if at a high price – as a token of special intellectual, moral, and emotional sensibility and distinction. And they permitted – at least in retrospect – the expression of mourning, anger, and other negative feelings in the form of physical symptoms, and this at a time when the public exhibition of intimate emotions was much more permissible than in former times but, at the same time, channeled by rigid norms which governed what counted as acceptable emotional expression.

These forms of adopting and 'embodying' the new model of nervous sensibility and nervous complaints overlapped with the medical view only to some extent. Physicians, based on their critique of civilization, held that nervous complaints were primarily caused by a misguided way of dealing with the body's natural needs, by bad nutrition, and by the all-too-sedentary lifestyle of the urban upper classes. In the predominantly individualistic accounts of patients and their relatives, however, this concept left hardly a trace. The new paradigm of 'nervous diseases' thus does not seem to have been particularly successful in disseminating norms of dietetic and emotional moderation. It was more effective in communicating other ideological and normative 'messages'. Contemporary accounts confirm the evidence provided by the patient letters: the large majority of people suffering from nervous complaints were women. The new paradigm may have served those women as a welcome means to comprehend and fight their suffering. It may have provided them with a symbolic language in which they could express their physical and emotional hardships or served as the distinguishing token of their individuality and exquisite sensibility. But the fact that women, more than men, adopted and embodied the new paradigm lent support to the physicians' notion of female nerves as more delicate, more irritable, and more flexible. Their own symptoms were a proof that women, due to their weak nervous system, were unable and ill-suited for demanding intellectual tasks or positions that entailed a certain amount of responsibility. With their complaints, female patients unintentionally justified the medical argument that, due to her very nature, the place for a woman was at home with her husband and children: precisely the limited life circumstances that some women, in reflecting on their situation, saw as a significant source of their suffering. In this manner, very much in line with Michel Foucault's analysis, a crucial pathway was opened for dominant discourse, in this case in support of patriarchal society and inegalitarian gender relations, which was all the more effective because it impacted on the very bodily constitution of the subject. With their disposition to symptoms of excessive nervous sensibility, women embodied the very patriarchal ideologies, power structures, and unequal opportunities for development which perhaps were making them ill in the first place, and which they simultaneously set in stone as the seemingly inevitable consequences of their natural physical condition.

Masturbation and Disease

The patient was only 20 years old but had already lived through years of serious suffering. According to his own account, it all began at the age of 13 or 14, when the nocturnal emissions began that would eventually lead to the young officer's present pitiable, weakened condition. Naturally, he initially did not even know what was happening to him. He had still been innocent as a lamb when he left collège. His father was more familiar with the wicked habits of students. He suspected his son of masturbation and was sure that this was the cause of his many ailments. And he was going to get to the bottom of the matter. 'He tortured me cruelly', the son related, 'to get me to admit to a habit that I had not become addicted to in any way.' When the son did not confess, his father began to spy on him, which was an equally inconclusive endeavor. Then he got the idea of inspecting his son's bedding and found what he was looking for: he discovered the evidence of nocturnal emissions or, as his son put it, 'the traces of a crime in which I was only passively involved, without my knowledge'. The father confronted his son about his allegedly sinful doing. Henceforth, the son was bombarded with sermons and threats. He was given 'all devotional and uplifting writings on this subject', with the result 'that I, for fear of being impure without my knowing it, began to observe myself very carefully'. He was now always afraid when he went to bed, 'hardly reassured by the holy water, by crossing myself and praying, which my confessor had recommended'. But his efforts were in vain.[185] 'Those watching me accused me of masturbation after inspecting my bed and this accusation for a crime of which I could not conceive alarmed me and occupied my thoughts without cease.' Only later, when he was more than 16 years old, was he to learn the cause of the events that happened in his dreams.[186]

Supported by the leading medical authorities and proven, it seemed, by countless empirical observations, the belief in the disastrous physical and mental effects of sexual self-gratification constituted, until the very recent past, one of the deeply ingrained, axiomatic certainties of our Western culture.[187] It was an established piece of common knowledge. Innumerable parents and educators warned and advised adolescents against it. Talk of the danger of 'spinal consumption' and of the beneficial effect of cold showers likely still sounds familiar to many readers. The harm which was done, the negative feelings of guilt, the experience of sinfulness and moral inferiority that this belief instilled in many adolescents can hardly be gauged – not to mention the consequences of more invasive measures including genital mutilation, which some 19th-century physicians promised would forestall the disastrous consequences of masturbation.

In the context of my analysis, the campaign against masturbation and the response it received among the population is of great interest for other reasons. The medical concepts and images that lay at the roots of and substantiated this campaign gave largely unveiled expression to central social values and norms and lent them a new, secular rationale. Not least because it no longer represents accepted medical knowledge, the anti-masturbation campaign is a particularly illustrative example of the social construction of medical truths. What is more, wide sectors of the population, at least among the educated classes, adopted the conviction that masturbation had disastrous effects on physical and mental health and interpreted their own bodily sensations and complaints in the light of this conviction. More impressively still than the rise of the nerve paradigm with its, in part, contradictory implicit messages, the success of the anti-masturbation campaign exemplifies the ability of the dominant medical discourse to impart the ruling elites' specific norms and values through new concepts and images of the human body.

A New Crusade

The moral-theological debate on masturbation has a long history. Medieval and early modern theologians and moralists were in agreement that it must be condemned as a sin against Nature.[188] Compared with adultery, fornication or sodomy, however, masturbation was for a long time regarded as a lesser sin. Only from the late 17th and early 18th centuries, particularly in England, where the – then still mostly moral-theological – debate on masturbation intensified, was masturbation declared one of the most salient threats to religious and spiritual welfare and one of the most serious sins of all. Not only did it violate the divine commandment 'Be fruitful and multiply',[189] but the man who spilled his semen in vain literally soiled his hands with the life of a child.[190] Masturbation was nothing less than a kind of 'clandestine murder'.[191] This view also found expression in newly coined terms such as 'Scelus ononiticum', 'the crime of Onan', 'Onanian' and soon 'onanism'.[192] These terms, from the late 17th century onward, were used as alternatives for or even took the place of older terms like 'mollities' (literally: 'softness'),[193] 'self-defilement' and, in medicine, 'masturbatio', 'masturpatio' or 'manustupratio'.[194] The Bible – as critics already pointed out at the time – actually said that Onan's sin was shirking his duty of giving a child to his brother's widow by performing coitus interruptus.[195] Associating masturbation with the biblical Onan, however, underlined impressively its criminal nature. As everybody knew, God had punished Onan with death.

Probably this theological reassessment of masturbation originated in large part in the new traducianist conceptions of ensoulment, which became established, initially among Protestant theologians, in the late 16th century. According to traducianists, and contrary to previous belief, God did not infuse the seed with the human soul several weeks after conception. Rather, the soul passed with the father's semen or from the parents directly to the child. If the semen already contained the soul of the child, this made the 'waste' of semen appear in an entirely different light.[196]

In early modern medical literature, masturbation, for a long time, played only a marginal role. It was discussed as a cause of 'gonorrhea', which was understood in very general terms as a pathological flow of semen. The discharge after intercourse with 'unclean' women, to which the term 'gonorrhea' exclusively refers today, was also taken for a seminal flow and thus constituted only a special form of 'gonorrhea'.[197] Roderigo da Castro, Baldassar Timaeus von Güldenklee, Michael Ettmüller, and others reported cases of excessive seminal flow due to masturbation and explained the pathophysiological processes. According to them, mechanical manipulation of the genitals brought about 'an excessive laxness and limpness' and damaged the closure of the urethra,[198] while the continual waste of valuable semen brought with it the generally acknowledged consequences of an excessive loss of balsamic substances, radical moisture, and innate heat, which traditional medicine also linked to excessive sex: 'Because those who indulge in it and are all too eager', wrote Johan van Beverwijk in his popular *Schat*, 'they drive out their natural warmth, chill and weaken their entire body, accumulate much coarseness, damage their sinews, make their body gouty and lame, and bring about a weakness of the mind, as well as the senses, indeed, as is often the case, even death itself.'[199]

This medical reasoning was in turn adopted by moralists and theologians to lend additional weight to their warnings. The physical consequences of masturbation were proof of its depravity. They demonstrated that masturbation not only contravened divine law but the laws of Nature as well. As early as the 1630s, Richard Capel warned that this 'self-defilement' deteriorated and weakened the body and rendered a person 'unfit for married life'.[200] In 1676, the *Letters of Advice*, the first known monograph treatise on masturbation and its dangers, explained how masturbation not only opposed the divine commandments but also led to a chronic, involuntary flow of semen and a limpness of the penis muscles, which ultimately made erections impossible.[201]

The synthesis of medical and moral-religious arguments against masturbation reached its apex with the famous *Onania, or the Heinous Sin of Self Pollution* from 1716.[202] 'Impurity' with oneself, as the reader learned from the publication, was even worse in God's eyes than impurity with others. Inevitably, young people in particular sometimes felt the desires of the flesh but these desires had to be stifled from the start because those who fell once for temptation would soon be overpowered by lustful thoughts.[203] And if the sinfulness of masturbation could not deter them from masturbation, many would doubtlessly not have done it 'had they only known about the physical suffering and the diseases' it could, and indeed often did, bring about. It seriously stunted one's growth. Paraphimosis, phimosis, urinary ailments, seminal flow, and other genital disturbances were to be feared. The semen became thin and watery and flowed forth 'unrefined' and without cause. Some suffered cramps and epileptic seizures, others consumption. And still others began to have excessive nocturnal emissions and 'a weakness of the penis and a loss of erection', just as if they had been castrated, or they became infertile. When they did procreate, their children were usually small weaklings who died soon after birth or were sickly their whole lives; these children were 'a misery to themselves, a dishonor to humane [sic] race and a scandal to their parents'.[204]

The author also described in great detail the devastating consequences of female masturbation. It made a woman's genitals limp, brought forth white discharge, and made her unable to retain male semen and thus to conceive. In the long term, this led to a pale and haggard complexion. Also, hysteric cramps appeared frequently and sometimes consumption, caused by the loss of radical moisture. Above all, however, infertility threatened, in the long run, to cause a complete inability to engage in the act of procreation. The women who experienced such detrimental effects availed themselves of 'all the cures Christianity could offer'. Nevertheless, they either suffered frequent, early miscarriages or failed to become pregnant at all.[205]

The work closed by offering readers of both sexes two special medications – at a steep price – which they could discreetly order from the publisher of *Onania*. In accordance with the views the author had presented before, both aimed predominantly at a treatment of the genitals. A 'Strength'ning Tincture' was supposed to treat the 'weaknesses' and the genital discharge in men and women, and a 'Prolifick Powder' to nourish the genitals and supply them with semen.[206]

Contrary to a widespread misinterpretation in historical studies on this subject, it should be emphasized that the harmfulness of masturbation was at the time not seen to lie primarily in the loss of semen during the act. It could become an issue when masturbation was practiced very often, but in this respect masturbation was not fundamentally different from other forms of sexual exhaustion. The harmful effects of masturbation were attributed above all to the unnatural, quasi-mechanical manipulation of the genitals. This manipulation was thought to cause a pathological relaxation the genital fibers, bringing about a dangerous, chronic loss of valuable semen or other balsamic, vital matter, independent of any sexual activity.

Onania was a bestseller. Within only a few years, numerous new editions and reprints followed the first edition and a supplementary volume went through several editions as well. Imitators and critics tried to profit from this success with their own publications.[207] One of them even composed declamatory poems which hauntingly warned 'masturbators' of the terrible trial that awaited them on Judgment Day.[208]

With its much-expanded later editions and the supplementary volume, *Onania* also marked the final point of transition from the predominance of the traditional moral-theological discourse to the predominance of medical arguments. In the later editions of *Onania*, a great number of letters (purportedly) written to the author can be found, in which patients described their ailments, allegedly caused by masturbation, and asked for medical advice. The author expressly offered such advice – for the appropriate fee, of course.

As I demonstrated several years ago,[209] the central if not exclusive source for the medical passages in *Onania* was the work of John Marten, a London specialist for venereal diseases who was well known at the time. Marten had not studied medicine but, it seems, originally worked as a pharmacist's assistant and later as a surgeon. Some years before the publication of *Onania*, he had written at length about the manifold dangers of masturbation in the sixth and seventh editions of his *Treatise on Venereal Diseases*.[210] In *Onania* his warnings are repeated almost word for word and combined with lengthy moral-theological passages. There is no clear evidence, however, to prove that Marten himself was the actual author and the lengthy moral-theological passages rather speak against that assumption. In his book *Solitary Sex*, Thomas Laqueur has, however, come forward with the astonishing claim that he has finally been able to identify *Onania*'s long sought author, namely none other than John Marten. Laqueur has found wide acclaim for this 'discovery'. Only readers of a note hidden at the back of his book will learn that Laqueur acknowledges that others before him, including myself, had already discussed Marten's possible authorship but had expressed reservations. Laqueur claims to have identified new evidence which 'clinches the case'. The only piece of new evidence he has actually found, however, is a short treatise, published in 1727 under the name 'Math. Rothos' and entitled *A whip for the quack: or, some remarks on M---n's Supplement to his Onania*, which he takes to prove that Marten was the author of *Onania*. Contrary to what Laqueur would like to make us believe, however, the text shows only that more than 10 years after the publication of *Onania* someone who lived, according to the dedication, far outside of London, in Exeter, thought that Marten was the author of *Onania*. Since 'Rothos' – or the author who used this name, which in German reads as 'red pants', as an alias – seems to have been a physician (he quotes Ettmüller and other author medical authors and harshly criticizes Marten as a quack) – he may simply have drawn his own conclusions from the striking similarities between Marten's treatise and the medical passages of *Onania*.[211] The issue of *Onania*'s authorship remains unresolved.

Though it is rather unlikely that Marten was the author of *Onania*, the crucial part which his *Treatise* played in the genesis of the medical campaign against masturbation nevertheless gives the whole story a remarkable, ironic twist: this campaign, with its massive moral undertones, owed its vital momentum to, of all people, a specialist in venereal diseases who was later even tried in court for disseminating obscene texts.[212]

Onania's success indicates that the message that masturbation was dangerous met with a receptive contemporary culture. At first without explicit reference to *Onania*, and presumably in ignorance of it, physicians on the European mainland began placing an increased emphasis on the health risks of masturbation. For example, Martin Schurig, in 1720, warned about weakness, spinal consumption, and seminal discharge as the consequences of deeds 'unworthy of a Christian'. He quoted Christian Franz Paullini, who had reported the case of a Frisian patient whose penis had a severe bend due to masturbation.[213] Soon afterward, in 1724, the editor of the German translation of Georg Ernst Stahl's gynecological treatises referred to 'the sinful tricks that lecherous women are in the habit of performing to appease their titillation' as one of the important causes of 'the whites'. Closely resembling male gonorrhea in this regard as well, the whites, it was said, robbed the female body of nutritious matter. Asthma, cachexia, hectic fever, hypochondria, dropsy, ulcers, hysteria, and other complaints were the result.[214] Friedrich Hoffmann also devoted himself to this subject.[215]

Gradually, *Onania* had an impact outside of England. In 1730, it was translated into Dutch and in 1736 into German.[216] Some of its passages found their way into other texts of popular medicine. For instance, the *Neue wohleingerichtete Frauenzimmerapotheke*,[217] in 1741, included an entire chapter taken from *Onania* on the dangers of female masturbation. Shortly afterward, Zedler's famous *Universallexikon* warned about the manifold dangers of 'self-abuse' for men and women, using almost identical words.[218] Particularly among German Philanthropists, such warnings attracted keen attention. Critics now had to fear that any doubt they voiced about the detrimental effects of self-gratification would be construed as an irresponsible endorsement of that practice.[219]

The medicalization of the anti-masturbation discourse culminated in Tissot's *Onanism* of 1760. Although Tissot claimed that he intended to concern himself exclusively with the medical aspects of the subject, his treatise was imbued with moral and religious condemnation. Later, he even refused to answer any letters asking for consultation on the matter so that he could make his help available to more 'deserving patients'.[220] The little treatise became a huge success and went through dozens if not hundreds of editions in a number of languages. With Tissot the moral and religious condemnation of masturbation had now received definitive scientific substantiation. Even clergymen conceded that Tissot's medically motivated warnings made a far more compelling appeal to the youth than did depictions of the eternal torment of hell in a distant future.[221]

Historians have often overrated the originality of Tissot's text. After all, *Onania*, with its many imitations and translations, had been on the market for decades. And despite his patronizing critique of *Onania* – which he believed to be the work of a physician named Bekker – Tissot's description of the various physical consequences of masturbation was largely based on *Onania* and on the letters of 'afflicted persons' that were included in its later editions and to which he added only a few further observations of his own.

The most influential contribution of Tissot's work to the medical discourse on masturbation was his insistence on the particularly dangerous effects of masturbation on the nervous system. He thus successfully linked the campaign against masturbation to the powerful new paradigm of the nerves. Tissot's very own creation, finally, was a series of arguments which served to prove that even moderate masturbation was far more harmful than equally frequent sex in marriage. He pointed especially to the common and particularly harmful vertical position masturbators tended to assume during the act, to the unhealthy consequences of the resulting, irrefutable feelings of guilt (which his text effectively fostered), to the development of a physical and mental dependency on such activity, and to the loss of balsamic substances which otherwise were balanced by the absorption of balsamic substances from the partner (a notion which few physicians at the time would have been ready to accept).

It seems, then that the outstanding success of Tissot's treatise on onanism can neither be attributed to its – not particularly original – ideas nor to its scientific quality. Here was a renowned physician, who was soon to become world famous as the author of *Advice to People* and other popular medical advice books, turning himself into the propagandist and spokesman for a campaign that had for decades been picking up speed and finding its audience. The style of his treatise aimed to convey an image of objective, factual science and scholarship, including numerous quotations and pieces of evidence from the works of older and more recent medical authorities (which only upon closer inspection often prove to be irrelevant). With the scientific format of his text and thanks to his great reputation, he lent the campaign an air of indisputable respectability. At the same time, and much in the same manner as the author of *Onania*, he gratified his readership's voyeurism with graphic medical horror stories describing the physical and mental havoc wreaked by masturbation.

Readers' Response

Contrary to a widespread preconception that has permeated writing on the subject, the idea that masturbation had devastating physical consequences was quite plausible within the context of the dominant medical concepts of the time. The claim that excessive loss of semen was harmful stood uncontested, and the slackening, excessively relaxing effect of mechanical overstraining of genital fibers had already been pointed out by earlier physicians when they attributed pathological seminal discharge, for example, to horseback riding.[222]

Alongside the fear of excessive loss of seminal fluid, however, the scholarly medical tradition had also long maintained the idea that serious diseases could be caused by impeding the expulsion of semen. Semen was thought to accumulate in the genitals and push its way to the outside. If it stayed in the body too long, it spoiled. As mentioned earlier, the Galenic tradition had likewise ascribed a special, though perhaps inferior, seminal fluid to women. People, including physicians, knew that it was not an exclusively male phenomenon that sexual arousal brought forth moisture. In women, retained semen was understood to be an important cause of hysteria and prompted advice to marry early. It was not harmful but healthy from this traditional viewpoint for unmarried people to have involuntary nocturnal emissions or, failing this, to masturbate in moderation. Dramatic rumors circulated to this effect, for example about a high clergyman who, against all warnings, refused to dispose of his unnecessary semen by means of artificial stimulation – and ultimately died.[223]

The prevalence of such ideas about the health benefits of seminal emissions among the general population is above all attested to by those who continued fighting them. The anonymous young gentleman, for example, who was addressed in the *Letters of Advice* of 1676, described his fate in plain, cautionary terms for those who believed that by masturbating they simply afforded their body the necessary 'relief of nature'. In his answers to the letters of afflicted people, the author of *Onania* also dealt extensively with the fears connected with an insufficient evacuation of semen.[224] He sought to refute the idea that seminal retention was harmful by underlining that negative effects of seminal retention had been claimed predominantly in the case of women who, according to the latest findings, did not have any semen of their own at all. To counter prevailing notions that evacuating semen was healthy, he also felt compelled to print a text by L. Salomon Schmieder in the supplementary volume to *Onania*, which put seminal retention into a new light. Based on Daniel Tauvry's *Nouvelle Anatomie* of 1690,[225] Schmieder opposed the fear of the harmful consequences of seminal retention with the notion of a healthy and productive circulation of retained semen in the body: if semen, after it had been refined in the genitals, was not expelled, it moved back into the body and helped bring forth and maintain the characteristic male features, which – significantly – were lacking in the semenless eunuch.[226] Among Tissot's patients, there were still those, however, who confessed their 'vice' but vowed, as we will see shortly, that they had not known of the harmfulness of their doing for a long time, until Tissot's treatise or another pamphlet on masturbation had fallen into their hands and opened their eyes.

Against the background of this originally converse understanding, the medical anti-masturbation campaign is also an excellent example of successful medical popularization. A new medical concept, new experiences and knowledge, were presented to the general public and gained acceptance. From today's point of view, the campaign's success can hardly be ascribed to the 'objective' validity of the new conceptions or to the overwhelming weight of empirical evidence – which raises the question: which particular circumstances, which cultural, social, and political context, made such a sweeping success possible?

I will begin exploring this issue through the testimonies of self-confessed victims of masturbation, that is, by looking into the attitudes and reactions of those who made the propositions of *Onania* and Tissot's work their own, using them to interpret their personal illnesses. Only very few 17th- and 18th-century personal testimonies about masturbation have come down to us.[227] The most famous exception is that of Jean-Jacques Rousseau, who admitted to this 'vice' (among others) in a hedged and stylized way in his *Confessions*.[228] Before him, English writers made occasional hints.[229] The fact that masturbation is mentioned only rarely suggests that it was closely linked to moral and religious feelings of guilt and sinfulness. That ordinary men and women were afraid of causing extensive physical or mental damage by masturbating cannot be demonstrated from autobiographies and similar personal testimonies.

Against this background, patient letters open the way to remarkable new insights into how warnings about the consequences of masturbation on health were received and, in individual cases, how they were harnessed for the interpretation of illness and personal life circumstances. The earliest reliable evidence of this kind which I have come across so far is from 1727 and appears in the letters to Geoffroy. As

mentioned above, earlier letters from self-confessed victims of masturbation can be found in John Marten's work and in *Onania*, but their
authenticity cannot be confirmed and was even called into question at
the time.[230]

The author of the letter to Geoffroy was a cloth merchant, around
35 years of age, in St Quentin. After a prolonged fever disease almost a
year earlier, he had temporarily made a full recovery. But then around
two months back, he said, the fever had returned and he was feeling so
weak that he was hardly able to climb a flight of 12 steps. He also suffered from attacks of dizziness and dramatic weight loss and was now
nothing but skin and bones, a pitiable sight. It had all started with a
serious cold, accompanied by a lot of thick sputum. Bloodletting helped
him get rid of this. He had never coughed up blood, but for 10 or 12 days
had been suffering from a dry cough and felt that his chest was very
weak. Unfortunately, his profession, he felt, was particularly harmful to
his condition because it forced him to talk all day.

The real cause of his complaints was not a matter of doubt to him. He
ascribed it to the fact that he had allowed himself to indulge numerous
times in 'liberties' with himself, which had weakened him seriously
and brought on his current state.[231] In a postscript, he explained the
circumstances: it had been over a year since he had last approached his
wife, after a difficult birth had brought her to the brink of death. She
had developed a great dislike for 'marriage' and they slept in separate
beds. In this situation he often gratified himself – in fact more often
than he had had intercourse with his wife – and he forced this upon
himself, which, he said, exhausted him a thousand times more than
sleeping with his wife.[232]

It was plain to see, even for outsiders, that the cloth merchant was
seriously ill. With masturbation he had found an explanation for this
serious disease which physicians – not knowing the real cause – had
been unable to treat successfully. A comparable case can be seen in one
of Tissot's patients, who had begun masturbating at the tender age of
10 and had coughed up blood for the first time when he was 13. Despite
numerous treatments, his life had been marked by disease and suffering ever since. He continued to have serious rashes and, more recently,
never went longer than two months without coughing blood.[233]

There were also patients, however, who were just as convinced of
their serious disease but displayed fewer clearly visible symptoms, so
that people around them were reluctant to believe their complaints
about physical ailments. For example, late in the fall of 1772, 43-year-
old Monsieur Belfontaine turned to Tissot. His health, he wrote, had

been impaired for 20 years. He suffered from massive flatulence that drove him to sheer despair, from constipation, from rashes and itching, sensations of coldness all over his body as well as sweaty feet, headaches and vapors. His digestion was laborious and his memory and eyesight, he thought, had weakened. But his physicians were not ready to believe in his complaints, the more so as he had a strong frame and muscular build and looked young for his age. Not surprisingly, their counsel had brought little improvement, though he had even, following medical advice, gone on a four-year trip around the world, crossing three continents. After reading of Tissot's *Onanism*, he hoped he had finally reached the bottom of his suffering. Could it be that 'this unfortunate pleasure, so widespread among the youth' was at the root of his suffering? He believed that, compared with others, he had probably not overdone it, but perhaps it had nevertheless been too much considering the natural weakness of his fibers. What was more, he had taken to it very early – before puberty. And later, for fear of exhausting himself, he had not allowed himself to ejaculate but had stopped just short of it. On top of that, he had demanded from the opposite sex to be gratified manually for fear of his strict father, who would never have forgiven him had he contracted a venereal disease that early in his life. His physician, he said, had explained to him that he was far from having spinal consumption or any of the other horrible conditions described by Tissot in his treatise, but this reassurance was not enough for M. Belfontaine. Also, he felt little desire for the opposite sex. His member was weak; it was hard work to make it erect and then the erection never lasted long. And when he did get together with a woman, which happened rarely enough, he ejaculated almost instantly – a sure sign, he thought, that his fibers had already become very slack and that it had become difficult for them to control his body openings.[234]

Among the patient letters to Geoffroy, Haller, and Tissot that I have examined, more than two dozen can be found in which either the patient or the people around him or her ascribe to masturbation a significant role in causing the respective disease. Those affected were almost exclusively men. Only Mme de Chastenay mentioned, alongside innumerable other issues, that she had sexually gratified herself even as a 12-year-old girl.[235] In most cases, the patients complained about local genital symptoms which they took to result from a pathological limpness and weakening of the genitals, assumedly caused by the unnatural 'manipulation'. Some suffered mainly from uncontrolled seminal flux or nocturnal emissions. One 28-year-old man who had masturbated occasionally during a three-month period complained that regardless of which side he slept on, he was overcome by nocturnal emissions. The ejaculation was not very strong. This led him to assume that his 'seminal organs' were lacking power, that they were 'irritated', and that it had become necessary to restore their 'tonicity'.[236] Another man described his nervous system as so irritable that, if he did not want to suffer a nocturnal emission, he could not read a single line in bed before going to sleep. He even attached an apparatus to his back that forced him to sleep on his side because, as he had learned from experience, the emissions occurred less often in this position.[237] With other patients, disturbances of their male sexual abilities took center stage. One 47-year-old, for whom Geoffroy was asked for advice, had been completely impotent for 20 years. He had started to masturbate when he was only 12 years old and had also often been with women later on.[238] Others suffered from premature ejaculation[239] or found that their weakened member was twisted towards one side.[240] Alongside those problems, most of the affected patients, like the aforementioned cloth merchant from Paris, complained about a dwindling of strength, a loss of appetite, tiredness, and similar signs of physical decay and waning vitality. An English law student, for instance, believed the nocturnal emissions he had brought upon himself with masturbation were enough to 'weaken and enfeeble [him] in every respect described by Tissot as the symptoms of such a habit'.[241] In individual cases, various other complaints came into the picture as well. For example, the patient of Geoffroy's just mentioned, who had been impotent for 20 years, also suffered from severe diarrhea for three years, combined with weakness, a pale face, and pangs that resembled rheumatic pains.[242] Another patient suffered from intense chest pains, shortness of breath, coughing, and purulent sputum.[243] And still another considered his whole body – his 'whole machine', as he put it – to be in disarray. Among other complaints, he counted serious weight loss, a pale, bluish complexion, and increasingly poor vision.[244] Breaking out in sweats, itching, and shortness of breath were likewise connected to masturbation, as were manifest edemas and rashes that took the shape of pustules or boils covering the face, the genital area, or even the entire body.[245]

Two symptoms featured prominently among the plethora of ailments, however: pains and nervous complaints. Monsieur Chillaud was hardly able to stand because of the pain caused by 'the lack of nervous substance' in his feet.[246] Monsieur de Roussany suffered from serious pain in his hips, a pain that worsened significantly after sex and even more so after sexual self-gratification.[247] Others complained of pain in their stomach, legs, joints, head, teeth, neck, or back.[248] The nervous complaints ranged from a nervous tickling and an excessive 'flexibility' of the nerves to a growing limpness of the body and mind, including a lack of concentration and memory, an inability to make decisions, and diminishing eyesight.[249] One patient, according to his surgeon's report, had even weakened his nervous system to such a degree that, every time he masturbated, he would remain motionless and without sensation for 20 minutes afterwards, finally going into convulsions and epileptic fits, during which he screamed and threw back his head and nearly choked on his own saliva.[250]

In addition to nervous symptoms in the stricter sense of the word, there were complaints about mood changes: about feeling 'despondent', 'sad', melancholic, or even suicidal.[251] 'I experienced that horrible state described in *Onanisme*', one patient wrote, looking back on the worst time of his suffering. 'It seemed to me as if my entire existence were suspended in nothingness; my taste and my feelings were almost completely numbed and thoughts of my own destruction pursued me day and night. [...] I was like a walking machine, dead to any kind of feeling.'[252]

These letters paint a vivid picture of how the warnings of the anti-masturbation campaign constituted a source of meaning and orientation for patients, who adopted them to interpret their own complaints. At the same time, their accounts stand as particularly illustrative examples of the aforementioned phenomenon of a narrative reconstruction of one's biography. To those who were willing to accept masturbation as the cause of their suffering, the teachings of the disastrous consequences of masturbation offered more than simply a conclusive explanation of their disease and guidance for its prevention and treatment in the future. Their entire histories now appeared in a different light, namely as the inescapable consequence of their 'sin', their 'criminal habit', of which they had become guilty and for which they were now being 'chastised'.[253] A number of patients described in great detail the long road from their discovery of masturbation to their late realization of its calamitous consequences on their health. Each story is different, but there are many similarities. Some discovered masturbation at the tender age of 7, 10 or 12 years, by chance, or misled by schoolmates or co-workers.[254] They soon fell prey to the vice and indulged in it as often as five or six times a day in individual cases,[255] and sometimes they worsened matters by having excessive intercourse with women. The first consequences on their health began to set in but they did not initially recognize them as such. Their complaints worsened over time. They consulted various physicians and surgeons, whose attempts at treating them through bloodletting, laxatives, or diets had, in retrospect, made their illness only worse, weakening the body even further. It often took many years before the patients finally learned – usually more or less by chance and often from reading one of the anti-onanism works – that they were suffering the disastrous consequences of their earlier wrongdoing. Their reading, or an enlightening conversation about the topic, became the turning point in their lives. They distanced themselves from their vice – and, through this, from their entire past – using a language that at times is reminiscent of accounts of religious conversion. They confessed to the abomination of their 'crime', their 'sin', and if they had become impotent, they saw this as the just punishment meted out to the 'tool of the crime'.[256] A good number of patients explicitly vowed to abstain from their vice or even renounced all sexual activity.

The Social Construction of the Anti-Masturbation Discourse

Absurd as it may seem to today's readers to assume a causal relation between sexual self-gratification and serious, even fatal, diseases, for some contemporaries of the period, the anti-onanism texts became the source of an irrevocable truth, a truth some had experienced first-hand. Even more sweepingly and outspokenly than other new medical theories of the time, the anti-masturbation discourse also communicated specific norms, values, and interests, however.

The moral and political 'message' of anti-onanism texts changed, of course, over time and varied according to particular (sub)cultural context: among English Puritans, for example, or German Philanthropists, certain aspects became more pronounced than elsewhere.[257] In late 17th- and early 18th-century England, where the campaign began, religious arguments initially held sway. The struggle against the vice of masturbation was not only rooted in the theological re-evaluation of semen, now considered by some to possess a soul, but was also an integral part of a sweeping campaign against all forms of moral 'uncleanliness' or 'defilement'.[258] Masturbation was prone to call forth particularly deep anxieties about the spiritual welfare of people. Since there was no partner present in masturbation, moralists warned, it required imagination and stimulated the imagination, in turn, to the utmost.[259] The place of a real, flesh-and-blood partner was taken by the imagined partner – and this partner was always available. In the end, thinking and desire were increasingly ruled by a lecherous imagination. The 'scum of sinful thoughts' threatened in an almost literal, physical sense to soil the mind or the brain, which was seen as the 'temple of the Holy Spirit'.[260] Once the 'uncleanliness' had taken hold, it was almost inevitable that the masturbator would become ever more entangled in the vice. 'Impurity' gained 'governance over his heart', keeping it forever in its grasp. The Holy Spirit could not dwell in such a person.[261] The masturbator was increasingly unable to fend off 'dirty' thoughts and ultimately became enslaved to his 'animal desires'.[262] He was overcome by lust even during religious exercises.[263] Sexual self-gratification was in this way a menace to central Protestant and particularly Puritan and Pietist ideals. The abandonment to the rule of animal desires took the place of a ceaseless pious self-reflection and a constant watchfulness over one's own thinking and doing.

Other, more worldly, values and ideals were in danger as well. As the 'solitary vice' par excellence, masturbation challenged the divine 'order of love'[264] as well as marriage as 'society's foundation'.[265] Masturbators learned to prefer their lonely vice to marriage and thus ultimately threatened to 'bring an immediate end to all of Nature's governance and order'.[266] Such concerns reflected the growing appreciation of marriage as companionship based on mutual respect and affection that had started in the 16th century, thanks to some degree to the Protestant Reformation. But political interest played its part as well. According to the prevailing political theories of the time, power and prosperity were founded upon a large and healthy population fit for work and military service. Those who refused to procreate in marriage, therefore, also harmed the interests of society.

Closely related to the enhanced status of marriage and the family, attention was also increasingly paid to the family's task of raising children to become orderly and productive citizens. Masturbation seemed to jeopardize this aim more than almost anything else. It was undisputed that masturbation was particularly prevalent among older children and adolescents.[267] Grammar schools, boarding schools, and colleges were considered veritable hotbeds of the vice[268] – the very institutions, that is, to which the educated classes sent their offspring to ensure them a successful future.[269] Those children who had already succumbed to the vice, it was feared, would sooner or later lead their innocent schoolmates to perdition. The consequences were all the more grave as the still tender mind of adolescents was less able to fend off lecherous fantasies. Having fallen prey to the vice this early, a person was essentially lost once and for all. As Adriaan Beverland put it in 1698: 'The Barrier that fenc'd their Chastity is broke, and the Enemy to Purity and Holiness makes daily Inroads, and ravages through every Passage of the conquer'd Soul.'[270] In Calvinist circles worries about the consequences of adolescent sexuality tended to be particularly pronounced since, based on the doctrine of original sin, children were not innocent but already carried the seeds of moral corruption.[271] In late 18th-century Germany, proponents of Philanthropic pedagogy, who were often rooted in Pietism, dominated the discussion.[272] All of this may have been a major reason why the campaign against masturbation took hold first in England, Holland, and (predominantly Protestant) northern Germany, and only slowly spread to the Catholic countries.

The harmful effects of masturbation on physical health brought welcome general support to the claim that masturbation was not only a sin against God's commandments but violated the natural order as well. The various arguments of the medical anti-onanism discourse, moreover, picked out individual aspects of the moral-theological discussion and provided scientific arguments to back them. For instance, the warning that masturbation would lead to an almost uncontrollable habit – even to addiction in the modern sense of the word – found its medical counterpart in physicians' references to the chronic genital irritation caused by masturbation, which led to a desire for gratification and thus to an endless repetition of the act. The concern that one would lose control over one's thoughts due to the influence of sexual fantasies was reflected in the medical warnings about an increasing loss of mental capacities and control over one's body, including cramps, convulsions and urinary incontinence. The concern that marriage was in jeopardy – marriage as the foundation of ordered social life and the means of nurturing a large population fit for work and military service – was further heightened by physicians' warnings of permanent damage to the genitals. Impotence and other sexual disturbances made masturbators unfit for marriage in the first place. And when they did have children, these children would be sickly due to the weakened, inferior semen from which they stemmed – to such a degree that they would be a burden to society rather than an asset.

One principal message was shared by all of the authors who, beginning in the late 17th century, denounced masturbation as one of the greatest threats to individual and collective welfare: masturbation was an evil that had to be battled against using all available means. It was not sexuality as such that was objectionable; in marriage, sexuality was even seen as part of a divine or natural order and as indispensable to the survival of humanity. Rather, the point was to distinguish forbidden sexual practices (most, we would say in retrospect) from permissible ones. In some ways, the history of the anti-masturbation campaign can be read as the history of a growing medicalization, naturalization, and (judged by the criteria of contemporary science) scientification of the discussion of sexuality and the desires of the flesh. Deviant forms of sexuality – and first and foremost masturbation, which was seen to have assumed epidemic proportions – would sooner or later result in damage to health.[273] At a time when religious arguments lost their influence on contemporary society, medicine provided a new justification for traditional norms of acceptable sexual behavior and sexual moderation.

Conclusion:
A New Bourgeois Habitus

The 'semantic networks' surrounding the two new concepts – the nervous complaints and the ailments due to masturbation – were connected only to a limited degree. The nervous paradigm revolved around dynamic, functional changes, around the effects of excessive stimulation of irritable nervous fibers on the orderly processes within the body. The medical anti-masturbation campaign in contrast focused, for decades, primarily on mechanical damage to genital fibers which harmed, first of all, sexual performance and only as a secondary consequence made the whole body suffer as well, due to the resulting excessive loss of valuable seminal matter. Overlaps between the two 'semantic networks' can be found above all in the notion that excessive semen loss damaged in particular the nervous system and, later, in Tissot's claim that the act of masturbation itself brought about an extreme over-stimulation of the nervous system, which could manifest itself in convulsions and epileptic fits.

As vehicles of implicit or explicit ideological 'messages' the discourse on nerves and the anti-masturbation discourse also functioned differently. The fears, values, norms, and interests communicated by the discourse on nerves were less obvious and in part even stood in conflict with one another – with the critique of the negative effects of urban civilization on the one hand and the positive ideal of moral sensibility on the other. The 'moral' message of the medical anti-masturbation campaign, by contrast, was much less ambiguous and, in the early publications, was often given more space than medical content proper.

On a more fundamental level, however, the new paradigm of nervous irritability and the anti-masturbation campaign both ultimately contributed to an overarching, long-term trend toward a new body concept, a new body ideal. They were part of a development which Norbert Elias in *The Civilizing Process* has described as the emergence of modern 'homo clausus'.[1] Traditionally, as we have seen, the body was characterized by a constant, health-preserving flow of fluids within and across its borders. Substantial excretion or discharge was all-important for health and was embraced wholeheartedly. People were warned against artificially hindering excretion and, if the need arose, it was assisted

213

using artificial means. Thus for a long time, drastic emetics and laxatives, bloodletting and other purging procedures formed the basis of
almost all medical treatment.

Though in different ways, the rise of the nerves and the anti-masturbation campaign both radically called this body conception into
question. The rise of the nerves played a crucial part in the long-term
trend away from mobile fluids and spirits toward solid and substantial
constituents: fibers and organs were increasingly seen as the principal
material substrate of human physiology and pathology. They rather
than humors and spirits were responsible for the body's functions and
malfunctions. The open and permeable body of the learned medical
tradition was superseded and replaced by a compact, internally firm
body mass that was largely sealed off from the outside. The vital basis
for maintaining good health was no longer the unobstructed flow of
humors but the integrity and orderly performance of the solid parts
and the strength of the 'life force' of the organism as a whole. Life force
became the pivotal concept in medical guidebooks, of which Hufeland's
Makrobiotik is the most famous example.[2] The strong response that early
homeopathy earned at the beginning of the 19th century, particularly
in genteel circles, is a good illustration of the positive lay response to
this new view. Homeopathy's great attractiveness, according to its followers, was the fact that it 'was based on the principle of temperance'
and that 'the small dosage of medication that it prescribes to patients
cannot have such a negative effect as the mass of mixtures often used in
allopathy, which, unfortunately, has frequently produced infelicitous
results.'[3] With homeopathic treatment, human nature was not so easily
deprived of the force it needed to fight the disease. The belief was that
homeopathy was easy on the body.[4] Accordingly, it made sense that
one should prefer 'the physician who spares the patient the drudgery of
allopathy' to the one who used it and that 'one would much rather take
the small dosage without the nasty taste' than take 'profuse, plentiful
medication of the most loathsome taste'.[5]

The general shift toward perceiving bodily excretions as dangerous rather than beneficial was most tangibly expressed by the anti-
masturbation discourse. As described above, a substantial and ongoing
loss of blood, semen, or other vital substances was also considered as
potentially harmful in earlier times, but the fundamentally positive
assessment of bodily evacuation nevertheless prevailed; well into the
19th century, physicians described this notion as the predominant
one among the less educated classes in towns and rural areas. Its per-
sisting strength can be glimpsed in the rural response to the 'mild',

'gentle' remedies of early homeopathy. According to one member of the Bavarian Parliament, only the future would tell whether homeopathy would find great favor 'in our dear Bavaria in particular, [...] because the strong Bavarian tribe desires big spoons and big steins'.[6] And in the view of a contemporary country doctor, the rural population did not hold homeopathy in particularly high esteem 'because this doctrine appears too long-winded to them, not strong enough and seems to lack, in a word, the necessary bombshell effect'.[7] Among the upper classes of the 18th and 19th centuries, on the other hand, we find ample evidence of growing concerns about *any* loss of bodily substance – even if only temporary and minimal. Instead of drastic purging procedures, many patients preferred the gentler diarrheic effect of whey and mineral water.

Seen from this perspective, both the discourse on nerves and the anti-masturbation campaign communicated the new ideal of a compact, closed-off, strong body, whose well-being depended on a limited and controlled exchange of matter across its borders. At the same time, the danger of losing control over one's body and one's morals and thinking alike was effectively communicated through the evocative imagery of the pathological consequences of masturbation and nervous overstimulation. The masturbator was not only under the dominion of 'impure thoughts' but, in the extreme case, lost all control over his body openings and emissions. He was bathed in his own excrement. And someone suffering from 'the nerves' had as little control over his thoughts and feelings as he had over the cramps and convulsions of his nerves and muscles.[8]

Furthermore, this new ideal body was, par excellence, a male body.[9] Women were subject to a considerable loss of substance every month, simply due to their nature. Women also were seen as tending much more toward uncontrolled behavior due to the heightened sensibility of their nervous systems. And men who destroyed their virility (and their sexual prowess) through self-gratification consequently came closer to the physical condition of women. They became effeminate 'weaklings'.

The discourse on nerves and the anti-masturbation discourse were thus closely connected to the development and manifestation of a body ideal which, in some respects, was tailored to the rising bourgeoisie. Through an increased esteem for the vital force of the natural body and a clear dissociation from idle courtly society, this class adopted the muscular, active body of the working man as an ideal, yet without embracing it fully: the nervous constitution of the bourgeoisie remained distinct, of a higher order. And by emphasizing a temperate, controlled manner of dealing with the body and its excretions, the bourgeoisie set itself off from the (alleged) excesses of both the noble classes and the mob.

This idea of a firm, delimited body that is controlled from the inside (and no longer dominated by, for example, a constant urge to disgorge) still shapes, to use Pierre Bourdieu's term, the prevalent 'habitus' in Western industrial societies today.[10] Of course, the close and fundamental correlation of the medical discourse on masturbation and the nerves with this modern Western habitus elicits the difficult issue of cause and effect: to what degree can we understand the medical discourse as a major driving force behind this new habitus? And to what degree did, vice versa, a more general change in habitus pave the road for the new medical understanding of the body as epitomized by the discourse on nerves and on the harmful consequences of masturbation? I will not try and provide a conclusive answer to these questions, but in closing I would like to highlight some important points.

First, it is clear that the development of medicine always has its own inner dynamic. It is not simply the product of social and cultural processes. Without the experimental research on nerve and muscle fibers, the medical doctrine of the nerves would no doubt have taken a different course. And this research, as far as we can tell, owed itself only very indirectly to the socio-cultural developments sketched out so far. It rather reflects a growing esteem for experimental studies performed under controlled conditions as a source of authoritative medical knowledge. This research played a crucial role also in other areas and in particular in the disputes that raged between the proponents of mechanistic conceptions of the body and those who attributed vital properties to the human body and its parts. The teaching of the irritability and sensibility of nerve and muscle fibers was one of the most powerful arguments in favor of a vitalist interpretation of the body.[11]

At the same time, there are certainly indications that the rise of the doctrine of the nerves within academic medicine was closely intertwined with the social and cultural developments of the day and owed much to them. On a rather hypothetical, speculative level, we can relate the renaissance experienced by vitalist conceptions to contemporaneous socio-cultural developments. For example, we might link the body's passivity within the mechanistic paradigm to the citizen's limited political role within Absolutism, while seeing the active, more self-determined, independent vitalist body as the expression of a new bourgeois self-confidence. There are, however, two much more concrete circumstances that I would like point out in this context. First, physicians were never simply physicians but also fellow citizens with a culturally shaped body of their own, a body that necessarily served as an important reference point for their new concepts and theories. Some leading proponents of the doctrine of the nerves even expressly declared themselves victims of nervous complaints. In other words, the general cultural changes that made the (educated) public receptive to the new body concept also had an influence on physicians. Second, as Nicholas Jewson pointed out years ago, fundamentally different mechanisms were at work in the development and implementation of medical innovations in the 18th century as compared to today.[12] While today it is almost exclusively the scientific community that decides on the success or failure of a medical innovation, in the past, the role of patients and the general lay public was at least equally influential. The reason for this is that medical development used to be furthered predominantly by physicians in their private practices and not by specialists at universities or in specialized research facilities. According to Jewson, the status, advancement opportunities, and income of those practicing physicians depended crucially on the benevolence of a small minority of genteel, rich, and influential patients.

As we have seen, Jewson underestimated the degree to which many physicians also treated patients who were of the same social standing as themselves or even of inferior standing. His basic argument nevertheless remains valid. On the relatively free and unregulated medical market of the early modern period, physicians had to adapt to their patients' wishes and expectations. This circumstance necessarily favored innovations that were suited to the patients' preconceptions and desires. The temporal relation between the spread of nervous complaints among the population and the increased importance of the 'nerves' in medicine as practiced by physicians seems to confirm this. Some physicians of the early 18th century explicitly described the vapors as a new phenomenon to which they were reacting and whose spread they were trying to explain after the fact. Or they admitted, at least, that their scientific interest in the subject had sprung first and foremost from cases of the illness they had encountered in their practice. This suggests that physicians who paid increased attention to vaporous or nervous complaints were responding to processes of somatization in their society, or to the fact that patients were now increasingly interpreting even vague malaises as illnesses and seeking the help of physicians. As a side-point, it is interesting to note that the new ideal of sensibility had emerged in French and English literature quite some time before sensibility was 'discovered' as a vital characteristic of the nervous system.

The success of the campaign against sexual self-gratification was likewise not simply the result of a new medical discourse. The essential initial boost experienced by the campaign during the late 17th and early 18th centuries in England was owed on the one hand to the campaign against *all* forms of 'uncleanliness' (and not to the medical discussion of it), and on the other to the tangible economical interests of John Marten and other members of the venereal trade whose claims – at first embedded in moral-theological argumentation – became widely disseminated with *Onania* and the publications that followed in its wake. Here as well, the question arises how changing bodily experience also affected the medical authors themselves. To what degree did their intensified warnings about the weakening consequences of self-gratification also express their personal bodily experience, their perception of their own culturally shaped bodies – especially if they allowed themselves such liberties? In other words, there is much to be said in support of the idea that the anti-masturbation discourse was simultaneously a driving force behind and a product of the new, more self-contained, self-controlled, and restrained habitus, a habitus which was also that of the physicians of the day, as men of their time.

Notes

Introduction

1. F. Loetz has claimed that the 'sick' (were) turned into 'patients' in the 19th century (Francisca Loetz, *Vom Kranken zum Patienten. 'Medikalisierung' und medizinische Vergesellschaftung am Beispiel Badens 1750–1850*. Stuttgart 1993). But her finding is based on a somewhat anachronistic modern understanding of the term 'patient' as a (largely passive) partner in the therapeutic interaction. The term 'patient' (or 'pacient', 'patiens') was quite common already in the late Middle Ages though it was associated more closely than it is today with the original meaning of 'patiens' (Latin for 'suffering').
2. See the programmatic plea by Roy Porter, 'The patient's view. Doing medical history from below', *Theory and society* 14 (1985), 175–98.
3. For a good overview of changing issues and approaches see Frank Huisman/ John Harley Warner (eds), *Locating medical history. The stories and their meanings*. Baltimore/London 2004.
4. Claudine Herzlich/Janine Pierret, *Malades d'hier et malades d'aujourd'hui: de la mort collective au devoir de guérison*. Paris 1984; Dorothy Porter/Roy Porter, *Patient's progress. Doctors and doctoring in eighteenth-century England*. Cambridge/ Oxford 1989; iidem, *In sickness and in health. The British experience, 1650–1850*. New York 1989; Barbara Duden, *The woman beneath the skin: a doctor's patients in eighteenth-century Germany*. Cambridge, MA 1991 (German orig. 1987); Lucinda McCray Beier, 'In sickness and in health: A seventeenth-century family's experience', in Roy Porter (ed.), *Patients and practitioners. Lay-perceptions of medicine in pre-industrial society*. London 1985, pp. 101–28; Robert Jütte, *Ärzte, Heiler und Patienten. Medizinischer Alltag in der frühen Neuzeit*. Munich/Zürich 1991; Jens Lachmund/Gunnar Stollberg, *Patientenwelten. Krankheit und Medizin vom späten 18. bis zum frühen 20. Jahrhundert im Spiegel von Autobiographien*. Opladen 1995; Christoph Lumme, *Höllenfleisch und Heiligtum. Der menschliche Körper im Spiegel autobiographischer Texte des 16. Jahrhunderts*. Frankfurt 1996. For surveys of work in this area see Katharina Ernst, 'Patientengeschichte. Die kulturhistorische Wende in der Medizinhistoriographie', in: Ralf Bröer (ed.), *Eine Wissenschaft emanzipiert sich: Die Medizinhistoriographie von der Aufklärung bis zur Postmoderne*. Pfaffenweiler 1999, pp. 97–108; Flurin Condrau, 'The patient's view meets the clinical gaze', *Social history of medicine* 20 (2007), 525–40; Eberhard Wolff, 'Perspektiven der Patientengeschichtsschreibung`, in Norbert Paul/Schlich (eds), *Medizingeschichte. Aufgaben, Probleme, Perspektiven*. Frankfurt/New York 1998, pp. 311–30; for recent case studies see: Miriam Schriewer, '*Kann der Körper genesen, wo die Seele so gewaltig krankt?*` *–Weibliche Gemüts- und Nervenleiden in der Patientenkorrespondenz Hahnemanns am Beispiel der Kantorstochter Friederike Lutze (1798–1878)*. Cologne 2011 [forthcoming] and Judith Oxfort, '*Meine Nerven tanzen*'. *Die Krankheiten der Madame de Graffigny (1695–1758)*. Cologne 2010.
5. Michael Stolberg, *Heilkunde zwischen Staat und Bevölkerung. Angebot und Annahme medizinischer Versorgung in Oberfranken im frühen 19. Jahrhundert*.

Unpubl. med. diss. Munich 1986; Loetz, *Vom Kranken* [see note 1]; Annemarie Kinzelbach, *Gesundbleiben, Krankwerden, Armsein in der frühneuzeitlichen Gesellschaft. Gesunde und Kranke in den Reichsstädten Überlingen und Ulm, 1500–1700*. Stuttgart 1995; Laurence W. B. Brockliss/Colin Jones, *The medical world of early modern France*. Oxford 1997.

6. Cf. for example Lumme's claim that laypeople in the 16th century attributed all their diseases essentially to pathological phlegm (Lumme, *Höllenfleisch* [see note 4]).

7. Gail Kern Paster, *The body embarrassed. Drama and the disciplines of shame in early modern England*. Ithaca, NY 1993; Michael C. Schoenfeldt, *Bodies and selves in early modern England. Physiology and inwardness in Spenser, Shakespeare, Herbert, and Milton*. Cambridge 1999.

8. This reservation is supported by the fact that the images and conceptions of the physiological and pathological processes in the female body that Duden finds in her source are remarkably similar to those advanced by physicians in Georg Ernst Stahl's ambit. Storch, as Duden knows, was an enthusiastic follower of Stahl.

9. For an overview see Michel Feher et al. (eds), *Fragments for a history of the human body*. New York 1989; David Hillman/Carla Mazzio, *The body in parts. Fantasies of corporeality in early modern Europe*. New York/London 1997; Michael Stolberg, 'Körpergeschichte und Medizingeschichte', in: Bröer, *Medizinhistoriographie* [see note 4], pp. 85–95; Heiko Stoff, 'Diskurse und Erfahrungen. Ein Rückblick auf die Körpergeschichte der neunziger Jahre', *Zeitschrift für Sozialgeschichte des 20. und 21. Jahrhunderts* 14 (1999), 142–60; for bibliographic surveys see Barbara Duden, 'A repertory of body history', in: Feher et al., *Fragments*, pp. 471–578, and Maren Lorenz, *Leibhaftige Vergangenheit. Einführung in die Körpergeschichte*. Tübingen 2000, pp. 174–238.

10. Jacques Revel/Jean-Pierre Peter, 'Le corps. L'homme malade et son histoire', in: Jacques LeGoff/Pierre Nora (eds), *Faire de l'histoire. Nouveaux objets*. Paris 1974, pp. 169–91.

11. Bryan S. Turner, *Regulating bodies. Essays in medical sociology*. London/New York 1992; Elisabeth List/Erwin Fiala (eds), *Leib Maschine Bild. Körperdiskurse der Moderne und Postmoderne*. Vienna 1997; Deborah Lupton, *Medicine as culture. Illness, disease and the body in Western societies*. London 1994.

12. With the three levels I primarily mean to distinguish between different levels of analysis. By contrast, there is the distinction between essentialist and non-essentialist positions which aims primarily at the respective underlying philosophical premises; cf. Lorenz, *Leibhaftige Vergangenheit* [see note 9].

13. This refers to a type of historical anthropology that takes its orientation more or less from cultural history; see for example Bob Scribner/Ronnie Po-Chia Hsia (eds), *Problems in the historical anthropology of early modern Europe*. Wiesbaden 1997. It does not refer to currents in philosophical anthropology that lean more toward an essentialist, trans-historical conception of the body, as for example in Christoph Wulf (ed.), *Vom Menschen. Handbuch Historische Anthropologie*. Weinheim/Basel 1997; see also Gert Dressel, *Historische Anthropologie. Eine Einführung*. Vienna 1996; Richard von Dülmen, *Historische Anthropologie. Entwicklung, Probleme, Aufgaben*. Cologne 2000.

14. Mary Douglas, *Natural symbols. Explorations in cosmology.* London 1996; eadem, *Purity and danger. An analysis of concepts of pollution and taboo.* Harmondsworth 1970.
15. For a critical survey see Ronald C. Simons/Charles C. Hughes, *The culture-bound syndromes. Folk illnesses of psychiatric and anthropological interest.* Dordrecht 1985.
16. Raymond Prince, 'The concept of culture-bound syndromes. Anorexia nervosa and brain-fag', *Social science and medicine* 21 (1985), 197–203; Cheryl Ritenbaugh, 'Obesity as a culture-bound syndrome', *Culture, medicine and psychiatry* 6 (1982), 347–61; Caroline Giles Banks, '"Culture" in culture-bound syndromes. The case of anorexia nervosa', *Social science and medicine* 34 (1992), 867–84; Mari Rodin, 'The social construction of premenstrual syndrome', *ibid.* 35 (1992), 49–56; Michael G. Kenny, 'Latah: the symbolism of a putative mental disorder', *Culture, medicine and psychiatry* 2 (1978), 209–31; Thomas W. Johnson, 'Premenstrual syndrome as a Western culture-specific disorder', *ibid.* 11 (1987), 337–56.
17. Cf. the ground-breaking work by Peter L. Berger/Thomas Luckmann, *The social construction of reality. A treatise in the sociology of knowledge.* London 1967; major early contributions to the study of social construction in medicine were P. Wright/A. Treacher (eds), *The problem of medical knowledge. Examining the social construction of medicine.* Edinburgh 1982; M. R. Bury, 'Social constructionism and the development of medical sociology', *Sociology of health & illness* 8 (1986), 137–69; on the concept of 'framing' see below.
18. Cf. Michel Foucault, 'La politique de la santé au XVIIIᵉ siècle', in: idem et al., *Les machines à guérir (aux origines de l'hôpital moderne).* Brussels/Liège 1979, pp. 11–21; idem, *The birth of the clinic. An archaeology of medical perception.* London 1973; idem, *Discipline and punish. The birth of the prison.* London 1977; idem, *Madness and civilization. A history of insanity in the age of reason.* New York 1965.
19. To name only a few works: Alphonso Lingis, *Foreign bodies.* New York/London 1994; David Armstrong, *Political anatomy of the body. Medical knowledge in Britain in the twentieth century.* Cambridge 1983; see also Turner, *Regulating bodies* [see note 11].
20. Along similar lines, Friedrich Nietzsche had already accused the idealist philosophy of his day of 'Leibverachtung', that is, of being contemptuous of the body; cf. Kogaku Arifuku, 'Der Leib als große Vernunft bei Nietzsche und das Problem des Leibes in der Zen-Theorie Dogens', *Geschichte und Gegenwart* 17 (1998), 156–79.
21. Michel Foucault, *The history of sexuality*, 3 vols. New York 1978–86.
22. Cf. Steven Shapin/Simon Schaffer, *Leviathan and the air pump. Hobbes, Boyle, and the experimental life.* Princeton 1985; Steven Shapin, *A social history of truth. Civility and science in seventeenth century England.* Chicago/London 1995; Bruno Latour, *Science in action. How to follow scientists and engineers through society.* Cambridge, MA 1987; David J. Hess, *Science studies. An advanced introduction.* New York/London 1997; Mario Biagioli (ed.), *The science studies reader.* New York/London 1999. For social-constructionist approaches in the history of medicine see for example Ludmilla Jordanova, 'The social construction of medical knowledge', *Social history of medicine* 8 (1995), 361–81; Jens Lachmund/Gunnar Stollberg (eds), *The social construction*

of illness. Illness and medical knowledge in past and present. Stuttgart 1992; Thomas Schlich, 'Wissenschaft: Die Herstellung wissenschaftlicher Fakten als Thema der Geschichtsforschung', in: Paul/Schlich, *Medizingeschichte* [see note 4], pp. 107–29; Philipp Sarasin/Jakob Tanner (eds), *Physiologie und industrielle Gesellschaft. Studien zur Verwissenschaftlichung des Körpers im 19. und 20. Jahrhundert.* Frankfurt 1998.

23. Cf. Antonio Gramsci, *Quaderni del carcere*, 4 vols. Ed. by V. Gerratana. Turin 1975.

24. This is, in my view, the major drawback of Philipp Sarasin's otherwise stimulating analysis of 19th- and early 20th-century hygienic and dietetic discourses (Philipp Sarasin, *Reizbare Maschinen. Eine Geschichte des Körpers 1765–1914.* Frankfurt 2001).

25. Cf. Joan Wallach Scott, *Gender and the politics of history.* Rev. edn. New York 1999; Judith Butler, *Gender trouble. Feminism and the subversion of identity.* New York 1990.

26. Cf. William M. Reddy, 'Against constructionism. The historical ethnography of emotions', *Current anthropology* 38 (June 1997), 326–51.

27. The aporia of the radical relativism inherent in some poststructuralist approaches in general historiography has rightly been pointed out. Not only do such approaches have far-reaching ethical and political implications, because there is no longer any criterion according to which some histories are more plausible than others; histories that deny the Holocaust could be written with the same claim to truth as those that describe and study it. Especially in their debates with supporters of differing positions, radical relativists also fall prey to what Habermas has termed 'performative contradiction': they intend to convince their readers or listeners of the superiority of their own position, but, by their own standards, the legitimacy of this position cannot be distinguished from that, of the biologist ideologies, such as violent racism or misogyny, which they are battling.

28. See for example Margaret Lock, 'Cultivating the body. Anthropology and epistemologies of bodily practice and knowledge', *Annual review of anthropology* 22 (1993), 133–55.

29. Turner, *Regulating bodies* [see note 11], pp. 16f.

30. Carol Bigwood, 'Renaturalizing the body (with a little help from Merleau-Ponty)', *Hypatia* 6 (1991), issue 3, 54–73; Laure Lee Downs, 'If "woman" is just an empty category, then why am I afraid to walk alone at night? Identity politics meets the postmodern subject', *Comparative studies in society and history* 35 (1993), 414–37.

31. Lyndal Roper, *Oedipus and the devil. Witchcraft, sexuality and religion in early modern Europe.* London/New York 1994, p. 17.

32. Charles E. Rosenberg/Janet Golden (eds), *Framing disease. Studies in cultural history.* New Brunswick 1992.

33. Thomas J. Csordas (ed.), *Embodiment and experience. The existential ground of culture and self.* Cambridge 1994; see also Robert B. Desjarlais, *Body and emotion. The aesthetics of illness and healing in the Nepal Himalayas.* Philadelphia 1992.

34. Marcel Mauss, 'Body techniques', in: idem, *Sociology and psychology. Essays.* London 1979, pp. 95–123.

35. Cf. Pierre Bourdieu, *Outline of a theory of practice.* Cambridge 1977.

36. Michael Stolberg, '"Mein äskulapisches Orakel!": Patientenbriefe als Quelle einer Kulturgeschichte der Krankheitserfahrung im 18. Jahrhundert', *Österreichische Zeitschrift für Geschichtswissenschaft* 7 (1996), 385–404; see also Séverine Pilloud/Micheline Louis-Courvoisier, 'The intimate experience of the body in the eighteenth century: between interiority and exteriority', *Medical history* 47 (2003), 451–72.

37. Guenter B. Risse, 'Doctor William Cullen, physician, Edinburgh. A consultation practice in the eighteenth century', *Bulletin of the history of medicine* 48 (1974), 338–51. Numerous other letters of consultation were written by local physicians and surgeons, usually at the patient's request; since they tend to reflect primarily the physician's or surgeon's point of view I will only occasionally make use of them, mainly when they contain an account of what the patient him- or herself said.

38. Cf. Michael Stolberg, 'Krankheitserfahrung und Arzt-Patienten-Beziehung in Samuel Hahnemanns Patientenkorrespondenz', *Medizin, Gesellschaft und Geschichte* 18 (1999), 101–20; Séverine Pilloud/Stefan Hächler/Vincent Barras, 'Consulter par lettre au XVIIIᵉ siècle', *Gesnerus* 61 (2004), 232–53.

39. Occasionally patients also resorted to the help of intermediaries who were more familiar with the physician and could perhaps better influence him to make a judgment or persuade him to visit in person; see for example Kantonbibliothek St Gallen 94, letter from the abbess Margaretha, Maggenau, 16 August 1616, who asked Schobinger for a personal visit to another patient who lived seven or eight hours further away; see also Micheline Louis-Courvoisier/Séverine Pilloud, 'Le malade et son entourage au XVIIIᵉ siècle. Les médiations dans les consultations épistolaires adressées au Dr Tissot', *Revue médicale de la Suisse romande* 120 (2000), 939–44; Séverine Pilloud, 'Mettre les maux en mots, médiations dans la consultation épistolaire au XVIIIᵉ siècle: Les malades du Dr Tissot (1728–1797)', *Canadian bulletin of medical history* 16 (1999), pp. 215–25.

40. SB Berlin, Ms.germ.fol. 99, 420a, 420b, 421a, 421b, 422a, 422b, 423a, 423b, 424, 425 and 426.

41. Cf. J. C. W. Moehsen, *Leben Leonhard Thurneissers zum Thurn. Ein Beitrag zur Geschichte der Alchemie wie auch der Wissenschaften und Künste in der Mark Brandenburg gegen Ende des 16. Jahrhunderts*. Berlin/Leipzig 1783.

42. I had already finished my analysis of the sources and published the first results (Stolberg, *Orakel* [see note 36]) and was working on this book when Vincent Barras, Séverine Pilloud, and Micheline Louis-Courvoisier in Geneva and Lausanne embarked on a major research project on Tissot's correspondence. In the process, they have also assigned new call numbers to the letters, which make it easier to identify individual letters. I have not found it necessary to check and change all my references according to the new and certainly useful system. Usually the letters can quite easily be found also on the basis of the chronological list of letters which I encountered in the library during my research there and which contains the date, name, age, and sex of the patient and/or the author or the person who forwarded the letter.

43. Tissot, *Avis*; idem, *De la santé des gens de lettres*, 3rd edn. Lausanne 1775; idem, *L'onanisme*; idem, *Essai sur les maladies des gens du monde*, 3rd edn. Paris 1771; cf. Charles Eynard, *Essai sur la vie de Tissot*. Lausanne 1839;

Antoinette Emch-Dériaz, *Tissot: physician of the Enlightenment*. New York 1992; for an introduction especially to the medical context see: Vincent Barras/Micheline Courvoisier (eds), *La médecine des lumières: tout autour de Tissot*. Chêne-Bourg 2001.

44. Moehsen, *Leben* [see note 41], p. 15; using a couple of letters to Thurneisser for her own analysis, Ulinka Rublack has already rightly underlined their importance for a history of the body (Ulinka Rublack, 'Körper, Geschlecht und Gefühl in der Frühen Neuzeit', in: Paul Münch (ed.), *'Erfahrung' als Kategorie der Frühneuzeitgeschichte* (Historische Zeitschrift, Beiheft 31, 2001), pp. 99–105; Paul Delaunay, *Le monde médical Parisien au dix-huitième siècle*. Paris 1906, p. 35; Laurence W. B. Brockliss, 'Consultation by letter in early eighteenth century Paris. The medical practice of Étienne François Geoffroy', in: Ann F. LaBerge (ed.), *French medical culture in the nineteenth century*. Amsterdam 1994, pp. 79–117; Emch-Dériaz, *Tissot* [see note 43], p. 48; a letter of consultation addressed to Tissot by Napoleon Bonaparte in 1787 had been mentioned by Charles Eynard (Eynard, *Essai* [see note 43], pp. 238–41); Daniel Teysseire has more recently published the particularly rich dossier of one of Tissot's patients, which, however, primarily reflects the physician's point of view (Daniel Teysseire, *Obèse et impuissant. Le dossier médical d'Élie-de-Beaumont 1765–1776*. Grenoble 1995).

45. Geoffroy's and Hoffmann's answers have survived. In the case of Tissot, it is only exceptionally that the complete response letter is available, although in many cases short, hand-written notes by Tissot on the diagnosis and therapy can be found written on the patient letter. They possibly served as the basis on which Tissot's secretary could compose a full-length response letter (Emch Dériaz, *Tissot* [see note 43], p. 48), or they were perhaps only an aide-memoire, like the log books some doctors kept of their practice. The responses of Thurneisser and Hahnemann are also only sporadically available, either as first drafts or, if patients held onto them, as part of other source collections.

46. IGM Stuttgart, HA, Best. B and C; cf. Jörg Meyer, '"… als wollte mein alter Zufall mich jetzt wieder unter kriegen." Die Patientenbriefe an Samuel Hahnemann im Homöopathie Archiv des Instituts für Geschichte der Medizin in Stuttgart', *Jahrbuch des Instituts für Geschichte der Medizin der Robert Bosch Stiftung* 3 (1984), 63–79; Walter Nachtmann, 'Les malades face à Hahnemann (d'après leur correspondance juin–octobre 1832)', in: Olivier Faure (ed.), *Praticiens, patients et militants de l'homéopathie aux XIXe et XXe siècles (1800-1940)*. Lyon 1992, pp. 139–53; Bettina Brockmeyer, *Selbstverständnisse: Dialoge über Körper und Gemüt im frühen 19. Jahrhundert*. Göttingen 2009.

47. IGM Stuttgart, HA, Best. C.

48. There is little reason to doubt the authenticity of these letters. But it is likely that Hoffmann only published a small portion of the letters addressed to him in the original; therefore we can presume that he made a purposeful choice. The patient letters to Lorenz Heister and other physicians, in the Trew collection in Erlangen, Germany, are likewise mostly from the first half of the 18th century. I have consulted them only selectively, as Marion M. Ruisinger intended to use them as the central source for her research about Lorenz Heister and his patients (in the meantime, she has published her

findings: Marion Maria Ruisinger, *Patientenwege. Die Konsiliarkorrespondenz Lorenz Heisters (1683–1758) in der Trew-Sammlung Erlangen.* Stuttgart 2008).

49. Cf. Hildegard Tanner, *Medizinische Konsultationsschreiben aus Albrecht von Hallers Briefsammlung 1750–1775. Inventar und Analyse.* Diss. med. Bern 1994.

50. BL London, Mss. Sloane; Jurin, *Correspondence*; UB Leiden, Mss. Marchand 3; AS Bologna, Studio.

51. Winfried Schulze (ed.), *Ego-Dokumente. Annäherung an den Menschen in der Geschichte.* Berlin 1996; I am very grateful to Brigitte Berger, who went through many of the German 16th- and 17th-century autobiographies on which I draw in this book.

52. Cf. Gabriele Jancke, *Autobiographie als soziale Praxis. Beziehungskonzepte in Selbstzeugnissen des 15. und 16. Jahrhunderts. Köln im deutschsprachigen Raum.* Cologne 2002; see also Klaus Arnold et al. (eds): *Das dargestellte Ich. Studien zu Selbstzeugnissen des späteren Mittelalters und der frühen Neuzeit.* Bochum 1999.

53. In terms of medical culture, early modern England differs from continental Europe in several respects, ranging from the somewhat different structure of the medical market, in which learned physicians were even less dominant than on the Continent, to the influence of Puritan traditions of self-scrutiny on patients' accounts of their diseases; cf. Porter/Porter, *Patient's progress* [see note 4]; iidem, *In sickness* [see note 4]; Andrew Wear, 'Puritan perceptions of illness in seventeenth century England', in: Roy Porter (ed.), *Patients and practitioners. Lay perceptions of medicine in pre-industrial society.* London 1985, pp. 55–99; McCray Beier [see note 4].

54. Even among the upper classes, spelling was, of course, not as standardized at the time as it is today. However, French spellings such as 'frecheur' (instead of 'fraîcheur'), 'oguementer' (instead of 'augmenter') and especially 'aigue-saquetement' (instead of 'exactement') make one presume that they were phonetic. Such spellings were at the time already highly unusual.

55. *Le miracle opéré dans la nouvelle Hemorrhoïsse par Jesus-Christ présent dans la Sainte Eucharistie à Paris le 31 May 1725.* Sine loco 1726; Anne Le Franc, 'Relation de la maladie que j'ai eûe pendant près de 28 ans, et dont j'ai été guerie par l'intercession du Bienheureux François de Paris Diacre', in: *Dissertation sur les miracles.* [Paris] 1731, pp. 33–9; *Recueil des miracles; Second recueil; Recueil des pièces*; 'Pièces justificatives du miracle arrivé à Moisy en la personne de Louise Tremasse veuve Mercier', in: *Réflexions importantes sur le miracle arrivé au mois d'octobre dernier au Bourg de Moisy en Beauce, Diocèse de Blis, en la personne de Louise Tremasse Veuve Mercier.* Sine loco 1738 (separate pagination).

56. For the context see Jean-Claude Pie, 'Anne Charlier, un miracle eucharistique dans le faubourg Saint-Antoine', in: Jacques Gélis/Odile Redon (eds), *Les miracles miroirs des corps.* Saint Denis 1983, pp. 161–90; Eliane Gabert-Boche, 'Les miraculés du cimetière Saint Médard à Paris (1727–1735)', in: *ibid.*, pp. 127–57; Ulrike Krampl, '"Par ordre des convulsions". Überlegungen zu Jansenismus, Schriftlichkeit und Geschlecht im Paris des 18. Jahrhunderts', *Historische Anthropologie* 6 (1998), 33–62; Daniel Vidal, *Miracles et convulsions jansénistes au XVIIIe siècle. La maladie et sa connaissance.* Paris 1987.

57. HStA Stuttgart A 213; AS Bologna, Studio 339–51 contains documents of several cases of conflict over medical fees, some of them including brief written statements by or records of interrogations of those involved; cf. Gianna Pomata, *La promessa di guarigione. Malati e curatori in antico regime. Bologna XVI–XVIII secolo.* Bari 1994.

58. Ulinka Rublack, 'Pregnancy, childbirth and the female body in early modern Germany', *Past & present* 150 (1996), 84–110; Eva Labouvie, *Andere Umstände. Eine Kulturgeschichte der Geburt.* Cologne 1998.

59. Storch, *Kranckheiten*; see also Gockel, *Consilia*; Fischer, *Consilia*; Clacius, *Consilia*.

60. SB München, Cgm 6874; StA München, Hist. Verein, Ms. 401; StA Bamberg K3FIII 1481.

61. Roger Chartier, 'Volkskultur und Gelehrtenkultur. Überprüfung einer Zweiteilung und einer Periodisierung', in: Hans U. Gumbrecht/Ursula Link-Heer (eds), *Epochenschwelle und Epochenstrukturen im Diskurs der Literatur- und Sprachhistorie.* Frankfurt 1985, pp. 376–90.

62. Cf. Michael Stolberg, 'Probleme und Perspektiven einer Geschichte der Volksmedizin', in: Thomas Schnalke/Claudia Wiesemann (eds), *Die Grenzen des Anderen. Medizingeschichte aus postmoderner Perspektive.* Vienna 1998, pp. 49–73; Eberhard Wolff, '"Volksmedizin" als historisches Konstrukt. Laienvorstellungen über die Ursachen der Pockenkrankheit im frühen 19. Jahrhundert und deren Verhältnis zu Erklärungsweisen in der akademischen Medizin', *Österreichische Zeitschrift für Geschichtswissenschaften* 7 (1996), 405–30.

63. Thus for early modern France Brockliss/Jones, *Medical world* [see note 5], p. 283, have found 'a basically unitary medical universe'.

64. See for example Doreen Evenden Nagy, *Popular medicine in seventeenth-century England.* Bowling Green, OH 1988.

65. Tissot, *Advice*, vol. 2, pp. 299–302; I have great doubts when Frédéric Sardet postulates that most patients resisted a 'philosophy of the subject that leaned toward monadological conceptions', basing his conclusion merely on the fact that the great majority of patients did not adhere to Tissot's list of questions, which supposedly represented this philosophy. Many patients presumably had not even read Tissot's *Advice*, certainly not in the extended edition, and even when they had read the list of questions, they would likely have experienced it as simply too confining for the most part, since they were used to telling doctors their stories in a way they saw fit. (cf. F. Sardet, 'Briefe in der Kommunikation zwischen Arzt und Patient im 18. Jahrhundert. Annäherung an das Subjekt', in: Kaspar von Greyerz/Hans Medick/Patrice Veit (eds), *Von der dargestellten Person zum erinnerten Ich. Europäische Selbstzeugnisse als historische Quellen (1500–1850).* Cologne 2001, pp. 231–58).

I Illness in Everyday Life

1. Pansa, *Extract*, p. 2; abbreviated titles, without cross-references to previous notes, refer to the lists of manuscript and printed sources at the end of this book.

2. Alfons Labisch, *Homo hygienicus. Gesundheit und Medizin in der Neuzeit.* Frankfurt/New York 1992.
3. Melitta Weiss Adamson, *Medieval dietetics. Food and drink in regimen sanitatis literature from 800 to 1400.* Frankfurt 1995; Lucinda McCray Beier, 'In sickness and in health: A seventeenth-century family's experience', in: Roy Porter (ed.), *Patients and practitioners. Lay perceptions of medicine in pre-industrial society.* London 1985, pp. 101–28; Andrew Wear, 'Puritan perceptions of illness in seventeenth century England', in: *ibid.*, pp. 55–99.
4. Cornaro/Lessius, *Hygiasticon;* cf. Klaus Bergdolt, *Leib und Seele. Eine Kulturgeschichte des gesunden Lebens.* Munich 1999; Holger Böning, 'Medizinische Volksaufklärung und Öffentlichkeit. Ein Beitrag zur Popularisierung aufklärerischen Gedankengutes und zur Entstehung einer Öffentlichkeit über Gesundheitsfragen. Mit einer Bibliographie medizinischer Volksschriften', *Internationales Archiv für Sozialgeschichte der deutschen Literatur* 15 (1990), 1–92; Karel Daněk, Uebersicht der Geschichte der Gesundheitserziehung und der Gesundheitsführung (unpubl. manuscript of a paper presented at the Medical Faculty in Rostock, September 1965); Weiss Adamson [see note 3].
5. Philipp Sarasin, *Reizbare Maschinen. Eine Geschichte des Körpers 1765–1914.* Frankfurt 2001, p. 19.
6. Thus, for example, Edward Shorter, 'The history of the doctor/patient relationship', in: W. F. Bynum/Roy Porter (eds), *Companion encyclopedia of the history of medicine.* London/New York 1993, pp. 783–800, here p. 787.
7. Cf. Michael Stolberg, 'Die wundersame Heilkraft von Abführmitteln. Erfolg und Scheitern vormoderner Krankheitsbehandlung aus der Patientensicht', *Würzburger medizinhistorische Mitteilungen* 22 (2003), 167–77. 'The system must, that is, have worked', Charles E. Rosenberg concludes along the same lines (Ch. E. Rosenberg, 'The therapeutic revolution. Medicine, meaning, and social change in nineteenth-century America', in: idem/Morris J. Vogel (eds), *The therapeutic revolution. Essays in the social history of American medicine.* Pittsburgh 1979, pp. 3–25, here p. 5); there is a growing interest in recent cultural anthropological studies into the reasons for the success of therapies in non-Western cultures, which according to the standards of modern Western medicine have no causal effect on the course of diseases; cf. Thomas J. Csordas/Arthur Kleinman, 'The therapeutic process', in: Carolyn F. Sargent/ Thomas M. Johnson (eds), *Handbook of medical anthropology. Contemporary theory and method*, 2nd edn. Westport, Conn./London 1996, pp. 3–20.
8. Judith Devlin, *The superstitious mind. French peasants and the supernatural in the nineteenth century.* New Haven/London 1987, pp. 43–71; Michael Stolberg, '"Volksfromme" Heilpraktiken und medikale Alltagskultur im Bayern des 19. Jahrhunderts', in: Michael Simon/Monika Kania-Schütz (eds), *Auf der Suche nach Heil und Heilung: religiöse Aspekte der medikalen Alltagskultur.* Dresden 2001, pp. 155–73.
9. Rudolf Behrens/Roland Galle (eds), *Leib-Zeichen. Körperbilder, Rhetorik und Anthropologie im 18. Jahrhundert.* Würzburg 1998, editors' preface, p. 7.
10. Cf. Michael Bury, 'Chronic illness as biographical disruption', *Sociology of health and illness* 4 (1982), 167–82; Kathy Charmaz, 'Loss of self: a fundamental form of suffering in the chronically ill', *ibid.* 5 (1983), 168–95; Byron J. Good, *Medicine, rationality, and experience.* Cambridge 1994, pp. 124–8; in societies

which primarily understand disease as a threat to the entire community this rupture may be experienced above all in social life.

11. Cf. Elaine Scarry, *The body in pain. The making and unmaking of the world.* New York/Oxford 1985.
12. BCU Lausanne, FT, letter from Mme de Nomis, Turin 16 April 1785.
13. Thus Khevenhüller, *Tagebuch*, p. 302; Staiger, *Tagebuch*, p. 271.
14. Guyon, *Vie*, p. 350.
15. BCU Lausanne, FT, letter from Monsieur Hartmann, 22 May 1792; *ibid.*, letter from Leutnant Roussany, 10 June 1774.
16. *Ibid.*, letter from Herr von Jungken, 24 January 1772; almost four years later he was still in Tissot's care (idem, 18 November 1775).
17. *Ibid.*, letter from the Chevalier de Peyrelongue, Strasbourg 7 September 1785.
18. *Ibid.*, letter from Mme Herrmann, regarding a lady from Vienne, 21 May 1790.
19. *Ibid.*, letter from Mme A. C. de Konauw, 26 January 1773; Julie de l'Espinasse, who suffered from melancholy, used almost the same expression (BB Bern, HC, De l'Espinasse, 4 October 1762): 'I am often a burden to myself and to my servant.'
20. *Ibid.*, letter from Sig. Piazza, 18 September 1781 (orig. Italian).
21. Cf. Oswei Temkin, 'The scientific approach to disease: specific entity and individual sickness', in: idem, *The double face of Janus and other essays in the history of medicine.* Baltimore/London 1977, pp. 441–55.
22. Much less often, the disease was described as a state with the help of the verb 'to be', as with statements like 'I was ill', 'I am feverish' or 'I am dropsical.' Comparable expressions described the illness as a 'tumble' or 'fall'. In this sense, French-speaking patients frequently used the term 'tomber malade'. In German letters we find the analogous expression 'in Krankheit gefallen' (SB Berlin, Ms.germ.fol. 421b, fol. 205r, letter from a certain 'Ursula', ca. 1577). A successful treatment accordingly had to 'pull' the patient 'back up'.
23. BCU Lausanne, FT, letter from Monsieur Gochuat, Bischoffsheim, 1 November 1785.
24. See also Dietlinde Goltz, 'Krankheit und Sprache', *Sudhoffs Archiv* 53 (1969), 225–69, here pp. 233–8; E. J. Cassell, 'Disease as an "it": Concepts of disease revealed by patients' presentation of symptoms', *Social science and medicine* 10 (1976), 143–6.
25. SB München, Cgm 6874.
26. Gillian Bennett, 'Bosom serpents and alimentary amphibians. A language for sickness', in: Marijke Gijswijt-Hofstra et al. (eds), *Illness and healing alternatives in Western Europe.* London/New York 1997, pp. 224–42; Rina Knoeff, 'Animals inside. Anatomy, interiority and virtue in the early modern Dutch Republic', *Medizinhistorisches Journal* 42 (2007), 1–19.
27. RA Utrecht, Ms. 2200, medical recipe book, 1604, foll. 50r–v; *ibid.*, Ms. 2201, fol. 24r and fol. 51v.
28. Similarly Robert Jütte, *Ärzte, Heiler und Patienten. Medizinischer Alltag in der frühen Neuzeit.* Munich/Zürich 1991, pp. 124f.
29. SB Berlin, Ms.germ.fol. 420b, foll. 112r–113v, letter with illegible signature.
30. RA Utrecht, Ms. 2200, fol. 54r.

31. On the paramount importance of bodily evacuations in many other cultures see J. B. Loudon, 'On body products', in: John Blacking (ed.), *The anthropology of the body*. London 1977, pp. 161–78.

32. Cf. Christian Probst, *Fahrende Heiler und Heilmittelhändler. Medizin von Marktplatz und Landstraße*. Rosenheim 1992.

33. SB Berlin, Ms.germ.fol. 420b, foll. 114r–v, Abbot Erhardus 16 December [1574].

34. Little research has so far been done on the history of the subjective experience of pain; see above all Roselyne Rey, *History of pain*. Paris 1993; for a survey of medical theories on pain see Daniel de Moulin, 'A historical-phenomenological study of bodily pain in Western man', *Bulletin of the history of medicine* 48 (1974), 540–70.

35. SB Berlin, Ms.germ.fol. 420a, foll. 258r–v, 3 May 1571.

36. Stampfer, *Hausbüchl*, p. 68.

37. BIM Paris, Ms. 5245, fol. 158, letter from an unidentified clergyman, no date given.

38. Jurin, *Correspondence*, pp. 396–405, letters from the Irish bishop Cary, 1733.

39. SB München, Cgm 6874/137, Parsberg; for a detailed analysis see Alexander Berg, *Der Krankheitskomplex der Kolik- und Gebärmutterleiden in Volksmedizin und Medizingeschichte unter bes. Berücksichtigung der Volksmedizin in Ostpreussen*. Berlin 1935.

40. Hoffmann, *Medicina consultatoria*, vol. 11, pp. 129f., appendix to a letter 13 June (ca. 1735).

41. Hoffmann, *Medicina consultatoria*, vol. 10, pp. 315–22, 'morbid state' of a rural clergy-man with the initials 'J. D. F.', no date given.

42. BIM Paris, Ms. 5245, foll. 59r–v, undated report, probably written by the abbot or a fellow monk.

43. BIM Paris, Ms. 5241, foll. 73r–75r, unsigned letter, 29 June 1724, probably written by an inmate of the same monastery.

44. SB Berlin, Ms.germ.fol. 420a, fol. 183r, letter from the husband, Plauen, 29 July 1571.

45. BCU Lausanne, FT, undated letter concerning a 55-year-old female patient, possibly written by a physician.

46. *Ibid.*, letter from Mme de Konauw , 26 January 1773.

47. *Ibid.*, letter from a clergyman by the name of Olivier, 2 March 1774.

48. *Ibid.*, letter from Mlle Kirchberger, 31 May 1790.

49. *Ibid.*, letter from the husband, Monsieur Faugeroux, 12 June 1787.

50. Hoffmann, *Medicina consultatoria*, vol. 5, pp. 326–33, undated.

51. IGM Stuttgart, HA, Best. B 321660, patient's diary, winter 1831/32.

52. BCU Lausanne, FT, letter from Monsieur Feger, 11 July 1772.

53. Hoffmann, *Medicina consultatoria*, vol. 1, pp. 237–40 [ca. 1715].

54. SB Berlin, Ms.germ.fol. 423b, foll. 133r–v, no date given.

55. Hoffmann, *Medicina consultatoria*, vol. 1, pp. 254–7, letter concerning a 46-year-old nobleman [ca. 1715].

56. BCU Lausanne, FT, patient's letter, 16 June 1776; *ibid.*, letter from Monsieur Bouju, 5 January 1774.

57. *Ibid.*, letter from Mme Niels, 15 August 1773.

58. *Ibid.*, undated letter concerning a 55-year-old female patient, possibly written by a physician.

59. *Ibid.*, letter from Mme Marianne Doxat de Champvent, 5 May 1790.

60. *Ibid.*, letter from Monsieur Feger, 11 July 1772.

61. Jütte, *Ärzte* [see note 28], pp. 36–8 has similarly underlined the rich vocubulary of pain in the early modern period.

62. Michael Stolberg, '"Cura palliativa". Begriff und Diskussion der palliativen Krankheitsbehandlung in der vormodernen Medizin (ca. 1500–1850)', *Medizinhistorisches Journal* 42 (2007), 7–29; idem, *Die Geschichte der Palliativmedizin. Medizinische Sterbebegleitung von 1500 bis heute.* Frankfurt 2011.

63. Ronald D. Mann (ed.), *The history of the management of pain: from early principles to present practice.* Carnforth (UK)/Park Ridge, NJ (USA) 1988.

64. Hildanus, Wund-Artzney, p. 1239, letter from Utenhoven, 8 September 1596; see also Brandis, *Diarium,* p. 540, on sick Dr Peter Hagen.

65. Montaigne, *Essais,* pp. 268–70.

66. Khevenhüller, 2, p. 282 (1603).

67. Zimmern, *Chronik,* vol. III, p. 248.

68. Vincentz, *Goldschmiede-Chronik,* p. 389.

69. *Ibid.*, 497f.

70. Zimmern, *Chronik,* vol. III, p. 284.

71. *Ibid.*, vol. II, pp. 59f.

72. For a useful overview see David B. Morris, *The culture of pain.* Berkeley 1991; on the historical contingence of experiencing pain see Jakob Tanner, 'Körpererfahrung, Schmerz und die Konstruktion des Kulturellen', *Archiv für Kulturgeschichte* 75 (1993), 489–502.

73. SB München, Cgm 6874; StAM Hist. Ver. Ms. 401.

74. SB München, Cgm 6874/137, Parsberg.

75. *Ibid.*, Cgm 6874/118 Neumarkt in der Oberpfalz.

76. Cf. part 3.

77. For a more general analysis see Pedro Laín Entralgo, 'Das Erlebnis der Krankheit als geschichtliches Problem', *Antaios* 2 (1961), 285–98.

78. Robert Jütte, *Geschichte der Alternativen Medizin. Von der Volksmedizin zu den unkonventionellen Therapien von heute.* Munich 1996, pp. 148–62; Wear, 'Puritan perceptions' [see note 3]; Lucinda McCray Beier, 'In sickness and in health' [see note 3].

79. UB Basel, Ms. Fr. Gr. I 6, Küngast Horklin, 25 May 1570.

80. SB Berlin, Ms.germ.fol. 420a, foll. 164r–165r, Georg Krakewitz, 28 December 1571.

81. SB Berlin, Ms.germ.fol. 421b, foll. 118r–119r, Caspar von Hobergk, 9 March 1577.

82. Thus for example Bösch, *Liber,* p. 91.

83. SB Berlin, Ms.germ.fol. 420a, foll. 180r–v, Georg von Königsmark, Laetare Sunday 1571; the patient thought that his disease may have been a punishment for his sins.

84. Andreä, *Johann Valentin Andreä,* p. 57.

85. SB Berlin, Ms.germ.fol. 420a, foll. 20r–v, Anthonius Billig (ca. 1571); *ibid.*, 420b, fol. 135r, Johan Schagern, no date given; *ibid.*, 422a, fol. 113r, David Frank [?], no date given; Schad, *Memorial- und Reisebuch,* p. 411 (1626).

86. Hirsch, *Familienaufzeichnungen,* pp. 49f.

87. Andreae, *Leben,* p. 66.

88. Passer, *Berichte*, p. 331 (1682).

89. Geizkofler, *Selbstbiographie*, p. 63.

90. Michael Stolberg, *'Volksfromme' Heilpraktiken* [see note 8]; Jütte, *Geschichte* [see note 78], pp. 69–90; Peter Assion, 'Geistliche und weltliche Heilkunst in Konkurrenz. Zur Interpretation der Heilslehren in der älteren Medizin- und Mirakelliteratur', *Bayerisches Jahrbuch für Volkskunde* 1976/77, 7–23.

91. Michael Vovelle, *Piété baroque et déchristianisation en Provence au XVIIIᵉ siècle. Les attitudes devant la mort d'après les clauses des testaments*. Paris 1973; Elinor G. Barber, *The bourgeoisie in 18th-century France*. Princeton 1955, pp. 38–54.

92. BCU Lausanne, FT, letter from Monsieur Goret, 20 August 1791.

93. Among the mentally ill – whose delusions frequently refer to central images and topics of their respective culture and society – religious ideas played an important role in the 18th and 19th centuries; cf. for example Doris Kaufmann, '"Irre und Wahnsinnige". Zum Problem sozialer Ausgrenzung von Geisteskranken in der ländlichen Gesellschaft des frühen 19. Jahrhunderts', in: Richard van Dülmen (ed.), *Verbrechen, Strafen und soziale Kontrolle*. Frankfurt 1990, pp. 178–214, here pp. 192–5.

94. BCU Lausanne, FT, letter from Monsieur Le Chartier, 19 January 1776.

95. *Ibid.*, letter from the Abbé Tinseau, no date given, with a 'history of the disease' written in the third person by the patient.

96. *Ibid.*, letter from Dr Esperandieu, 27 February 1790; Tissot recommended among other things that he shave his head, presumably in order to facilitate the evacuation of morbid matter.

97. *Ibid.*, history of the disease of a 19-year-old woman, 10 February 1772.

98. *Ibid.*, letter from Mme de Chastenay, winter 1784/85.

99. *Recueil des pièces*, pp. 78–80, testimony of the patient's mother.

100. Mme Guyon's (b. 1648) autobiography (Guyon, *Vie*) was unusual in this respect; she ranked her manifold diseases among the major 'crosses' or trials which God had imposed on her; on Mme Guyon see also Elizabeth C. Goldsmith, *Publishing women's life stories in France, 1647–1720*. Aldershot 2001, pp. 71–97.

101. Christa Habrich, 'Pathographische und ätiologische Versuche medizinischer Laien', in: Wolfgang Eckart/Johanna Geyer-Kordesch (eds), *Heilberufe und Kranke im 17. und 18. Jahrhundert: Die Quellen- und Forschungssituation*. Münster 1982, pp. 99–123, with further references.

102. Hoffmann, *Medicina consultatoria*, vol. 5, pp. 274f., J. H. P., 6 May 1726.

103. Christa Habrich, 'Characteristic features of eighteenth-century therapeutics in Germany', in: William F. Bynum/Vivian Nutton, *Essays in the history of therapeutics*. Amsterdam/Atlanta 1991, pp. 39–49; Johanna Geyer-Kordesch, 'Cultural habits of illness. The enlightened and the pious in eighteenth-century Germany', in: Porter, *Patients and practitioners* [see note 3], pp. 177–204; Katharina Ernst, *Krankheit und Heilung. Die medikale Kultur württembergischer Pietisten im 18. Jahrhundert*. Stuttgart 2003.

104. Stephan Baron von Koskull, *Wunderglaube und Medizin. Die religiösen Heilungsversuche des Fürsten Alexander von Hohenlohe in Franken*. Bamberg 1988; *Dankende Geheilte in Bamberg auf das Gebeth des Fürsten A. v. Hohenlohe*. Würzburg 1821.

105. Eva Labouvie, *Verbotene Künste. Volksmagie und ländlicher Aberglaube in den Dorfgemeinden des Saarraumes (16.–19. Jahrhundert).* St Ingbert 1992; Stuart Clark, *Thinking with demons. The idea of witchcraft in early modern Europe.* Oxford 1997.
106. SB Berlin, Ms.germ.fol. 420a, fol. 198r, letter from the husband Ludolf, Wednesday after Exaudi 1571.
107. *Ibid.*, 200r, Saturday after Exaudi 1571.
108. Vetter, *Gesichten*, p. 74; later she interpreted her disease as a sign that she was among the elected.
109. Fleck, *Chronik*, pp. 219–21.
110. Bösch, *Liber*, pp. 78f. (ca. 1630).
111. Stampfer, *Hausbüchl*, pp. 12f.
112. In German the term *'angetane Krankheit'* was commonly used.
113. Stadtarchiv Konstanz H IX 48, cit. in Martin Burkhardt et al., *Konstanz in der frühen Neuzeit. Reformation, Verlust der Reichsfreiheit, österreichische Zeit.* Konstanz 1991, pp. 282–4.
114. Hartung, *Aufzeichnungen*, pp. 114f.; see also Hans de Waardt, 'Van exorcisten tot doctores medicinae. Geestelijken als gidsen naar genezing in de Republiek, met name in Holland, in de zestiende en de zeventiende eeuw', in: Willem de Blécourt et al. (eds), *Grenzen van genezing: Gezondheid, ziekte en genezen in Nederland, zestiende tot begin twintigste eeuw.* Hilversum 1993, pp. 88–114.
115. Cf. Willem de Blécourt, *Termen van toverij. De veranderende betekenis van toverij in Noordoost-Nederland tussen de 16de en 20ste eeuw.* Nijmegen 1990.
116. Hoffmann, *Medicina consultatoria*, vol. 5, pp. 274f., J. H. P., 6 May 1726; the medical faculty of Halle could find no evidence that his suspicions were justified (*ibid.*, pp. 282–4, 31 May 1726).
117. HStA Stuttgart A 213 Bü. 8416, testimonies and report of the chancellery, 16 November 1748; cf. Stolberg, Michael: *Die Harnschau. Eine Kultur- und Alltagsgeschichte.* Cologne/Weimar 2009, pp. 131–5.
118. Cf. Michael Stolberg, 'Alternative medicine, irregular healers and the medical market in nineteenth-century Bavaria', in: Robert Jütte et al. (eds), *Historical aspects of unconventional medicine. Approaches, concepts, case studies.* Sheffield 2001, pp. 139–62; on the other hand, in her research on the Duchy of Braunschweig-Wolfenbüttel in the 18th century, Mary Lindemann has hardly found a trace of such notions and practices (Mary Lindemann, *Health and healing in eighteenth-century Germany.* Baltimore/London 1996); it is possible, however, that this reflects only a lack of archival documentation.
119. Cf. for example Ranzovius, *Exempla*.
120. Wolf-Dieter Müller-Jahncke, *Astrologisch-magische Theorie und Praxis in der Heilkunde der frühen Neuzeit.* Stuttgart 1985, esp. pp. 135–75.
121. Vincentz, *Goldschmiede-Chronik*, p. 504.
122. Ficino, *Three books*; Fernel, *Universa medicina*.
123. SB Berlin, Ms.germ.fol. 424, foll. 187r–188r, Ladislaus Cubinius [?], 16 March 1582.
124. Michael MacDonald, *Mystical Bedlam. Madness, anxiety and healing in seventeenth-century England.* Cambridge 1981; see also Barbara Howard Traister, *The notorious astrological physician of London. Works and days of*

Simon Forman. Chicago/London 2001; Lauren Kassell, *Medicine and magic in Elizabethan London: Simon Forman – astrologer, alchemist, and physician.* Oxford 2005.

125. Michael MacDonald, 'The career of astrological medicine in England', in: Ole Peter Grell/Andrew Cunningham (eds), *Religio medici. Medicine and religion in seventeenth-century England.* Aldershot 1996, pp. 62–90.

126. Sachiko Kusukawa, *The transformation of natural philosophy. The case of Philip Melanchthon.* Cambridge 1995, esp. pp. 129–44; Claudia Brosseder, *Im Bann der Sterne. Caspar Peucer, Philipp Melanchthon und andere Wittenberger Astrologen.* Berlin 2004.

127. Müller-Jahncke, Astrologisch-magische Theorie [see note 120], pp. 175–93.

128. SB Berlin, Ms.germ.fol. 422b, fol. 285r, letter [1579].

129. SB Berlin, Ms.germ.fol. 420b, foll. 458r–459r, Simon Roter, Sunday after Misericordia 1575.

130. Wolf, *Commentariolus*, p. 31.

131. SUB Hamburg, Sup. ep. 4 30, fol. 140, copy of a letter, 26 February 1579 (orig. in French).

132. BIM Paris, Ms. 2037, foll. 25–8, letter from an unidentified patient, no date given; Helvetius's response is dated 24 January 1728.

133. UB Amsterdam, Ms. 14 CD 1, natal chart, 1 May 1745. On Ludeman see Hans de Waardt, 'Breaking the boundaries. Irregular healers in eighteenth-century Holland', in: Gijswijt-Hofstra, *Illness* [see note 26], pp. 141–60.

134. For a good contemporary overview see the entry 'Régime' in: *Encyclopédie,* vol. 14, 1767, pp. 11–16; see also William Coleman, 'Health and hygiene in the Encyclopédie. A medical doctrine for the bourgeoisie', *Journal of the history of medicine* 26 (1974), 399–421.

135. In learned medical writing, dietetics were traditionally dealt with in the context of the six 'non-naturals' (cf. Peter H. Niebyl, 'The non-naturals', *Bulletin of the history of medicine* 45 (1971), 486–92; L. J. Rather, 'The "six things non-natural"', *Clio medica* 3 (1968), 337–47; Saul Jarcho, 'Galen's six non-naturals', *Bulletin of the history of medicine* 44 (1970), 372–7); the number six results from the fact that the proper amount of physical activity and the right ratio of sleeping to waking were treated as two separate points; the sixth traditional non-natural, expulsion and retention, was not an aspect of lifestyle but referred primarily to a disorder inside the body, which at best could be influenced with medication and dietary measures.

136. Bösch, *Liber*, p. 102.

137. Geizkofler, *Selbstbiographie*, pp. 103f.

138. Andreä, *Johann Valentin Andreä*, p. 134.

139. Hummel, *Histori*, p. 40; Wolf, *Commentariolus*, p. 74; cf. Sabine Vogel, *Kulturtransfer in der Frühen Neuzeit. Die Vorworte der Lyoner Drucke des 16. Jahrhunderts.* Tübingen 1999, pp. 34–6.

140. UB Leiden, Mss. Marchand 3, G. van Boetzelaar, 13 October 1591 (orig. in Latin).

141. Vincentz, *Goldschmiede-Chronik*, p. 52.

142. Hoffmann, *Medicina consultatoria*, vol. 5, pp. 300f., 'v. D.', 10 March 1727.

143. SB Berlin, Ms.germ.fol. 421b, ff. 354r–355v, 19 October 1576.

234 *Notes*

144. BCU Lausanne, FT, undated letter from the unidentified patient's mother; at the time she consulted Tissot the son was 14 and suffered from convulsive fits.
145. Brandis, *Diarium*, pp. 451f.
146. Thus Ludwig von Diesbach thought a 'cold fever' to have been caused by corrupted food (Diesbach, *Aufzeichnungen*, p. 41).
147. Cf. Michael Stolberg, '"Zorn, Wein und Weiber verderben unsere Leiber". Krankheit und Affekt in der frühneuzeitlichen Medizin', in: Johann Anselm Steiger (ed.), *Passion, Affekt und Leidenschaft*. Wolfenbüttel 2005, pp. 1033–59.
148. SB Berlin, Ms.germ.fol. 421a, Sunday after Mary's conception, 1576.
149. Weinsberg, *Leben*, p. 261.
150. BIM Paris, Ms. 5241, unsigned letter, written in the third person, probably by a layperson, 15 September 1723.
151. BCU Lausanne, FT, undated letter from a 43-year-old female patient.
152. Hoffmann, *Medicina consultatoria*, vol. 5, pp. 300f., 'v. D.', 10 March 1727.
153. BCU Lausanne, FT, letter from Monsieur Reverdil, concerning a 30- to 40-year-old acquaintance, 17 July 1791.
154. Pirckheimer, *Autobiographie*, p. 132; see also Emil Reicke, 'Willibald Pirckheimer und sein Podagra', in: W. P. Eckert/Christoph von Imhoff (eds), *Willibald Pirckheimer. Dürers Freund im Spiegel seines Lebens, seiner Werke und seiner Umwelt*. Cologne 1971, pp. 184–202.
155. UB Erlangen, Ms. 1029/I, foll. 658f., 1 October 1714, signature illegible.
156. UB Basel, Ms. Fr. Gr. I 6, Hans Augustin Reich von Reichenstein, 6 May 1570.
157. For example, KB St Gallen, Ms. 94, fol. 23r, friar Placidus about a sick neighbor, 20 April 1605.
158. Orléans, *Leben*, p. 261 and pp. 267–9; Kristin Rönck has undertaken a detailed analysis of Liselotte's critical stance towards court medicine and the court (Kristin Rönck, *Gesundheit und Krankheit um 1700 in den Briefen Liselottes von der Pfalz*. Berlin, Humboldt-Universität 2000). Unfortunately, she has never published it and I am grateful to her for allowing me to see this work; see also Elborg Forster, 'From the patient's point of view. Illness and health in the letters of Liselotte von der Pfalz (1652–1722)', *Bulletin of the history of medicine* 60 (1986), 297–320.
159. Orléans, *Leben*, p. 205.
160. BCU Lausanne, FT, letter transmitted by Monsieur Bruna from an anonymous 37-year-old patient, 26 April 1773.
161. La Tour du Pin, *Journal*, p. 202.
162. Bövingh, *Lebensbeschreibung*, p. 146.
163. BB Bern, HC, A 94, letter from an unidentified patient, no date given.
164. BIM Paris, Ms. 5245 foll. 148r–149v, letter from the father of Mlle Herbolin, 23 April 1730.
165. Glückel von Hameln, *Denkwürdigkeiten*, pp. 171–5.
166. Hummel, *Histori*, p. 47.
167. Michael Stolberg, 'Aktive Euthanasie und Eugenik vor 1850', in: Ignacio Czeguhn/Eric Hilgendorf/Jürgen Weitzel (eds): *Eugenik und Euthanasie 1850–1945. Frühformen, Ursachen, Entwicklungen, Folgen*. Würzburg 2009, pp. 9–26.

168. Gareth Williams, 'The genesis of chronic illness: narrative reconstruction', *Sociology of health and illness* 6 (1984), 175–200; Catherine Kohler Riessman, 'Strategic uses of narrative in the presentation of self and illness. A research note', *Social science and medicine* 30 (1990), 1195–1200.

169. Ian Robinson, 'Personal narratives, social careers and medical courses. Analysing life trajectories in autobiographies of people with multiple sclerosis', *ibid.*, 1173–86.

170. Tissot, *Onanisme*.

171. For a more detailed account see Michael Stolberg, 'An unmanly vice: Self-pollution, anxiety, and the body in the eighteenth century', *Social History of Medicine* 13 (2000), 1–21.

172. Andreä, *Johann Valentin Andreä*, p. 37.

173. BCU Lausanne, FT, letter from Mme de Merande, 4 October 1783.

174. Cf. Leon Eisenberg, 'The physician as interpreter. Ascribing meaning to the illness experience', *Comprehensive psychiatry* 22 (1981), 239–48.

175. Thus a survey among a group of 100 rheumatic patients found that lay theories only 'in the fewest cases' matched medico-scientific theories (Hans-Eckhard Langer/Helmut Bormann, Krankheits-Bild als Krankheits-(Be-)Deutung. Versuch einer Meta-Theorie der Laientheorien bei rheumatischen Erkrankungen', in: Heinz-Dieter Basler et al. (eds), *Psychologie in der Rheumatologie*. Berlin 1992, pp. 55–82.

176. *Ibid.*; Rosemarie Welter-Enderlin, *Krankheitsverständnis und Alltagsbewältigung in Familien mit chronischer Polyarthritis*. Munich 1989.

177. Langer/Bormann, *Krankheits-Bild* [see note 175], pp. 75f.

178. Hermann Faller, 'Das Krankheitsbild von Herz-Kreislauf-Kranken. Ein Gruppen- und Methodenvergleich', in: Claus Bischoff/Helmuth Zenz (eds), *Patientenkonzepte von Körper und Krankheit*. Bern 1989, pp. 86–102; Welter Enderlin, *Krankheitsverständnis* [see note 176].

179. BCU Lausanne, FT, letter from Monsieur Goret, 20 August 1791.

180. *Ibid.*, letter from Monsieur de Char(r)itte, no date given.

181. Michael Stolberg, *Die Cholera im Grossherzogtum Toskana: Ängste, Deutungen, und Reaktionen im Angesicht einer tödlichen Seuche*. Landsberg 1995; Patrice Bourdelais/Jean-Yves Raulot, *Une peur bleue. Histoire du choléra en France, 1832–1854*. Paris 1987; the idea is much older, going back to times of plague in the Middle Ages.

182. From the perspective of modern medicine, the term may have referred to quite different diseases, ranging from cerebral hemorrhage and occlusion of the carotid or cerebral arteries to myocardial infarction and pulmonary embolism.

183. Brokes, *Tagebuch*, pp. 372f., note.

184. Michel Vovelle, *Mourir autrefois. Attitudes collectives devant la mort aux XVIIe et XVIIIe siècles*. Paris 1974, pp. 193f.

185. *Conversations-Lexicon*, vol. 1, p. 752.

186. BUP Geneva, Ms. suppl. 1908, foll. 260r–261v, Marquise de Contades, 13 July 1772.

187. SB Berlin, Ms.germ.fol. 420b, foll. 50r–52v, Johannes Theobaldus on the case of Simon Braschevius, Wednesday before Ascension Day 1574.

188. Cf. the detailed case study in Daniel Teysseire, *Obèse et impuissant. Le dossier médical d'Élie-de-Beaumont 1765–1776*. Grenoble 1995.

189. Hunter, *Treatise*, p. 202; Hunter himself underlined the importance of emotional influences.
190. BCU Lausanne, FT, letter from Schwitzer de Buonas, Lucerne, 2 October 1793.
191. *Ibid.*, letter from an anonymous English patient, 15 May 1792.
192. *Ibid.*, letter concerning an anonymous male patient, written by a friend and probably dictated by the patient, 16 August 1772; he asked for the reply to be sent to a clergyman in Geneva.
193. Cf. Pierre Darmon, *Le tribunal de l'impuissance*. Paris 1979; and for the preceding period, based on testimonies from trials for marriage annulment, Daniela Hacke, 'Gendering men in early modern Venice', *Acta Histriae* 8 (2000), n. 1, 49–68.
194. *Recueil des miracles*, pp. 7f., testimony of 57-year-old Marie Jeanne Orget.
195. Glückel von Hameln, *Denkwürdigkeiten*, pp. 171–5.
196. Platter, *Tagebuch*, p. 427.
197. Weinsberg, *Denkwürdigkeiten*, vol. V, p. 50.
198. SB Berlin, Ms.germ.fol. 421b, foll. 161r–162v, Chr. von Falckenberg, Friday after Pentecost 1577.
199. Bövingh, *Lebensbeschreibung*, p. 121.
200. Staal Delaunay, *Mémoires*, pp. 28f.
201. Guyon, *Vie*, pp. 124f.
202. Hoffmann, *Medicina consultatoria*, vol. 5, pp. 305–9, undated letter from an unnamed man concerning his sick wife.
203. HStA Wiesbaden, Abt. 369 Ms. 328, foll. 56r–57r, testimony in a case of suspected witchcraft, 5 February 1635.
204. BCU Lausanne, FT, letter from Monsieur Baville, 14 May 1774; see also *ibid.*, letter from a clergyman by the name of Olivier, 2 March 1774.
205. Cf. Anke Bennholdt-Thomsen/Alfredo Guzzoni, 'Zur Theorie des Versehens im 18. Jahrhundert. Ansätze einer pränatalen Psychologie', in: Thomas Kornbichler (ed.), *Klio und Psyche*. Pfaffenweiler 1990, pp. 112–25.
206. *Second recueil*, pp. 21–4.
207. BCU Lausanne, FT, letter from the mother, Mme de Bouchet, 12 October 1769.
208. Storch, *Kranckheiten*, vol. 2, pp. 183f., E. L. v. C., 12 September 1723; this is one of the very few patient letters in Storch's collection.
209. Brokes, *Tagebuch*, pp. 25f., note.
210. This is suggested, for example, by sources from the Nuremberg area which Andrea Reiter, Würzburg, has analyzed as part of her work on a forthcoming MD thesis; see also Kaufmann, *Irre* [see note 93].
211. RA Arnhem, Archive of the Counts of Coulembourg, Ms. 403, collection of letters from times of disease, early 17th century.
212. Stolberg, *Cholera* [see note 181].
213. In the testimonies on disputes with unauthorized healers, which have survived from Bologna (AS Bologna, Studio 339–350), this is a striking element; often the patients – who, in the end, were disappointed with the cure – had learned of the healer in question from relatives or intermediaries.
214. UB Leiden, Mss. Marchand 3, letter (in Latin) from an unnamed patient, 29 July 1591; friends had recommended various remedies for his violent anal pain.

215. She had let herself be convinced by some good-hearted people to send Thurneisser a sample of her urine, wrote, for example, Judith Daniels (SB Berlin, Ms.germ.fol. 423b, fol. 139r, ca. 1581).

216. SB Berlin, Ms.germ.fol. 420a, fol. 215r, Balzer von Barfuß (or Barwes), Wednesday after Lambertus 1571.

217. SB Berlin, Nachlaß Francke, 10.1/4: 48, Frid. Wilh. Louyse Solms (early 18th cent.).

218. BCU Lausanne, FT, letter from the father, a priest by the name of Cart, 21 June 1785.

219. BL London, Ms. Sloane 4075, foll. 94–6, letter from John Evelyn, 28 July 1703.

220. BCU Lausanne, FT, letter from the father, Monsieur Cart, 8 May 1785.

221. BL London, Ms. Sloane 4075, fol. 124, V. Ferguson, 13 November 1699.

222. Most works on the history of nursing have focused on the more recent past and/or on institutions and normative texts such as rules of conduct for hospital nurses; for a somewhat dated overview see Adelaide Nutting/ Lavinia A. Dock, *A history of nursing*, 2 vols. (orig. 1907). Tokyo/Bristol 2000; see also Barbara Mortimer/Susan McGann (eds), *New directions in the history of nursing. International perspectives.* London/New York 2005; Sylvelyn Hähner-Rombach (ed.), *Quellen zur Geschichte der Krankenpflege, mit Einführungen und Kommentaren.* Frankfurt 2008.

223. Guenter B. Risse, *Mending bodies, saving souls. A history of hospitals.* New York/Oxford 1999; Ulrich Knefelkamp. 'Über die Pflege und medizinische Behandlung von Kranken in Spitälern vom 14. bis 16. Jahrhundert', in Michael Matheus (ed.), *Funktions- und Strukturwandel spätmittelalterlicher Hospitäler im europäischen Vergleich.* Stuttgart 2005, pp. 175–94; Robert Jütte, 'Vom Hospital zum Krankenhaus: 16.–19. Jahrhundert', in Alfons Labisch/Reinhard Spree (eds), *'Einem jeden Kranken in einem Hospitale sein eigenes Bett'. Zur Sozialgeschichte des Allgemeinen Krankenhauses in Deutschland im 19. Jahrhundert.* Frankfurt 2005, pp. 31–50; John Henderson, *The Renaissance hospital. Healing the body and healing the soul.* New Haven 2006.

224. Bismarck, *Tagebuch*, p. 74.

225. Guyon, *Vie*, p. 95.

226. BL London, Ms. Sloane 4075, foll. 91–3, letter from John Evelyn, concerning his wife, who suffered from pulmonary disease, 1702.

227. BCU Lausanne, FT, letter from Mme Develay, 21 May 1791; *ibid.*, letter from the husband of Mme de Faugeroux, 12 June 1787; see also Jens Lachmund/Gunnar Stollberg, *Patientenwelten. Krankheit und Medizin vom späten 18. bis zum frühen 20. Jahrhundert im Spiegel von Autobiographien.* Opladen 1995, p. 54.

228. Vincentz, *Goldschmiede-Chronik*, p. 455.

229. Questel, *De pulvinari*; cf. Michael Stolberg, 'Active euthanasia in early modern society. Learned debates and popular practices', *Social history of medicine* 20 (2007), 205–21.

230. StA Nürnberg, B 14 II 9, fol. 52v, 1516.

231. StA Nürnberg B 14 II 18, foll. 163r–v and fol. 166v.

232. StA Nürnberg B 14 II 15, fol. 29r.

233. StA Nürnberg, D1 209, report of the urban poor administration (Stadtalmosenamt), 22 July 1769.

234. Bövingh, *Lebensbeschreibung*, p. 121; as a student Hermann Weinsberg, too, was nursed by his landlady (Weinsberg, *Denkwürdigkeiten*, vol. I, pp. 88f).
235. Scultetus, *Selbstbiographie*, p. 35.
236. Graffigny, *Correspondance*, vol. III, 262, October 1741; also in other places (for example *ibid.*, p. 341) she called herself a 'nurse' ('garde malade').
237. HStA Stuttgart A 209, Bü. 1547, report of the Faculty of Law in Tübingen on the witchcraft trial against the midwife Barbara Wildermuth, 24 October 1662.
238. *Ibid.*, Bü. 1921, testimony of the neighbor of an unnamed female patient who had allegedly died from excessive bloodletting, 2 January 1664.
239. Guyon, *Vie*, p. 41 and p. 58.
240. Hummel, *Histori*, p. 52.
241. SB Berlin, Ms.germ.fol. 421b, foll. 299r–300v, H. Birckholtz, 9 March 1577.
242. StA Nürnberg B 14 II 12, foll. 136r–v, 1519.
243. StA Köln, Testamente 2/B/848, 1530; see also Jütte, *Ärzte* [see note 28], pp. 202f.
244. Wolf, *Leben*, pp. 40f and pp. 44f.
245. UB Leiden, Mss. Marchand 3, G. van Boetzelaar, 13 October 1591.
246. BCU Lausanne, FT, letter from Mme de Marnais, 22 December 1784.
247. *Ibid.*, letter from the Marquise d'Agrain, 4 July 1785; her letter was occasioned by her wish not to see her 18-year-old daughter burdened with nursing the said older husband, in case his bad leg got worse.
248. *Ibid.*, letter from the Chevalier d'Alberey, 13 February 1790.
249. *Ibid.*, letter from Mme de Merande, 4 October 1783.
250. BCU Lausanne, FT, letter from the mother, Mme de la Roche, 19 January 1792.
251. According to Lachmund/Stollberg, *Patientenwelten* [see note 228], p. 56, the fact that the nursing attendant was male was underlined as something extraordinary, however.
252. Fabricius, *Lebensbeschreibung*, p. 40.
253. Hutten, *Brief*, pp. 337f.; cf. Michael Peschke, *Ulrich von Hutten (1488–1523) als Kranker und als medizinischer Schriftsteller*. Cologne 1985.
254. Staël-Holstein, *Mémoires*, p. 94.
255. Graffigny, *Correspondance*, vol. III, p. 330, editors' note to a letter from Devaux, August 1742.
256. *Second recueil*, testimony concerning Jean Baptiste le Doulx, pp. 14–17.
257. Christina Vanja, *Das Irrenhaus: Zur Kulturgeschichte des 'Wahnsinns'*. Hamburg 2007.
258. BCU Lausanne, FT, letter from Mme Develay, 21 May 1791.
259. BL London, Ms. Sloane 4075, pp. 61–7, letters from the spring of 1714.
260. BCU Lausanne, FT, undated report about an unnamed patient, probably written by a relative or a close acquaintance.
261. See among others Laurence W. B. Brockliss/Colin Jones, *The medical world of early modern France*. Oxford 1997; François Lebrun, *Se soigner autrefois. Médecins, saints et sorciers aux 17e et 18e siècles*. Paris 1983; Sabine Sander, *Handwerkschirurgen. Sozialgeschichte einer verdrängten Berufsgruppe*. Göttingen 1989; Margaret Pelling, *The common lot. Sickness, medical occupations and the urban poor in early modern England*. London 1998; Maria Conforti, 'Healing practices and medical professions in early modern

Europe', *Nuncius* 20 (2005), 219–29; Francisca Loetz, *Vom Kranken zum Patienten. 'Medikalisierung' und medizinische Vergesellschaftung am Beispiel Badens 1750–1850.* Stuttgart 1993; Jütte, *Ärzte* [see note 28]; Michael Stolberg, *Heilkunde zwischen Staat und Bevölkerung. Angebot und Annahme medizinischer Versorgung in Oberfranken im frühen 19. Jahrhundert.* Unpubl. med. diss. Munich 1986.

262. Stolberg, *Harnschau* [see note 117].
263. Graffigny, *Correspondance*, vol. III, p. 234.
264. Jurin, *Correspondence*, p. 398, 9 June 1733.
265. *Ibid.*
266. *Ibid.*, 328–30, 20 February 1726.
267. Cf. the excellent work by Andrew Wear, *Knowledge & practice in English medicine, 1550–1680.* Cambridge 2000; Wear focuses almost exclusively on the English situation, however.
268. BCU Lausanne, FT, letter from Mme Develay, 21 May 1791
269. *Ibid.*, letter from the Comtesse de Wedel, Copenhagen, 12 November 1784.
270. Hoffmann, *Medicina consultatoria*, vol. 5, pp. 326–33, letter from an unnamed vice-rector.
271. SB München, Cgm 6874.
272. BCU Lausanne, FT, letter from Major Bouju, 1773.
273. Cf. Probst, *Fahrende Heiler* [see note 32], esp. pp. 77–127; Alfred Franklin, La vie privée d'autrefois. Les médicaments. Paris 1891, pp. 207–37, based on a collection of contemporary advertisements.
274. Cf. Ailhaud, *Médecine universelle.*
275. BCU Lausanne, FT, letter from Monsieur Gringet, Chambery, 4 January 1784.
276. Franklin, *Vie privée* [see note 274], pp. 162–99; Laurence W. B. Brockliss, 'The development of the spa in seventeenth-century France', in: Roy Porter (ed.), *The medical history of waters and spas.* London 1990, pp. 23–47; Pascale Cosma-Muller, 'Entre science et commerce. Les eaux minérales en France à la fin de l'Ancien Régime', in: Jean-Pierre Goubert (ed.), *La médicalisation de la société française 1770–1830.* Waterloo, Ont. 1982, pp. 254–63.
277. There are even occasional hints that the employer undertook the bloodletting personally; cf. *Second recueil*, pp. 21–4, Marie-Anna Couronneau.
278. IGM Stuttgart, HA, Best. C 2, 27, 1837.
279. SB München, Cgm 6874; in the early 19th century, Mme Monteaux, for example, according to her own account, still 'quite frequently' underwent bloodletting (IGM Stuttgart, HA, Best. C 2, 14, Mme Monteaux, Paris 29 May 1837).
280. Especially in France, from the late 17th century, educated surgeons achieved a relatively high status after they had separated themselves from the barber-surgeons (Toby Gelfand, *Professionalizing medicine. Paris surgeons and medical science and institutions in the eighteenth century.* Westport, CT 1980).
281. Overview in Jütte, *Geschichte* [see note 78].
282. Henry Tronchin, *Un médecin du XVIIIᵉ siècle. Théodore Tronchin (1709–1781).* Paris/Geneva 1906.
283. Pomme, *Traité.*

284. The degree of repression varied considerably from state to state; the development was particularly marked for example in the 19th-century kingdom of Bavaria (Stolberg, *Heilkunde* [see note 262]).

285. Matthew Ramsey, *Professional and popular medicine in France, 1770–1830. The social world of medical practice.* Cambridge 1988; Gianna Pomata, *La promessa di guarigione. Malati e curatori in antico regime. Bologna XVI–XVIII secolo.* Bari 1994; Probst, *Fahrende Heiler* [see note 32]; Stolberg, *Heilkunde* [see note 262]; David Gentilcore, *Medical charlatanism in early modern Italy.* Oxford 2006.

286. Brockliss/Jones, *Medical world* [see note 262], pp. 288f.; Lachmund/Stollberg, *Patientenwelten* [see note 228]; Dorothy Porter/Roy Porter, *Patient's progress. Doctors and doctoring in eighteenth-century England.* Cambridge/Oxford 1989, p. 53 and pp. 209f.

287. Brockliss/Jones, *Medical world* [see note 262], pp. 288–90.

288. Cf. Pierre Bourdieu, *A social critique of the judgment of taste.* Cambridge, MA 1984.

289. According to his own account, Monsieur Dauphin, a 31-year-old barrister, had already spent between 28,000 and 30,000 livres on his treatment before turning to Tissot. At that time, an ordinary worker could expect to earn about 200 livres per year (BCU Lausanne, FT, 5 June 1772).

290. Caylus, *Souvenirs*, p. 42f.

291. François Lebrun, 'Médecins et empiriques à la cour de Louis XIV', *Histoire, économie, société* 3 (1984), 557–66; Brockliss/Jones, *Medical world* [see note 262], pp. 291f.

292. Geizkofler, *Selbstbiographie*, pp. 90–2.

293. Chastenay, *Mémoires*, vol. II, p. 36.

294. Deffand, *Lettres*, vol. II, pp. 14–16, 1 February 1770.

295. Cf. Camille Vieillard, 'Un uromante au XVIIIᵉ siècle. Michel Schuppach', *Bulletin de la Société Française d'Histoire de la Médecine* 2 (1903), 146–64; Eugen Wehren, 'Das medizinische Werk des Wundarztes Michel Schüppach (1707–1781) an Hand seiner Rezept- und Ordinationsbücher', *Berner Zeitschrift für Geschichte und Heimatkunde* 47 (1985), 85–166; Genlis, *Souvenirs*, pp. 194–6.

296. Tissot, *Advice*, vol. 2, pp. 275–98, 'Of mountebanks, quacks and conjurers'.

297. BCU Lausanne, FT, physician's letter, in Latin, concerning the sick Domenican friar Schinz; *ibid.*, letter concerning a nine-year-old girl with 'absences', written in the third person, probably by her parents; Schuppach prescribed a plaster on the head to draw out the morbid matter, but with no success.

298. Brockliss/Jones, *Medical world* [see note 262], pp. 284f and pp. 537–40.

299. Michael Stolberg, 'The decline of uroscopy in early modern learned medicine, 1500–1650', *Early science and medicine* 12 (2007), 313–36; idem, *Die Harnschau. Eine Kultur- und Alltagsgeschichte.* Cologne/Weimar 2009.

300. Alfred Franklin, *La vie privée d'autrefois. Les médecins.* Paris 1892, p. 146.

301. Brockliss/Jones, *Medical world* [see note 262], p. 535; Stolberg, *Heilkunde* [see note 262], pp. 259–76.

302. Chastenay, *Mémoires*, vol. II, p. 35.

303. Cf. Michael Stolberg, 'Krankheitserfahrung und Arzt-Patienten-Beziehung in Samuel Hahnemanns Patientenkorrespondenz', *Medizin, Gesellschaft und Geschichte* 18 (1999), 101–20.

304. BCU Lausanne, FT, letter from Monsieur Gringet, 4 January 1784; the consultation was made possible through the generous help of the Comtesse de Perron; the patient was probably employed by her and impressing the countess may well have been Villermoz's principal aim.

305. This research is currently being undertaken in a joint research venture funded by the Deutsche Forschungsgemeinschaft, with partners in Berlin, Ingolstadt, Stuttgart, Würzburg, Zürich, Bern, Innsbruck, and Bolzano.

306. Chastenay, *Mémoires*, vol. I, p. 43 and p. 294; *ibid.*, vol. II, pp. 17–19.

307. *Ibid.*, vol. I, p. 43 and vol. II, p. 13.

308. *Ibid.*, vol. I, p. 154 and p. 205.

309. BCU Lausanne, FT, history of Monsieur Larrei's disease, written in the third person but clearly by himself, 5 January 1784.

310. *Ibid.*, letter from Monsieur Marcard, Livorno/Florence, 6 December 1785.

311. SB Berlin, Ms.germ.fol. 421b, foll. 118r–119r, Caspar von Hobergk, 9 March 1577.

312. Hoffmann, *Medicina consultatoria*, vol. 10, pp. 279–82, letter from a 44-year-old female patient.

313. BCU Lausanne, FT, letter from the husband, 21 June 1776.

314. *Ibid.*, illness report from 40-year-old Monsieur Torchon de Lihu, 26 April 1785.

315. *Ibid.*, letter from Mme Develay concerning her sick husband, 21 May 1791; as M. Louis-Courvoisier and A. Mauron have rightly underlined, such accounts cast substantial doubts on the common assumption that tensions and conflicts between patients and physicians are primarily the result of modern medical technology with its reliance on 'objective' diagnosis; see Micheline Louis-Courvoisier/A. Mauron, '"He found me very well; for me, I was still feeling sick". The strange worlds of physcans and patients in the 18th and 21st centuries', *Medical humanities* 28 (2002), 9–13.

316. Lindsay Wilson, *Women and medicine in the French enlightenment. The debate over maladies des femmes*. Baltimore/London 1993, pp. 94f.

317. Levis, *Souvenirs*, pp. 237f.

318. *Ibid.*, 238.

319. *Ibid.*, 241; Anne-Charles Lorry (1726–83).

320. Burney, *Journals*, p. 603.

321. Levis, *Souvenirs*, pp. 240f.; Michel Philippe Bouvart (1711–87).

322. Marmontel, *Mémoires*, p. 304.

323. See Jens Lachmund/Gunnar Stollberg, 'The doctor, his audience, and the meaning of illness: The drama of medical practice in the late 18th and early 19th centuries', in: iidem (eds), *The social construction of illness*. Stuttgart 1992, pp. 53–66; I disagree with their claim, however, that the reality of the disease and its 'meaning' were constituted only in the course of this 'drama'; in my view, the kind of antecedent, culturally and biographically formed patterns of perception and interpretation analyzed in this book had a decisive influence, too.

324. N. D. Jewson, 'Medical knowledge and the patronage system in 18th-century England', *Sociology* 8 (1974), 369–85; idem, 'The disappearance

of the sick man from medical cosmology, 1770–1870', *Sociology* 10 (1976), 225–44.

325. For a more detailed analysis see Michael Stolberg, 'La négociation de la thérapie dans la pratique médicale du XVIIIe siècle', in: Olivier Faure (ed.), *Les thérapeutiques: savoirs et usages.* Lyon 1999, pp. 357–68.

326. BCU Lausanne, FT, letter from Mme Viard d'Arnay, 6 July 1778; *ibid.*, letter from Monsieur Gringet, 4 January 1784.

327. KB St Gallen, Ms. 94, foll. 119r–120r, B. B[odmer], 15 March 1616.

328. *Ibid.*, foll. 91r–v, Sister Afra, 4 May 1613.

329. *Ibid.*, fol. 125r, letter from an unnamed abbot, 1 August 1616.

330. UB Basel, Ms. Fr. Gr. I 6, Hans Georg Reinach, May 1570.

331. BCU Lausanne, FT, letter from Monsieur de Condé, 7 October 1773.

332. UB Basel, Ms. Fr. Gr. I 6, Hans Augustin Reich von Reichenstein, 15 May 1570.

333. Thus SB Berlin, Ms.germ.fol. 421b, Caspar von Hobergk, 9 March 1577.

334. SB Berlin, Ms.germ.fol. 420b, foll. 25r–26r, Anthonius Kholl, Königsberg, 14 August 1574.

335. WL London, Western Manuscripts, Ms. 6114, letter from an unnamed patient, Aubonne, 28 July 1801.

336. BL London, Ms. Sloane 4039, foll. 188f., Lord Hatton, letter from the brother, 20 September 1703.

337. BCU Lausanne, FT, letter from Mme de Chastenay, 21 February 1785.

338. SB Berlin, Ms.germ.fol. 420b, foll. 14r–v, A. von Bradowa, Sunday after Dr Bartholomew 1574.

339. BCU Lausanne, FT, letter from the Freifrau von Werthern, 3 February 1792 (in French).

340. WL London, Western Manuscripts, Ms. 6114, letters from an unidentified patient, 12 July and 10 September 1801.

341. BCU Lausanne, FT, undated letter from the sick sister of Mme Butex; in the patient's idiosyncratic spellling: 'l'on ma seigné et cela tres mal a propos et encor une fort grosse saignee qui ma entièrement epuisé et fort deranger toutes les organes de mon corps'; she suffered from indigestions and vapors and she also was unhappy about the 'very bad pills' which a previous physician had given her.

342. Boerhaave, *Correspondence*, p. 196, no date given.

343. BCU Lausanne, FT, letter from Monsieur Bardin [?], ca. 1781.

344. *Ibid.*, letter from Mme d'Arthaud, letter from the daughter, 16 July 1768.

345. WL London, Western Manuscripts, Ms. 6114, letter from M. Raiss, summer 1801.

346. BL London, Ms. Sloane 4036, fol. 155, John Ray, 16 October 1693.

347. BL London, Ms. Sloane 4075, fol. 46, letter from an unidentified patient, Kensington, 1 February 1707.

348. Paul Joseph Barthez (1734–1806), for example, was said to have acquired his fame thanks to the successful cure of the the Comte de Périgord (Levis, *Souvenirs*, p. 240).

349. Ferrières, *Mémoires*, pp. 296–301, report about Mirabeau's death in 1791.

350. Malboissière, *Lettres*, pp. 294f., 25 October 1765.

351. Orléans, *Leben*, pp. 206f.

352. BL London, Ms. Sloane 4038, 23, Monsieur Cheyne [?], 25 June 1700.

353. BL London, Ms. Sloane 4039, foll. 188f., 20 September 1703.
354. BCU Lausanne, FT, letter concerning a 43-year-old clergyman from the Great Saint Bernhard, probably written by a fellow clergyman, 18 October 1789.
355. Hummel, *Histori*, p. 40.
356. BCU Lausanne, FT, letter from Monsieur Decheppe de Morville, one of a series of letters, 1782.
357. RA Arnhem, Archive of the Counts of Coulembourg, Ms. 403, P. D. Steelant, p. 17, September 1614.
358. Rem, *Tagebuch*, p. 23.
359. BCU Lausanne, FT, letter from Monsieur de Walmoden (ca. 1781).
360. *Ibid.*, letter from Mme de Konauw, 26 January 1773.
361. SB Berlin, Ms.germ.fol. 422a, foll. 120r–121r, undated and unsigned letter, which the patient probably gave to Thurneisser.
362. UB Leiden, Mss. Marchand 3, G. von Boetzelaar, 14 December 1591.
363. Cf. Michael Stolberg, 'Therapeutic pluralism and medical conflict in the 18th century. The patient's view', in: Jürgen Helm/Renate Wilson (eds), *Medical theory and therapeutic practice in the eighteenth century. A transatlantic perspective.* Stuttgart 2008, pp. 95–112. For overviews of early modern medical theory see Wear, *Knowledge* [see note 268]; Lawrence I. Conrad et al., *The Western medical tradition 800 BC to AD 1800.* Cambridge 1995; Andrew Wear/Roger French/Ian M. Lonie (eds), *The medical renaissance of the sixteenth century.* Cambridge 1985; Lester S. King, *The road to medical Enlightenment 1650–1695.* London/New York 1970; idem, *The medical world of the eighteenth century.* Chicago 1958; Andrew Cunningham (ed.), *The medical enlightenment of the eighteenth century.* Cambridge 1990; Guenter B. Risse, 'Medicine in the age of Enlightenment', in: Andrew Wear (ed.): *Medicine in society. Historical essays.* Cambridge 1992, pp. 149–95; Peter Elmer (ed.), *The healing arts. Health, disease and society in Europe, 1500–1800.* Manchester 2004.
364. SB Berlin, Ms.germ.fol. 420a, foll. 293r–294r, G. Stange, undated. 'Quintessence' refers to the product from isolation – usually by distillation – of the specifically efficacious parts, for example of a medicinal plant.
365. Hoffmann, *Medicina consultatoria*, vol. 4, pp. 12–14, unnamed 34-year-old patient, 25 February 1723.
366. Mary Fissell ('The disappearance of the patient's narrative and the invention of hospital medicine', in: Roger French/Andrew Wear (eds), *British medicine in an age of reform.* London/New York 1991, pp. 92–109) comes to the conclusion that the 'patient narrative' lost much of its former importance in medical practice around 1800; her claim is based principally on her analysis of the very specific situation of the predominantly poor and underprivileged inmates in hospitals in contemporary Bristol, however. Her findings about the physician–patient relationship in this particular setting cannot easily be applied to the relationship between physician and the more affluent and educated patients in general practice.
367. Graffigny, *Correspondance*, vol. IV, p. 214, 18 March 1743; for a detailed analysis of Mme de Graffigny's relationship with her physicians see Judith Oxfort, *'Meine Nerven tanzen'. Die Krankheiten der Madame de Graffigny (1695–1758).* Cologne 2010.

368. Stolberg, *Harnschau* [see note 117].
369. StA Köln A 459, foll. 8r–9r, 11 February 1750.
370. StA Köln 459, foll. 21r–22r, 4 April 1791; cf. AS Bologna, Studio 339–51; based on about half a dozen well documented cases of this kind, Gianna Pomata (Pomata, *La promessa* [see note 287]) has underlined the importance of this notion of a healing contract; see also Jütte, *Ärzte* [see note 28], p. 140, p. 147 and pp. 221f.; it is very difficult to assess, however, how widespread this practice actually was and how often learned physicians – and not, as in the Bologna cases, unauthorized healers – were confronted with this notion.

II Perceptions and Interpretations

1. Roy Porter (ed.), *The popularization of medicine 1650–1850*. London/New York 1992; Corinne Verry-Jolivet, 'Les livres de médecine des pauvres aux XVIIᵉ et XVIIIᵉ siècles. Les débuts de la vulgarisation médicale', in: *Maladies médecines et sociétés. Approches historiques pour le présent*, vol. 1. Paris 1993, pp. 51–66; Roselyne Rey, 'La vulgarisation médicale au XVIIIe siècle: le cas des dictionnaires portatifs de santé', *Revue d'histoire des sciences* 44 (1991), 413–33; Mary Lindemann, '"Aufklärung" and the health of the people. "Volksschriften" and medical advice in Braunschweig-Wolfenbüttel, 1756–1803', in: Rudolf Vierhaus (ed.), *Kultur und Gesellschaft in Nordwestdeutschland zur Zeit der Aufklärung*. Tübingen 1992, pp. 101–20; Mireille Laget, 'Les livrets de santé pour les pauvres aux XVIIᵉ et XVIIIᵉ siècles', *Histoire, économie, société* 3 (1984), 567–82; see also Roger Cooter/Stephen Pumfrey, 'Separate spheres and public places. Reflections on the history of science popularization and science in popular culture', *History of science* 32 (1994), 237–67; in what follows I will use the term 'popular' to refer to medical writing which was directed primarily at lay readers rather than professional healers; this does not necessarily imply that these texts aimed at the wider population, let alone succeeded in reaching it.
2. For a good survey of German language writing in this area see Holger Böning, 'Medizinische Volksaufklärung und Öffentlichkeit. Ein Beitrag zur Popularisierung aufklärerischen Gedankengutes und zur Entstehung einer Öffentlichkeit über Gesundheitsfragen. Mit einer Bibliographie medizinischer Volksschriften', *Internationales Archiv für Sozialgeschichte der deutschen Literatur* 15 (1990), 1–92.
3. I have pleaded elsewhere in more detail for an approach to the history of medical popularization which systematically takes the lay audience's reading practices and responses into account (Michael Stolberg, 'Medical popularization and the patient in the 18th century', in: Willem de Blécourt/Cornelie Usborne (eds), *Cultural approaches to the history of medicine. Mediating medicine in early modern and modern Europe*. Basingstoke/New York 2004, pp. 89–107). Mary Fissell has presented some useful remarks on the potential impact of readers' responses on the way popularizing authors conceived their works but has not tested her assumptions empirically; see Mary E. Fissell, 'Readers, texts, and contexts', in: Porter, *Popularization* [see note 1], pp. 72–96.

4. BCU Lausanne, FT, letter from Monsieur de Pollet, 20 April 1772.
5. *Ibid.*, letter from Monsieur Le Chartier, 19 January 1776.
6. Tissot, *De la santé.*
7. BCU Lausanne, FT, letter from Monsieur d'Eyrand, 6 August 1776.
8. Maraise, *Correspondance*, p. 119; the same is true for Mme de Graffigny (Graffigny, *Correspondance*).
9. BCU Lausanne, FT, letter from the Comtesse de Vougy, 3 April 1785; with similar arguments the impoverished Mme de Boubers justified her request for a free copy of Tissot's 'Avis' (*ibid.*, letter from Mme de Boubers, 30 December 1792).
10. Tissot, *Avis*, copy in UB München, shelfmark Med. 621, p. 132 and p. 136.
11. Tissot, *Anleitung*, 1772 edn, copy in SB München, München, shelfmark Path. 1258, pp. 159–68.
12. Especially in the libraries in Munich, Paris, and London.
13. Porter, 'Introduction', in: idem, *Popularization* [see note 1], p. 9.
14. Hoffmann, *Medicina consultatoria*, vol. 5, pp. 326–33, letter from an unidentified vice-rector, no date given, quoting Georg W. Wedel's *Tractatus de medicamentorum facultatibus cognoscendis et applicandis* (Jena 1678); some further examples in Stolberg, 'Medical popularization' [see note 3].
15. Cf. Robert A. Aronowitz, *Making sense of illness. Science, society, and disease.* Cambridge 1998.
16. See the ground-breaking work by Ruth Benedict, *Patterns of culture.* New York 1946; cf. Erwin H. Ackerknecht, 'Primitive medicine and culture pattern', *Bulletin of the history of medicine* 12 (1942), 545–74; Arthur Kleinman, 'Medicine's symbolic reality. On a central problem in the philosophy of medicine', *Inquiry* 16 (1973), 206–13.
17. Byron J. Good, 'The heart of what's the matter. The semantics of illness in Iran', *Culture, medicine and psychiatry* 1 (1977), 25–58.
18. I owe this image to Hans-Eckhard Langer/Helmut Bormann, 'Krankheits-Bild als Krankheits-(Be-)Deutung. Versuch einer Meta-Theorie der Laientheorien bei rheumatischen Erkrankungen', in: Heinz-Dieter Basler et al. (eds), *Psychologie in der Rheumatologie.* Berlin 1992, pp. 55–82.
19. Thomas Lux, 'Semantische Netzwerke bei Laien und Spezialisten. Eine Studie zum Hypertonus in den USA', in: idem (ed.), *Krankheit als semantisches Netzwerk. Ein Modell zur Analyse der Kulturabhängigkeit von Krankheit.* Berlin 1999, pp. 73–98, cit. p. 80.
20. Arthur Kleinman, *Patients and healers in the context of culture. An exploration of the borderland between anthropology, medicine, and psychiatry.* London 1980, pp. 109f., referring, in particular, to Evans Pritchard's work.
21. Langer/Bormann, *Krankheits-Bild* [see note 18].
22. See Kathrin Zedlitz, *Pica. Die Geschichte einer vergessenen Essstörung.* Cologne 2010.
23. Klaus Brunnert, *Nostalgie in der Geschichte der Medizin.* Düsseldorf 1984.
24. Mary Lindemann, *Medicine and society in early modern Europe.* Cambridge 1999, p. 17.
25. See E. Schöner, *Das Viererschema in der antiken Humoralpathologie.* Wiesbaden 1964.
26. By the 18th century, learned physicians generally doubted that black bile ranked among the four natural humors.

27. SB Berlin, Ms.germ.fol. 422a, foll. 101r–102r, Ludwig Albricht, Thursday after Jubilate 1578.
28. For example SB Berlin, Ms.germ.fol. 423a , foll. 245r-248v, Remich Rechwerm, 13 February 1580.
29. SB Berlin, Ms.germ.fol. 424, foll. 16r–v, Jorgen Eckhard, 22 March 1582; *ibid.* foll. 187r–188r.
30. Hoffmann, *Medicina consultatoria*, vol. 10, pp. 279–82, letter from an unidentified 44-year-old patient, ca. 1732.
31. SB Berlin, Ms.germ.fol. 421a, fol. 62r, spring 1576.
32. SB Berlin, Ms.germ.fol. 420b, foll. 479r–v, Matthäus Cuno [?], 7 February 1575.
33. Johann Christoff Amberg, Stadtschreiber in Feldkirch, for example wrote about his 'person or complexion' ('*Persohn oder Complexion*') as if the two terms were synonymous (KB St Gallen, Ms. 94, foll. 241r–243r, 29 October 1624).
34. UB Erlangen, Ms. 948, 342r–344v, letter from an unidentified patient, who suffered from pulmonary disease and impotence, apparently addressed to Johann Georg Fabricius (1593–1668), no date given.
35. BCU Lausanne, FT, letter from Monsieur Goyen, 30, 10 December 1772; BIM Paris, Ms. 5245, foll. 61r–62r, letter concerning an unidentified 38-year-old woman, no date given.
36. BIM Paris, Ms. 5242, foll. 215r–216r, letter from a sick nun to her sister, 10 September 1724.
37. BCU Lausanne, FT, letter from Mme de Villeterque, no date given.
38. *Ibid.*, letter from Monsieur Decheppe de Morville, 20 July 1783; *ibid.*, letter from Mme de Chastenay, fall 1784.
39. *Ibid.*, letter from the Comtesse de Wedel, 21 November 1784.
40. *Ibid.*, letter from Jacques Le Meilleur, medical student, 26 March 1770.
41. *Ibid.*, letter from the surgeon Rouvière, 26 May 1783.
42. Andreä, *Johann Valentin Andreä*, p. 134.
43. BCU Lausanne, FT, letter from Ouvrard de Liniere, 12 August 1772; *ibid.*, history of Monsieur Larrei's disease, written in the third person but clearly by himself, 5 January 1784.
44. Thus, for example, in several of her letters, Mme de Sévigné (Sévigné, *Lettres*). On the long history of humors and passions see Noga Arikha, *Passions and tempers. A history of the humours*. New York 2007.
45. One patient wrote explicitly that he was of a 'lively' and indeed restless 'humor', like all people of biliary temperament. He had a medical training, however (BCU Lausanne, FT, letter from Monsieur Lantrac, 10 September 1789).
46. BIM Paris, Ms. 2075, foll. 25–8, letter from an unidentified 45-year-old sick man, no date given (Helvetius answered on 24 January 1728).
47. FTTZ Regensburg, Haus- und Familiensachen, 3786, letter, in French, to the hereditary prince, 18 January 1803.
48. BCU Lausanne, FT, letter from the Baronesse von Vrentz [?], 8 April 1771.
49. *Ibid.*, letter from the brother of a 19-year female patient, 10 February 1772.
50. *Ibid.*, letter from Monsieur de Diesbach, no date given.
51. UB Leiden, Mss. Marchand 3, Jacob Baldwein von Zweybruch, 30 June 1593 (in Latin).
52. Harvey, *Exercitatio*.

53. Tissot, *Anleitung*, p. 578.
54. Many authors referred to John Freind's detailed analysis of the effects of plethora, especially in amenorrhoic women, which at the same time served as evidence for Freind's mechanistic-hydraulic understanding of human physiology in general (Freind, *Emmenologia*. London 1729, pp. 77–110; cf. Jacqueline Albrecht-Chanton, *Die Menstruationslehre von John Freind (1675–1728)*. Diss. med. Bern 1997).
55. BCU Lausanne, FT, letter from Mme Mieg, Mulhouse 9 May 1790.
56. Thus KB St Gallen, Ms. 94, fol. 115r, letter from the Abbot Augustin von Einsiedeln, 4 March 1616.
57. BCU Lausanne, FT, letter from the Chevalier de Virieu, 28 July 1772.
58. For example Hoffmann, *Medicina consultatoria*, vol. 5, pp. 93–8, letter from a 31-year-old patient, written first in the third, then in the first person, no date given.
59. Michael Stolberg, 'Erfahrungen und Deutungen der weiblichen Monatsblutung in der Frühen Neuzeit, in: Barbara Mahlmann-Bauer (ed.), *Scientiae et artes. Vermittlung alten und neuen Wissens in Literatur, Kunst und Musik*. Wolfenbüttel 2004, pp. 913–31; see also Patricia Crawford, 'Attitudes to menstruation in seventeenth-century England', *Past & present* 91 (1981), 47–73.
60. Roussel, *Système*, pp. 196–202.
61. In early modern medical 'observationes' we find numerous cases of vicarious menstrual bleeding in women and regular bleeding, perceived as analogous to menstruation, in men; cf. Schurig, *Parthenologia*, pp. 83–127; Gianna Pomata, 'Uomini mestruanti. Somiglianza e differenza fra i sessi in Europa in età moderna', *Quaderni storici* 79 (1992), 51–103; Michael Stolberg, 'Menstruation and sexual difference in early modern medicine', in: Gillian Howie/Andrew Shail (eds): *The history of menstruation*. Basingstoke 2005, pp. 90–101.
62. BIM Paris, Ms. 5242, foll. 134r–135r, Monsieur Cerrier, no date given.
63. Khevenhüller, *Tagebuch*, p. 3.
64. WL London, Western Manuscripts, Ms. 6114, letter with illegible signature, 5 February 1801.
65. KB St Gallen, Ms. 94, foll. 241r–243r, letter from Johann Christoff Amberg, 29 October 1624.
66. Clacius, *Consilia*, pp. 11–14.
67. BCU Lausanne, FT, letter from the father, Monsieur Cart, 8 May 1785.
68. Cf. for example Flamant, *Kunst*, pp. 63–7.
69. Tissot, *Anleitung*, p. 168; without any attempt to reconcile the two explanations, Tissot shortly afterward also claimed, however, that the pressure exerted by the excessive amounts of blood which rushed in, slowed the 'movement of the nerves' (*ibid.*, p. 172).
70. *Ibid.*, pp. 173–5 he also distinguished a second type of apoplexy in patients with 'watery' and 'slimy' blood; presumably he was looking for a way to explain cases of apoplexy in patients who showed no sign of of plethora.
71. *Ibid.*, p. 168; see also Ettmüller, *Opera*, vol. II, pp. 296–312.
72. Brandis, *Diarium*, p. 500; Spener also mentioned speaking difficulties in his account of the fatal apoplexy of his godmother (Spener, *Lebens-Lauf*, p. 857); paralysis was, in part, associated with a loss of vital heat; sensations

of cold were therefore an important warning sign (BIM Paris, Ms. 5245, foll. 106r–v, account of Mme de Chambery's disease, written in the third person but probably by the patient herself, ca. 1730).

73. SB Berlin, Ms.germ.fol. 420a, foll. 88r–v, C. Morgriesser, 31 May 1571.
74. Hoffmann, *Medicina consultatoria*, vol. 5, pp. 45–7, letter, signed 'G. S. V.', 23 April 1729, addressed to an unknown physician, asking him to pass the letter on to Hoffmann.
75. *Second recueil*, pp. 21–4, Marie-Anna Courronneau.
76. BCU Lausanne, FT, letter from Mme de Brackel, 24 February 1793.
77. Graffigny, *Correspondance*, vol. III, p. 37, December 1740; see also *ibid.*, p. 66, letter to the physician John Clephane; the patient was treated with bloodletting and emetics.
78. Mettrie, *Traité*, pp. 8–10; on La Mettrie see Kathleen Wellman, *La Mettrie. Medicine, philosophy, and Enlightenment.* Durham/London 1992.
79. Hoffmann, *Medicina consultatoria*, vol. 5, pp. 290–2, letter from the patient.
80. BIM Paris, Ms. 5241, foll. 22r–23r, letter concerning an unnamed person, written in the third person, with an accompanying letter from the physician in charge, 18 May 1730.
81. BCU Lausanne, FT, letter from Monsieur Faugeroux, 12 June 1787.
82. Among laypeople the notion of a disturbed motion of the 'animal spirits' does not seem to have been current; cf. for example Ettmüller, *Opera*, vol. II, pp. 296–312.
83. KB St Gallen, Ms. 94, foll. 241r–243r, 29 October 1624.
84. BCU Lausanne, FT, letter from Monsieur Cheroz, Troyes, 16 January 1785.
85. BIM Paris, Ms. 5241, foll. 113r–v, letter, no date given, from an unidentified patient, probably written or at least dictated by himself. He suffered from, among other things, feelings of numbness, weakness in his legs, trembling, convulsive movements of his head, and an insurmountable melancholia; somewhat surprisingly he also suggested that he might be suffering from 'dropsy'.
86. Today, when a patient, in certain diseases, produces excessive amounts of red blood cells, bloodletting is occasionally still the preferred treatment.
87. BCU Lausanne, FT, letter from the Comtesse de Beauharnois, 20 May 1790.
88. KB St Gallen, Ms. 94, foll. 241r–243r, letter from Joh. Chr. Amberg, 29 October 1624; BCU Lausanne, FT, letter concerning a lady from Auxerre, written in the third person, 6 April 1790.
89. Staiger, *Tagebuch*, pp. 45f. (1623); a 25-year-old patient of Haller's similarly reported that he was subjected to bloodletting about a dozen times but that this was still not enough to stop his bloody expectoration; he also described his blood as increasingly pathological, however (BB Bern, HC, A 143, letter from an unidentified patient, no date given).
90. *Second recueil*, pp. 17–20.
91. For example BCU Lausanne, FT, letter, probably from Mme de Virazel, starting in the third person and then switching to the first, 12 September 1790.
92. SB Berlin, Ms.germ.fol. 420a, foll. 87r–v, G. Traupitz, 7 October 1571.

93. *Ibid.*, foll. 73r–v, G. Traupitz, Wednesday after Assumption Day, 1571; see also *ibid.*, foll. 175r–176r, Niclas von der Linde, no date given.

94. The most important alternative explanation was a 'tooth worm' or, much more rarely, caries.

95. BCU Lausanne, FT, letter from Monsieur Budé de Boisy, 23 April 1793.

96. *Ibid.*, letter from Mme du Bochage la Vallette, 11 January 1771; a dentist found only caries, however, and advised her to have the tooth extracted.

97. *Ibid.*, letter from the Chevalier d'Albarey, about the Comtesse de Mouroux, 14 August 1793.

98. SB Berlin, Ms.germ.fol. 420a, foll. 190r–v, 21 February 1572.

99. SB Berlin, Ms.germ.fol. 421a, fol. 62r, I. Keller, 24 April 1576.

100. KB St Gallen, Ms. 94, foll. 241r–243r, letter from Johann Christoff Amberg, 29 October 1624.

101. SB Berlin, Ms.germ.fol. 420a, foll. 177r–v, Petrus Groß, 29 March 1571.

102. A term which nicely illustrates this semantic connection is *'gutta serena'* (in French *'goutte sereine'*) which, at the time, was used for a massive deterioration of patients' eyesight without any visible changes in the eye; one of Tissot's patients took the term from a contemporary medical textbook (BCU Lausanne, FT, letter from Monsieur Bruckner, 29 November 1789).

103. Pauli, *Volksheilmittel*, p. 1.

104. Jurin, *Correspondence*, pp. 396f. (1733).

105. SB Berlin, Ms.germ.fol. 420b, fol. 284r, 25 August 1574.

106. SB Berlin, Ms.germ.fol. 420b, foll. 186r–v, 18 May 1574.

107. BCU Lausanne, FT, letter from Monsieur de Schilden, Lausanne, 5 May 1790.

108. *Ibid.*, letter concerning Monsieur Gilbert, written in the third person, no date given.

109. Hummel, *Histori*, pp. 47f.

110. *Ibid.*, letter from B. Duclaux, writing about a sick 'philosopher'; Duclaux was skeptical about the physician's diagnosis of 'rheumatism', because the knee was painful but not red and swollen.

111. *Ibid.*, letter from the daughter, 9 August 1784; today so-called 'tophi', hard deposits of uric acid beneath the skin, are considered as typical signs of gout; in rheumatoid arthritis and arthrosis the bones themselves and the surrounding connective tissue can exhibit nodular growth.

112. *Ibid.*, letter from the Baron de Beaucourse, 25 September 1783.

113. *Ibid.*, letter from 'M. R.', Ludwigslust, 19 February 1790.

114. BIM Paris, Ms. 5245, foll. 148r–149v, Monsieur Herbolin, 23 April 1730, with an undated account of the disease.

115. BCU Lausanne, FT, letter from Monsieur de Schilden, Lausanne, 5 May 1790.

116. *Ibid.*, letter from Mme de Wisne, concerning Monsieur de Croix, no date given.

117. SLUB Dresden C 457, fol. 8v.

118. BCU Lausanne, FT, letter from Monsieur Goyen, 10 December 1772.

119. Pauli, *Volksheilmittel*, pp. 1–5; Langer/Bormann, *Krankheits-Bild* [see note 18].

120. BCU Lausanne, FT, letter from the brother and the patient himself, 26 April 1785; cf. Boscius, *Podagra*, pp. 7f.

121. Blancardus, *Abhandlung*, p. 14; distillation was taken to show that the gouty knots contained 'volatile salt'.
122. BCU Lausanne, FT, letter from Lieutenant Roussany, Paris, 10 June 1774.
123. *Ibid.*, letter from an unidentified patient, apparently a member of the army, born in 1736, ca. 1775.
124. *Ibid.*, letter from the Baron de Beaucourse, 25 September 1783.
125. One of the most popular purgatives in the 18th century was Ailhaud's powder, which is also sometimes mentioned in the patient letters; cf. *L'ami des malades ou discours historique & apologétique sur la poudre purgative de M Ailhaud*. Carpentras 1770.
126. Pauli, *Volksheilmittel*, pp. 3f.
127. BIM Paris, Ms. 5245, foll. 164r–165v, concerning an unidentified patient, no date given.
128. BIM Paris, Ms. 5241, foll. 93r–94v, account of the disease of a 37-year-old nun, no date given (response dated 22 November 1722); by that time she had been suffering for 13 months from dyspnea and coughing and was 'without a sound'.
129. BCU Lausanne, FT, letter from Monsieur Cart, 8 May 1785.
130. *Ibid.*, letter from the Baron de Beaucourse, 25 September 1783.
131. *Ibid.*
132. *Ibid.*, letter from Monsieur de Schilden, Lausanne, 5 May 1790; *ibid.*, letter from J. B. Grimaldi, Genova, 12 May 1793, asking for advice about the best source.
133. BIM Paris, Ms. 5241, foll. 166r–167v, letter from Geoffroy to his colleague Dubois in Rennes, 3 May 1729.
134. Huldrych M. Koelbing, *Die ärztliche Therapie. Grundzüge ihrer Geschichte*. Darmstadt 1985, p. 103.
135. See for example Fuchs, *De curandis morbis*, p. 220.
136. Anhart, *Consilium podagricum*, p. 19.
137. BCU Lausanne, FT, letter from Mme de Wisne, writing about Monsieur de Croix, no date given.
138. Tissot, *Anleitung*, pp. 418–24 'Von den Gichtern'; see also *ibid.* 553–5; when applied to infants and young children, the term 'Gichter' had basically the same meaning as 'Fraisen' ('convulsions') and in the 19th century both terms were still used interchangeably even in official statistics on causes of death (HStAM, MInn 61717, annual reports of the district physicians, 1833–38). Especially in French medicine, very similar symptoms were associated with the 'vapeurs' and some physicians began to interpret these 'gouts' as a kind of nervous disease. Around 1813, Christoph Wilhelm Hufeland even claimed that, due to the process of civilization, diseases in general had become 'more refined and decorporalized'. The 'material forms of gout, podagra, etc.' could no longer be found. The morbid humor, Hufeland believed, had changed 'into a volatile matter which attacks the nerves'; it 'has turned into nervous gout'. (Hufeland, *Geschichte der Gesundheit*, p. 39); Venel similarly found a shift away from local gout in feet and hands towards 'nervous gout' (Venel, *Observations*, p. 2).
139. StAM, Hist. Ver. Ms 401/37Traunstein; SB München, Cgm 6874/190 Wassertrüdingen; *ibid.*, Cgm 6874/180 Vilsbiburg; *ibid.*, Cgm 6874/173 Sulzbach; *ibid.*, Cgm 6874/148 Regenstauf.

140. SB München, Cgm 6874/44 Erlangen Land.
141. Otto Münsterer, 'Grundlagen, Gültigkeit und Grenzen der volksmedizinischen Heilverfahren', *Bayerisches Jahrbuch für Volkskunde* 1950, pp. 9–20, here p. 11; SB München, Cgm 6874/118 Neumarkt/Opf.
142. SB München, Cgm 6874/174, about the 'wise men' from Liebenstein and Waltershof.
143. SB München, Cgm 6874/154 Roth.
144. *Ibid.*, Cgm 6874/115 Nabburg; *ibid.*, Cgm 6874/183 Vohenstrauß; *ibid.*, Cgm 6874/118 Neumarkt/Opf.
145. In French: 'acretés', 'humeurs acres'; in English 'acrimonies'; in Italian 'acrimonie'.
146. Gustav Jungbauer, *Deutsche Volksmedizin. Ein Grundriß*. Berlin/Leipzig 1934, p. 20; Carly Seyfarth, *Aberglaube und Zauberei in der Volksmedizin Sachsens. Ein Beitrag zur Volkskunde des Königreichs Sachsen*. Leipzig 1913, p. 66; Pauli, *Volksheilmittel*, p. 26; SB München, Cgm 6874/142 Riedenburg.
147. See in particular de la Boë, *Idea*, book 1, pp. 338.
148. 'Peau, maladies de', in: *Encyclopédie*, vol. 12, 1765, pp. 217–19.
149. BCU Lausanne, FT, letter from a female patient in her 20s, no date given.
150. BIM Paris, Ms. 5241, foll. 146r–147v, letter from Monsieur de Cullembourg, by all appearances a layperson himself, concerning a 51-year-old Ursuline nun, spring 1714.
151. SB Berlin, Ms.germ.fol. 420a, foll. 152r–v, Hans von Löben, 1571.
152. BCU Lausanne, FT, letter from Monsieur Bouju, 5 January 1774.
153. BIM Paris, Ms. 5241, foll. 236r–239v, letter from Monsieur Gouët's daughter to Mme Geoffroy, 26 May 1728.
154. BCU Lausanne, FT, letter from the Baron de Beaucourse, 25 September 1783.
155. BIM Paris, Ms. 5241, foll. 151r–v, physician's advice to Mme Fleury.
156. SB München, Cgm 6141, notice book of a physician in Coburg, Germany, 1739, fol. 217.
157. BCU Lausanne, FT, letter from the Chevalier d'Albarey, writing about the Comtesse de Mouroux, Turin, 13 February 1790; later she suffered from a white vaginal discharge; her physician diagnosed a strong acrimony of her humors with an active ferment, due to spicy, salty and hot food (*ibid.*, accompanying note).
158. BCU Lausanne, FT, letter from Mme Rostaing, Fribourg, 6 October 1793.
159. BB Bern, HC A 94, letter from a 44-year-old patient, ca. 1760.
160. Hoffmann, *Medicina consultatoria*, vol. 1, pp. 237–40, letter from a 63-year-old patient.
161. BCU Lausanne, FT, letter from the Chevalier de Peyrelongue, Strasbourg, 7 September 1785.
162. *Ibid.*, letter from Mme au Bavement, 17 December 1783; *ibid.*, account of an unidentified patient's disease, sent to Tissot by Monsieur Lindner and possibly referring to his own sickness, 12 April 1774.
163. BCU Lausanne, FT, letter concerning a 43-year-old clergyman from the Great Saint Bernhard, probably written by a fellow clergyman, 18 October 1789.
164. SB Berlin, Ms.germ.fol. 420b, foll. 460r–461r, M. Kühne, 12 January 1575.
165. BCU Lausanne, FT, letter from Mme Herrmann writing about an unidentified female patient, spring 1790.

166. *Ibid.*, letter from Conseiller Gualtieri, ca. 1773.

167. BIM Paris, Ms. 5241, foll. 111r–112v, physician's advice to a more than 80-year-old patient with massive trembling.

168. Hunauld, *Vapeurs*, pp. 88f.; similarly Pauli, *Volksheilmittel*, p. 26, about contemporary folk beliefs.

169. BIM Paris, Ms. 5242, foll. 134r–135v, Monsieur Cerrier, concerning his newly wed daughter, no date given.

170. BIM Paris, Ms. 5241, foll. 194r–196v, letter concerning Monsieur Collart, written, probably by a layperson, in the third person, no date given.

171. Andreas Johannes Jäckel, 'Aphorismen über Volkssitte, Aberglauben und Volksmedizin in Franken, mit besonderer Rücksicht auf Oberfranken', *Abhandlungen der naturhistorischen Gesellschaft zu Nürnberg* 2 (1861), 148–258, here p. 151.

172. SB Berlin, Ms.germ.fol. 422a, foll. 101r–102v, L. Albricht, Thursday after Jubilate 1578; a physician's diagnosis of a 'sharp lymph' is also mentioned in an undated report, probably written by a physician, about a 30-year-old female patient with migraines (BIM Paris, Ms. 5245, foll. 155r–156r).

173. BCU Lausanne, FT, letter concerning a female patient in her 20s, no date given.

174. SB Berlin, Ms.germ.fol. 424, foll. 365r–367v, 14 May 1582.

175. BCU Lausanne, FT, letter from Monsieur Magelli, referring to the pimples his 52-year-old sister had suffered from for years, Fribourg 28 May 1793.

176. *Ibid.*, 25 March 1775; similarly *ibid.*, letter concerning Mlle Dupré, who was suffering from a nervous disease and a rash on her forehead, 11 May 1793.

177. *Ibid.*, letter from Mme de Constable, Yverdon, 15 March 1792.

178. *Ibid.*, letter from Dr Risler, writing about a 55-year-old female patient, 17 March 1790.

179. *Ibid.*, letter from Mme Battel, concerning her sick husband, 24 December 1773.

180. Cf. Oliver König, 'Haut', in: Christoph Wulf (ed.), *Handbuch historische Anthropologie*. Weinheim 1997, pp. 436–45.

181. See the ground-breaking analysis by Terence S. Turner, *The social skin*. Berkeley 1980.

182. Claudia Benthien has published a nuanced study (*Im Leibe wohnen. Literarische Imagologie und historische Anthropologie der Haut*. Berlin 1998) on the historical anthropology of the skin, based on an analysis of literary works, popular and scientific accounts, and visual sources; see below for critical remarks on some of her results.

183. Cf. *ibid.*, pp. 88–100; for the role of the Sisamnes story in late medieval and early modern culture see Valentin Groebner, *Gefährliche Geschenke. Ritual, Politik und die Sprache der Korruption in der Eidgenossenschaft im späten Mittelalter und am Beginn der Neuzeit*. Konstanz 2000, pp. 139–41, with further references.

184. The famous 'Encyclopédie' gave macabre expression to this notion in a story about shoes that a Parisian surgeon had made out of human skin (see 'Peau humaine passée', in: *Encyclopédie*, vol. 12, 1765, p. 220).

185. See Edmund Wießner (ed.), *Heinrich Wittenwilers Ring*. Leipzig 1931, coll. 26d–27a, for the advice to use 'sweating baths' in order to get rid of superfluities 'between the flesh and also the skin' ('zwüschen Flaisch und auch

der Haut'); Benthien, *Haut* [see note 182] does not even mention this important concept.

186. The notion has survived in modern French phrases like 'entre cuir et chair', meaning 'beneath the skin' or 'under the skin'. Native French speakers no longer seem to be aware, however, of the original meaning of this phrase, which once referred literally to a space between the flesh and the skin.

187. SB Berlin, Ms.germ.fol. 421a, foll. 313r–314v, Wulf von Closter (ca. 1576); see also Krafft, *Reisen*, pp. 222f. ('zwischen Hautt vnd Flaisch').

188. BCU Lausanne, FT, letter from Mme Monget, 2 November 1776.

189. *Ibid.*, letter from Mme Arthaud, 1 September 1768; *ibid.*, letter from Mme Chenou, 26 June 1793.

190. SB Berlin, Ms.germ.fol. 420b, foll. 118r–119r, Caspar von Hobergk, 9 March 1577.

191. BCU Lausanne, FT, account of Monsieur de Lihu's disease, no date given, accompanied by a letter from his brother, 26 April 1785.

192. *Ibid.*, letter from Mme Reverdil, 17 July 1791, concerning an unidentified 30- to 40-year-old acquaintance.

193. Flügel, *Volksmedizin*, p. 52; Brenner-Schaeffer, *Darstellung*, pp. 25f. and p. 58.

194. Benthien, *Haut* [see note 82], p. 84.

195. Guyon, *Vie*, p. 121; when Guyon was sick with smallpox, she underwent bloodletting and the desired rash appeared immediately; Maraise, *Correspondance*, pp. 118–22, 14 July 1783.

196. 'Peau, maladies de', in: *Encyclopédie*, vol. 12 (1765), pp. 217–19.

197. BCU Lausanne, FT, letter from Mme de Chastenay, 8 November 1784; *ibid.*, letter from Françoise de Valerien, 4 October 1783.

198. *Ibid.*, letter from Marianne Nomis de Pollon, 2 June 1776; the child continued to suffer from episodes of pain and fever, however, and complained about a viscous, slimy expectoration which sometimes occasioned a feeling of suffocation.

199. For example BIM Paris, Ms. 2075, foll. 25–8, copy of a letter from an unidentified patient, no date given (Helvetius replied on 24 January 1728).

200. Guyon, *Vie*, p. 122.

201. In the patient letters there is occasionally talk of a 'lichenous humor' ('humeur dartreuse').

202. BCU Lausanne, FT, letter from Mme au Bavement, 17 December 1783.

203. *Ibid.*, letter from Mme Monget, 2 November 1776.

204. *Relation de la maladie*, p. 4.

205. Marmontel, *Mémoires*, p. 255.

206. *Recueil des pièces*, pp. 9–11.

207. BCU Lausanne, FT, letter from Monsieur Muroz de la Borde, 15 December 1772; skin-changes massively disfigured his face.

208. Zwinger, *Theatrum*, pp. 284–7; Musitano, *Opera*, pp. 25–9.

209. Retif, *Monsieur Nicolas*, vol. II, pp. 147f.

210. *Recueil des miracles*, pp. 7f., testimony of the patient.

211. BCU Lausanne, FT, letter from an unidentified female patient, by then in her mid-30s, no date given; in her later years, she suffered from a 'crust' (in French: 'galle') in her hair, which she attributed to the evacuation of a morbid humor.

212. BIM Paris, Ms. 2075, foll. 25–8, letter from a 45-year-old female patient, ca. January 1728.
213. BCU Lausanne, FT, letter from Mme de Menou, 26 June 1793.
214. See the forthcoming dissertation by Maximilian Mayer, *Der Skorbut in der Medizin des 16. und 17. Jahrhunderts.* Würzburg 2011.
215. Helmont, *Aufgang*, p. 560.
216. For a good historical overview see Lind, *Scurvy*; cf. also Wierus, *Medicarum observationum*, pp. 7–35; Sennert, *Opera*, vol. 2, pp. 504–35; Alberti, *Introductio*, pp. 343–5.
217. Zedler, *Universal-Lexikon*, vol. 34, p. 880, drawing on G. E. Stahl.
218. StA Bremen 2 S 7 a.12, testimony, 1 October 1647.
219. James Lind [see note 216] is hailed as the first physician to recognize the beneficial effects of citrus fruits in the prevention of scurvy.
220. Hoffmann, *Medicina consultatoria*, vol. 4, pp. 172–7, letter from Baron C. G. v. L., 26 September 1723; Hoffmann confirmed that the patient suffered the effects of a 'penetrating scurvious acrimony of the serum', which resulted from an obstructed perspiration and a disturbed humoral flow through the liver and had deposited itself on the membranous and sinewy parts (*ibid.* pp. 177–85, letter of reply, 20 October 1723).
221. Hoffmann, *Medicina consultatoria*, vol. 1, pp. 277–91, I. E. F., 5 October 1720.
222. Sennert, *Opera*, vol. 2, pp. 504–35; according to van Helmont, apart from the gums, the respiration and the thighs were primarily affected.
223. SB Berlin, Ms.germ.fol. 421a, foll. 148r–v, Peckatel, Warsaw, no date given.
224. BCU Lausanne, FT, physician's advice to Mme de Launay, 1790.
225. Zedler, *Universal-Lexikon*, vol. 34 (1742), p. 881.
226. *Recueil des pièces*, pp. 9–21.
227. BCU Lausanne, FT, letter from an unidentified 28-year-old patient, no date given.
228. *Ibid.*, letter from the Gräfin von Wedel, 28 January [1790]; similarly BB Bern, HC, A 94, letter from an unidentified 44-year-old female patient, ca. 1760.
229. BL London, Ms. Sloane 4075, foll. 73r–74r, letter from S. Downing, 19 July 1726.
230. *Ibid.*
231. SB Berlin, Ms.germ.fol. 420b, foll. 152r–153r, letter from L. von Closter, 25 October 1574.
232. BCU Lausanne, FT, letter from the Comtesse de Champagne, 11 April 1773.
233. *Ibid.*, letter from Mme de Disse, 13 June 1774.
234. *Ibid.*, letter from the Comtesse de Vougy, 3 April 1785.
235. SB Berlin, Ms.germ.fol. 421b, foll. 299r–300v, letter from H. Birckholz, 9 March 1577.
236. Wierus, *Medicarum observationum* [see note 221], pp. 7–35.
237. Though the concept is closely related to the imagery of the werewolf, melancholy lycanthropes did not usually kill and eat people or animals; according to the physicians' case histories they preferred solitude, roamed around graveyards and carried around parts of corpses; cf. Michael Stolberg, 'Lykanthropie', in: *Der Neue Pauly*, vol. 15/1: *Wissenschafts- und*

Rezeptionsgeschichte, part La-Ot. Ed. by Manfred Landfester. Stuttgart/ Weimar, 2001, pp. 243–6.

238. SLUB Dresden, C 466, fol. 16, on the scurvy of Frau Marschallin.
239. BIM Paris, Ms. 5242, foll. 18r–v, physician's advice, no date given.
240. *Ibid.*, foll. 150r–151v, Geoffroy's advice to Mlle Herbollin, 27 April 1730.
241. Brunner, *Podagra*, p. 10.
242. BIM Paris, Ms. 5242, foll. 79r–80v, prescription for an unidentified 33-year-old female patient in response to a letter of consultation, 20 April 1727.
243. Hoffmann, *Medicina consultatoria*, vol. 1, pp. 254–7, account of the disease of a 46-year-old aristocratic male patient.
244. Zedler, *Universal-Lexikon* 34 (1742), p. 886.
245. BCU Lausanne, FT, letter from the mother of postmaster Briante, 25 November 1791.
246. *Ibid.*, Mme de Corselles, 17 January 1790; similarly, though probably written by a physician: *ibid.*, letter concerning Mme de Diesbach, 18 March 1790.
247. Fleck, *Chronik*, pp. 219–21.
248. BCU Lausanne, FT, letter from the patient, 15 February 1772.
249. BIM Paris, Ms. 5241, foll. 236r–239v, letter apparently addressed to Mme Geoffroy, 26 May 1728.
250. BIM Paris, Ms. 5241, fol. 240, medical advice, 14 February 1728.
251. BCU Lausanne, FT, letter from the Chevalier d'Albarey writing about the Comtesse de Mouroux, 13 February 1790; BIM Paris, Ms. 5241, fol. 240r, prescription for Monsieur Gouet, 14 February 1728, for the patient's 'sharp' blood, which was mixed with 'bilious particles'.
252. *Ibid.*, letter from the Chevalier de Belfontaine, 25 November 1772.
253. Marmontel, *Mémoires*, pp. 243f.
254. For an overview see Charles Edward A. Winslow, *The conquest of epidemic disease. A chapter in the history of ideas.* New York/London 1967, pp. 117–43; Karl-Heinz Leven, *Die Geschichte der Infektionskrankheiten von der Antike bis ins 20. Jahrhundert.* Landsberg 1997.
255. Kiechel, *Reisen*, p. 300.
256. According to Sebastian Schertlin, almost 5000 men died of 'pestilence' outside the walls of Rome in 1526/27, 'because of the dead bodies which had not been buried' (Schertlin, *Leben*, p. 13).
257. The term 'infection' (Latin: 'infectio') was originally used by dyers in reference to the fact that tiny amounts of dye can color large volumes of fluid (Oswei Temkin, 'An historical analysis of the concept of infection', in: idem, *Studies in intellectual history.* Baltimore 1968, pp. 123–47).
258. UB Leiden, Mss. Marchand 3, Gideo van Boetzelaar, 13 October 1591 (in Latin).
259. Vincentz, *Goldschmiede-Chronik*, p. 369.
260. Passer, *Berichte*, p. 291 (April 1680).
261. See the detailed account by Carlo Cipolla, *Miasmi ed umori. Ecologia e condizioni sanitarie in Toscana nel Seicento.* Bologna 1989; Alain Corbin, *The foul and the fragrant. Odor and the French social imagination.* Leamington Spa 1986.
262. Explicit reference to 'contagium' or 'miasms' is rare in patient letters and personal testimonies; see for example Bismarck, *Tagebuch*, pp. 74f., September 1636 ('contagium').

263. Wolff, *Leben*, p. 7.
264. BCU Lausanne, FT, letter from Monsieur Marcard, 6 December 1785.
265. *Ibid.*
266. *Ibid.*, letter concerning Mme de Diesbach, 18 March 1790, probably written by a physician; BIM Paris, Ms. 5242, foll. 75r–77r, prescription; *ibid.*, letter from Mme au Bavement, 17 December 1783 (she suffered from vaginal discharge); see also *ibid.*, letter from Monsieur de Walmoden (ca. 1781), who mentions the previous diagnosis of a 'virus'; Peyrilhe, *Cancer*; Becane, *Observations*.
267. For a good overview see Andrew Wear, *Knowledge & practice in English medicine, 1550–1680*. Cambridge 2000, pp. 275–349; on literary/artistic representations see Margaret Healy, *Fictions of disease in early modern England*. Basingstoke/New York 2001, esp. pp. 50–122.
268. Especially rich are, in this respect: Pellikan, *Hauschronik*; Vincentz, *Goldschmiede-Chronik*; Brandis, *Diarium*; Weinsberg, *Denkwürdigkeiten*.
269. Weinsberg, *Denkwürdigkeiten*, vol. II, pp. 30f.; see also Bozenhart, *Schicksale*, p. 226.
270. Vincentz, *Goldschmiede-Chronik*, p. 53.
271. Glückel von Hameln, *Denkwürdigkeiten*, p. 67.
272. Weinsberg, *Denkwürdigkeiten*, vol. I, p. 156; *ibid.*, vol. II, p. 31 and p. 321.
273. Vincentz, *Goldschmiede-Chronik*, p. 369.
274. UB Leiden, Mss. Marchand 3, letter from Friedrich Heiming, Bremen, ca. 1594 (in Latin).
275. Weinsberg, *Denkwürdigkeiten*, vol. II, pp. 31f.
276. *Ibid.*, p. 132.
277. Bösch, *Liber*, p. 75.
278. Ryff, *Selbstbiographie*, pp. 56f.
279. Weinsberg, *Denkwürdigkeiten*, vol. I, p. 156.
280. Vincentz, *Goldschmiede-Chronik*, p. 12.
281. Weinsberg, *Denkwürdigkeiten*, vol. I, pp. 175f.
282. Pellikan, *Hauschronik*, pp. 12f.
283. See Jon Arizabalaga et al., *The great pox. The French disease in Renaissance Europe*. New Haven 1997.
284. Zimmern, *Chronik*, vol. II, p. 71.
285. Fröschel, *Hauschronik*, p. 17.
286. Weinsberg, *Denkwürdigkeiten*, vol. I, p. 120.
287. Zimmern, *Chronik*, vol. II, pp. 59f.
288. Bövingh, *Lebensbeschreibung*, p. 121.
289. Rem, *Tagebuch*, p. 26.
290. Oldecop, *Chronik*, pp. 164f.
291. *Ibid.*; Vincentz, *Goldschmiede-Chronik*, p. 456; Dreytwein, *Esslingische Chronik*, pp. 50f.
292. Weinsberg, *Denkwürdigkeiten*, vol. I, p. 64.
293. *Ibid.*
294. Vincentz, *Goldschmiede-Chronik*, pp. 455f.
295. Betrand, *Relation*; BM Avignon, Ms. 3192, foll. 115r–120r, copy of a letter from Monsieur de Raymon Jr to Dr Bellier in Carpentras.
296. Michael Stolberg, *Die Cholera im Grossherzogtum Toskana: Ängste, Deutungen, und Reaktionen im Angesicht einer tödlichen Seuche*. Landsberg 1995; idem,

'Public health and popular resistance. Cholera in the Grand Duchy of Tuscany', *Bulletin of the history of medicine* 68 (1994), 254–77.

297. Bövingh, *Lebensbeschreibung*, p. 145.

298. WL London, Western Manuscripts, Ms. 6114, letter from M. Fasnacht [?], 4 May 1801.

299. BCU Lausanne, FT, letter from Mme Despens [?], 28 October 1771.

300. Holsten, *Kriegsabenteuer*, pp. 11f.

301. LHA Schwerin 2.12.-1/22, 80, letter from Sophia, 29 June 1593.

302. BIM Paris, Ms. 5245, fol. 176r, letter from the brother of a 48-year-old female patient, 10 January 1724.

303. BCU Lausanne, FT, letter from Monsieur Piciet [?], Geneva, 17 July 1767.

304. WL London, Western Manuscripts, Ms. 6114, letter from the nephew of Monsieur Bourgeois, 28 May 1801.

305. BCU Lausanne, FT, letter from Monsieur Claret, Geneva, 7 April 1790.

306. Brenner-Schaeffer, *Darstellung* [see note 193], p. 27; Flügel, *Volksmedizin* [see note 193], p. 54 and p. 66; Max Höfler, *Deutsches Krankheitsnamenbuch*. Hildesheim/New York 1970 (repr. of the 1899 edn), p. 827; StA Bamberg, K3FIII, Gefrees 1828.

307. KB St Gallen, Ms. 94, fol. 117r, letter from Agatha Zimermann, 11 January 1616.

308. BIM Paris, Ms. 5242, foll. 150r–151r, letter from an old Dominican friar, 3 December 1703.

309. Some physicians assumed a further, fourth step in which the various parts of the body produced semen.

310. IGM Stuttgart, HA, Best. B 321548 and B 331062, letters from the Vicomte de Bonneval, 30 October 1832 and 18 August 1833.

311. BCU Lausanne, FT, letter from an unidentified patient, 26 April 1773.

312. SB Berlin, Ms.germ.fol. 422b, fol. 201r, Sigmund von Daberstich [?], 8 May 1579.

313. BIM Paris, Ms. 5245, foll. 139r–v, treasurer Monsieur de Soulas, no date given (reply 28 February 1728).

314. BCU Lausanne, FT, letter from Monsieur de Pollet, 20 April 1772.

315. SB Frankfurt, Ms 334, 17 July 1665.

316. SB Berlin, Ms.germ.fol. 423a, foll. 311r–314r, J. Lindt, Easter 1580.

317. Ledel, *Breviarium*, p. 43, letter from A. Hahn.

318. StA Bamberg K3FIII, Höchstadt 1828; similarly SB München, Cgm 6874/199, Wertingen and *ibid.*, Cgm 6874/144, Pottenstein.

319. Wittich, *Praeservator*, p. 172; cf. Michael Stolberg, 'Der gesunde und saubere Körper', in: Richard van Dülmen (ed.), *Die Erfindung des Menschen*. Vienna 1998, pp. 295–306.

320. As one patient reported, even sleeping with an open mouth could have serious consequences – in his case an immediately increasing occurence of nightmares (BIM Paris, Ms. 5245, foll. 5r–6r, Monsieur Aubriot, 30 January 1724).

321. Andrew Wear, 'The history of personal hygiene', in: W. F. Bynum/Roy Porter (eds), *Companion encyclopedia of the history of medicine*. London/New York 1993, pp. 1283–1308, here p. 1301.

322. Cf. Norbert Elias, *The civilizing process*. New York 1978.

323. Hoffmann, *Medicina consultatoria*, vol. 1, pp. 287–91, 'I. E. F.', 5 October 1729.

324. Krafft, *Reisen*, pp. 324f.
325. Hoffmann, *Medicina consultatoria*, vol. 5, pp. 300f., letter from 61-year-old 'v. D.', 10 March 1727.
326. KB St Gallen, Ms. 94, letter from Friar Placidus, 27 April 1605.
327. SB Berlin, Ms.germ.fol. 420b, foll. 137r–v, Saturday after Annunciation 1574.
328. SB Berlin, Ms.germ.fol. 421b, foll. 199r–200v, H. Birckholtz, 9 March 1577.
329. BCU Lausanne, FT, letter from the court advocate Dauphin, 5 June 1772.
330. BIM Paris, Ms. 5245, foll. 139r–v, letter from treasurer Monsieur de Soulas, no date given (reply 28 February 1728).
331. BIM Paris, Ms. 5245, foll. 59r–60v, letter concerning a 78-year-old Benedictine friar, referring to the opinion the patient expressed himself, no date given (reply 3 November 1719).
332. Hoffmann, *Medicina consultatoria*, vol. 11, pp. 129f., Graf von 'N.', 22 June [1731?].
333. *Ibid.*, pp. 294–6, 'Historia morbi', ca. 1726/27, written by the 63-year-old patient himself.
334. Bösch, *Liber*, p. 106.
335. SB Berlin, Ms.germ.fol. 421b, foll. 146r–148r, Ph. von Farnrode, 22 April 1577.
336. For example Krafft, *Reisen*, pp. 324–6.
337. Bösch, *Liber*, p. 78, for example, described himself as 'full of mucus and obstructed' (*'häfftig verschlimt und verstopfft'*).
338. BCU Lausanne, FT, letter from Jean Colomb, 17 January 1792.
339. Krafft, *Reisen*, p. 325.
340. BCU Lausanne, FT, letter concerning a 33-year-old patient, probably Mme de Virazel, written initially in the third and then in the first person, 12 September 1790.
341. BIM Paris, Ms. 5241, foll. 233r–234r, Mme de Pont Farcy, 27 January 1730.
342. BIM Paris, Ms. 5245, foll. 187r–188r, account of Mlle Darmenon's disease, written in the third person but probably by herself, ca. 1727.
343. Barbara Duden (*The woman beneath the skin: a doctor's patients in eighteenth-century Germany*. Cambridge, MA 1991) has justly emphasized the significance of the early modern notion of obstructions; due to the marked Stahlian orientation of her source and its gynecological focus – in which menstrual disturbances took priority – she has exaggerated the importance of this notion, however, and has not paid sufficient attention to concepts and explanatory models which were crucial for the early modern understanding of other diseases.
344. BCU Lausanne, FT, account of Monsieur Larrei's sickness, written in the third person but clearly by himself, 5 January 1784.
345. UB Erlangen, Trew collection, letter from the bailiff Stambke, 1750.
346. Jacquin, *Abhandlung*, pp. 405–10, notes by the German translator, Georg Neuhofer; see also Alfred Franklin, *La vie privée d'autrefois. Les médicaments*. Paris 1891, pp. 71–4; Elborg Forster, 'From the patient's point of view. Illness and health in the letters of Liselotte von der Pfalz (1652–1722)', *Bulletin of the history of medicine* 60 (1986), 297–320, here pp. 309f.
347. Essich (*Gesundheitswörterbuch*, pp. 99–104) ranked the rejection of clysters among the prevailing 'prejudices'.

348. Pellikan, *Hauschronik*, pp. 163f.
349. BIM Paris, Ms. 5244, foll. 85r–86v, letter concerning Monsieur Laval, written in the third person, probably by a layperson, 10 July 1729.
350. BCU Lausanne, FT, letter from Mme de Konauw, 26 January 1773.
351. *Ibid.*, letter from an unidentified patient, no date given.
352. *Ibid.*, letter from the Comtesse de Vury, 2 April 1774; *ibid.*, letter concerning a 43-year-old clergyman from the Great Saint Bernhard, probably written by a fellow clergyman, 18 October 1789.
353. BIM Paris, Ms. 5244, foll. 85r–86v, letter concerning sick Monsieur Laval, written in the third person, probably by a layperson, 10 July 1729.
354. SB Berlin, Ms.germ.fol. 422a, fol. 5r, letter from A. Trotzberg, 15 June 1578.
355. BCU Lausanne, FT, letter from Monsieur Bruckner, 29 November 1789.
356. *Ibid.*, letter from Monsieur Marcard, 6 December 1785.
357. Spener, *Lebenslauf*, p. 863 (appendix).
358. Her physician sought to guide the morbid matter towards the urine (BIM Paris, Ms. 5244, foll. 79r–v, letter from Mlle de Refuge, ca. 1729, with advice by Monsieur Dubois.
359. BCU Lausanne, FT, letter from Lieutenant Roussany, 10 June 1774.
360. *Ibid.*, 'report' by Monsieur Lavergue, 19 May 1772.
361. *Ibid.*, letter from the Baron from Bamberg, illegible signature, 30 May 1792.
362. Georges Vigarello, *Concepts of cleanliness: Changing attitudes in France since the Middle Ages.* Cambridge/Paris 1988.
363. Hufeland, *Makrobiotik*, pp. 199–215.
364. BCU Lausanne, FT, letter from Monsieur Thiebaud, Evian, 3 February 1792.
365. Cf. Wittich, *Praeservator*, p. 181.
366. BCU Lausanne, FT, letter from Mme Neider, The Hague , 30 September 1780.
367. StA Bamberg K3FIII 1481, Kirchenlamitz 1828.
368. SB München, Cgm 6974/154 Roth; StA Bamberg K3FIII, Kirchenlamitz 1828; Schmitt, *Sitten*, p. 363.
369. SB München, Cgm 6874/118 Neumarkt/Opf.
370. Santorio, *De statica medicina*.
371. In the case of 78-year-old Monsieur Martine, for example, the physicians hoped to evacuate the excess fluid in his lungs and limbs via, among other methods, 'insensible transpiration' (BCU Lausanne, FT, physicians' advice, probably written by Dres Vilet, Petetin and Gilibert, no date given, and letter from the patient's niece, 11 February 1784).
372. BCU Lausanne, FT, letter from Mme La Millère, 1 March 1767.
373. Stolberg, *Erfahrungen* [see note 59].
374. The major source is Pliny, *Historia naturalis*, vol. 7, ch. 15 and vol. 28, ch. 20–23.
375. Commonly used terms were in German *'Zeit'*, *'Monatszeit'*, and *'das Monatliche'*, and in French *'règles'*, *'époques'*, *'menstrues'*, and *'ordinaires'*.
376. BB Bern, HC A 167, letter from an unidentified patient, Neufchâtel, 15 December 1773.
377. Massia, *Âge critique*, p. 19.

378. Fothergill, *Conseils*; Armand, *Considérations*, p. 14; Garnier, *Considérations*, p. 368; cf. Michael Stolberg, 'A woman's hell? Medical perceptions of menopause in pre industrial Europe', *Bulletin of the history of medicine* 73 (1999), 408–28.

379. Pierre Bourdieu, *Outline of a theory of practice*. Cambridge 1977.

380. Yvonne Verdier, *Façons de dire, façons de faire. La laveuse, la couturière, la cuisinière*. Paris 1979, pp. 20–41; Vieda Skultans, 'Menstrual symbolism in South Wales', in: Thomas Buckley/Alma Gottlieb (eds), Blood magic. The *anthropology of menstruation*. Berkeley 1988, pp. 137–60.

381. SB Berlin, Ms.germ.fol. 423b, foll. 133r–v, 27 May 1581.

382. BCU Lausanne, FT, letter from the Chevalier d'Albarey about the Comtesse de Mouroux, 14 August 1793.

383. BCU Lausanne, FT, letter from Monsieur Bruckner, 29 November 1789.

384. *Ibid.*, letter from Monsieur Claret, 7 April 1790.

385. Bösch, *Liber*, p. 93.

386. Thus a patient's husband suspected an 'obstructed liver and spleen' ('vorstopffte Leber vnd Miltz'); see SB Berlin, Ms.germ.fol. 421b, foll. 245r–v, no date given.

387. Vincentz, *Goldschmiede-Chronik*, p. 391.

388. Galen, 'Ad Glauconem de medendi methodo', in: idem, *Opera XI*, pp. 139–44; according to Galen, a breast cancer, for example, from which the vessels stretched out visibly into the surrounding tissue, resembled a living crab – in Latin 'cancer'; on the history of medical theories about cancer see Jacob Wolf, *Die Lehre von der Krebskrankheit von den ältesten Zeiten bis zur Gegenwart*, vol. 1. Jena 1907; L. J. Rather, *The genesis of cancer. A study in the history of ideas*. Baltimore/London 1978; Carl G. Kardinal/John W. Yarbro, 'A conceptual history of cancer', *Seminars in oncology* 6 (1979), 396–408; Luke Demaitre, 'Medieval notions of cancer. Malignancy and metaphor', *Bulletin of the history of medicine* 72 (1998), 609–37; bibliographical survey in James S. Olson, *The history of cancer. An annotated bibliography*. New York 1989.

389. Boerhaave, *Abhandlung*, pp. 1f.

390. Juraj Körbler, *Geschichte der Krebskrankheit. Schicksale der Kranken, der Ärzte, der Forscher. Der Werdegang einer Wissenschaft*. Vienna 1973, pp. 29–49; Milow, *Ich will aber nicht murren*, pp. 289–311.

391. Rivière, 'Praxis medica', in: idem, *Opera*, pp. 262f.

392. Astruc, *Treatise*, p. 243.

393. Boerhaave, *Abhandlung*, p. 3; Gendron, *Recherches*, pp. 15f.

394. BCU Lausanne, FT, letter from the husband of an unidentified 30-year-old patient.

395. Galen, *Opera XI*, p. 139.

396. BIM Paris, Ms. 5245, foll. 106r–v, account of Mme de Chambery's disease, written in the third person but probably by the patient herself, ca. 1730; among learned physicians this notion was increasingly criticized; cf. Gendron, *Recherches*, pp. 37–9.

397. Cf. Mary Douglas, *Purity and danger. An analysis of concepts of pollution and taboo*. Harmondworth 1970.

398. William A. Cooper, 'The history of the radical mastectomy', *Annals of medical history*, 3rd series 3 (1941), 36–54.

399. Boerhaave, *Abhandlung*, p. 1.
400. Houppeville, *Guérison*, p. 49.
401. *Ibid.*, 52.
402. *Recueil des miracles*, pp. 20f., testimony, 1 July 1728.
403. Burney, *Journals*, pp. 596–615, 30 September 1811.
404. Milow, *Ich will aber nicht murren*.
405. BCU Lausanne, FT, letter from the husband of an unidentified 30-year-old patient.
406. Foreest, *Observationum*, pp. 482–94; see also, however, *Recueil des miracles*, pp. 20f., with testimony by Elisabeth de la Loë, 1 July 1728, about the treatment of a breast tumor by the surgeon Vasseur.
407. BCU Lausanne, FT, letter from Monsieur Cheroz, 16 January 1785.
408. *Ibid.*, letter from the Chevalier de Peyrelongue, Strasbourg, 7 September 1785.
409. *Ibid.*, letter from Mlle de Maltzan, Florence, 12 May 1776.
410. *Ibid.*, letter from the Comtesse de Wedel, Copenhagen, 12 November 1784.
411. *Ibid.*; the countess had strong reservations about the advice of her local physician.
412. BL London, Ms. Sloane 4039, fol. 198r, letter from the brother of Lord Hatton, 4 October 1703.
413. BCU Lausanne, FT, letter from Mme Meyster de Verthamon, Bordeaux, 6 January 1777.
414. *Ibid.*, letter from Schwitzer de Buonas, Lucerne, 2 October 1793
415. BIM Paris, Ms. 5242, foll. 215r–216r, letter which a 50-year-old nun dictated to a friend, 10 September 1724.
416. BCU Lausanne, FT, letter from the Marquise de St Innocent, ca. 1785.
417. Hoffmann, *Medicina consultatoria*, vol. 4, pp. 12–14, letter from an unidentified patient, 25 February 1723.
418. Hoffmann, *Medicina consultatoria*, vol. 5, pp. 93–8, letter from a 31-year-old patient, written initially in the third and later in the first person.
419. BCU Lausanne, FT, letter from the patient's daughter, 9 August 1784.
420. *Ibid.*, letter concerning Mme le Pin, probably written by an acquaintance of hers, ca. 1784.
421. Cf. Stolberg, *Woman's hell* [see note 380].
422. BCU Lausanne, FT, Fonds Tissot, letter from Mme Viard d'Arnay, 6 July 1778.
423. BIM Paris, Ms. 5241, foll. 45r–48v, letter from the brother, 10 June 1729, and Geoffroy's reply, 3 July 1729.
424. Margaret Lock, *Encounters with aging. Mythologies of menopause in Japan and North America*. Berkeley 1993.
425. Andrew Wear, 'Explorations in Renaissance writings on the practice of medicine', in: idem/Roger French/Ian M. Lonie (eds): *The medical Renaissance of the sixteenth century*. Cambridge 1985, pp. 118–45.
426. On early modern scientific demonology see Stuart Clark, *Thinking with demons. The idea of witchcraft in early modern Europe*. Oxford 1997; Philipp Melanchthon produced a highly influential account of the concept of 'spirits' and 'vapors' in his *Liber de anima* (Melanchthon, *Liber*).
427. BIM Paris, Ms. 5241, foll. 289r–v, account of the disease of an unidentified 43-year-old female patient, no date given.

428. KB St Gallen, Ms. 94, fol. 18r, Christ. Schmidholzer writing about Barth. Schobinger, 23 May 1604.
429. SB Berlin, Ms.germ.fol. 420b, fol. 75r, letter from A. von Bradowa, 26 October 1574.
430. Vincentz, *Goldschmiede-Chronik*, p. 291.
431. BIM Paris, Ms. 5241, foll. 64r–65v, Monsieur Darmenon, 1 September 1717.
432. BCU Lausanne, FT, letter from Mme Priolo de Baliviere, 14 October 1791; BIM Paris, Ms. 5245, foll. 263r–266v, letter concerning a 17-year-old female patient, probably written by a relative, 11 November 1729.
433. BCU Lausanne, FT, letter from the Comte de Favét de Bosses, Turin, 7 November 1792.
434. Cf. Volker Hess, 'Die Normierung der Eigenwärme. Fiebermessen als kulturelle Praktik', in: idem (ed.), *Normierung der Gesundheit. Messende Verfahren der Medizin als kulturelle Praktik um 1900*. Husum 1997, pp. 169–88.
435. 'As if he had a fever' was how a worried father described the trembling hands and feet of his nine-year-old son (HStA Stuttgart A 209, Bü. 1668, petition from Hans Martin Grätzinger, 1742).
436. KB St Gallen, Ms. 94, fol. 23r, Fr. Placidus Heller, 20 April 1605.
437. BCU Lausanne, FT, letter from Mme Priolo de Baliviere, 14 October [1791?]
438. *Ibid.*, letter from Monsieur Gochuat, 1 November 1785.
439. BIM Paris, Ms. 5245, foll. 71r–72v, letter concerning Mme du Bourgneuf, initially written in the third person, no date given.
440. BCU Lausanne, FT, letter concerning Mme de Verdun, 27 October 1789, written in the third person but apparently dictated by the patient herself.
441. *Ibid.*, letter from Monsieur Decheppe de Morville, 24 February 1784.
442. *Ibid.*, letter concerning Mme Mackinnen [?], apparently written by a male acquaintance, 27 September 1786.
443. BIM Paris, Ms. 5241, foll. 64r–65v, letter from Monsieur Darmenon, 1 September 1717.
444. BIM Paris, Ms. 5425, foll. 5r–6r, letter from Monsieur Aubriot, 30 January 1724.
445. BCU Lausanne, FT, letter from Mme au Bavement, 17 December 1783.
446. BIM Paris, Ms. 5241, foll. 64r–65v, letter from Monsieur Darmenon, 1 September 1717 ('nulle fievre au poux').
447. BCU Lausanne, FT, letter from Major Bouju, 1773.
448. For example BIM Paris, Ms. 5241, foll. 308r–v, letter from Monsieur Darmenon, 27 October 1717; *ibid.*, letter from Mme Monget, 2 November 1776.
449. *Ibid.*, letter from the Comte de Favét de Bosses, Turin, 7 November 1792.
450. Mario Vegetti, *Tra Edipo e Euclide. Forme del sapere antico*. Milan 1983, pp. 41–58.
451. BCU Lausanne, FT, letter from Monsieur Boissière, 3 January 1793.
452. RA Utrecht, Archive of the family Geuns 46, letter from Wm. van Tuil, March 1802 (in Dutch).
453. BCU Lausanne, FT, letter from the Baron de Vigneule, 26 April 1785; *ibid.*, letter concerning a 43-year-old clergyman from the Great Saint Bernhard, probably written by a fellow clergyman, 18 October 1789.
454. BCU Lausanne, FT, letter from Monsieur Bruckner, 29 November 1789; he felt that the cinchona he had taken had successfully fought the fever

which he had developed when he had accidentally suppressed the habitual sweating of his feet by rubbing fat upon them.

455. Butzbach, *Odeporicon*, p. 205.
456. SB München, Cgm 6874/35, Donauwörth.
457. SB München, Cgm 6874/155, Rothenburg ob der Tauber.
458. BCU Lausanne, FT, letter from the lawyer Gayot, 14 September 1779; he complained that his physicians had only laughed at him.
459. Marmontel, *Mémoires*, p. 56.
460. WL London, Western Manuscripts, Ms. 6114, letter from Monsieur Fasnacht [?], 4 May 1801; he suspected a 'pneumonia' (*'poulmonie'*); in spite of the patient's bloody sputum, Verdeil, in contrast, diagnosed a 'nervous hypochondria' (*ibid.*, 18 May 1801).
461. Buchoz, *Traité*, p. 1.
462. Historians have hailed François de la Boë as the first physician to describe the 'tubercles', in 1679 (cf. Jean Dubos/René Dubos, *The white plague. Tuberculosis, man, and society*. New Brunswick 1987, p. 73), but Jean Fernel had already in the 16th centrury mentioned a local cheeselike hardening in the lungs (Fernel, *Universa medicina*, pp. 297–300).
463. BIM Paris, Ms. 5245, foll. 1r–2v, advice to a Barnabite novice, no date given; similarly, BCU Lausanne, FT, advice by Carlo Allione, September 1791.
464. BCU Lausanne, FT, letter from Monsieur Boissière, 3 January 1793. He attributed his fever, coughing, and loss of vigor to either a humor in his body or 'lung tubercles'.
465. On the history of medical debates about consumption see Dubos/Dubos, *White plague* [see note 462]; see also Barbara Gutmann Rosenkrantz (ed.), *From consumption to tuberculosis. A documentary history*. New York/London 1994.
466. Marmontel, *Mémoires*, p. 57.
467. Peter H. Niebyl, 'Old age, fever, and the lamp metaphor', *Journal of the history of medicine* 26 (1971), 351–68; Thomas S. Hall, 'Life, death and the radical moisture', *Clio medica* 6 (1971), 3–23.
468. SB Berlin, Ms.germ.fol. 420b, fol. 27r, letter from J. Breckow, 25 August 1574.
469. BCU Lausanne, FT, letter from 43-year-old patient's husband, 12 June 1787.
470. *Ibid.*, letter from a 33-year-old female patient, 25 June 1772.
471. *Ibid.*, letter from Mme de Launay, 1 June 1790.
472. Boerhaave, *Correspondence*, p. 132.
473. BCU Lausanne, FT, letter from Mme La Millière, 1 March 1767.
474. Cf. Lord Hatton's complaint: 'my Brother takes too much purging physick' (BL London, Ms. Sloane 4039, foll. 188f., 20 September 1703).
475. Hummel, *Histori*, p. 53.
476. *Recueil des miracles*, pp. 7–18.
477. BCU Lausanne, FT, letter from Mme de Sucher, 2 March 1782.
478. *Ibid.*, letter from Nancy de Brackel, 9 May 1790.
479. *Ibid.*, account of the disease of an unidentified patient with a post-scriptum by Monsieur Jungken.
480. *Ibid.*, letter from Mme de Konauw, 26 January 1773.
481. *Ibid.*, letter from the Marquise de Louvois, 29 October 1784.

482. Zimmern, *Chronik*, vol. II, p. 204.
483. UB Erlangen 948/III, foll. 503r–504r, excerpt from a letter which Vogel had written on 2 May 1659. Some patients resolutely rejected the diagnosis of 'dropsy' because their face, legs, and arms had never been swollen; German physicians distinguished 'Bauchwassersucht' ('abdominal dropsy' or 'ascites') – in which the limbs were often thin – from 'Gliederwassersucht' ('dropsy of the limbs') and sometimes also considered a belly filled with wind ('tympanites') as a form of dropsy (Wirsung, *Artzney Buch*, pp. 351–8).
484. UB Erlangen 948/III, foll. 519r–523r, 25 June 1661.
485. BCU Lausanne, FT, letter from F. von Jungken, 24 January 1772.
486. SB Berlin, Ms.germ.fol. 420b, foll. 400r–v, Moritz von Nischwitz, 11 May 1575.
487. SB Berlin, Ms.germ.fol. 424, foll. 214r–v, Opitz von Hacke, 28 March 1582.
488. Hummel, *Histori*, p. 47.
489. SB Berlin, Ms.germ.fol. 421b, foll. 77r–v, Georg von Doberck [?].
490. Zimmern, *Chronik*, vol. II, p. 204.
491. Weinsberg, *Leben*, pp. 402–5.
492. SUB Hamburg Sup. ep. 109, foll. 41r–v, advice from Konerding, 2 April 1647 (in Latin), mentioning the patient's own opinion.
493. BCU Lausanne, FT, letter from the Baron de Beaucourse, 25 September 1783.
494. *Ibid.*, letter from an unidentified patient, Lille, 2 June 1793, requesting that the answer be sent to Monsieur Warembourg.
495. *Ibid.*, letter from the friend of an impotent man, 16 August 1772.
496. Wedel, *De pollutione*; Castro, *Universa mulierum medicina*, pp. 95–108; Ettmüller, *Opera*, vol II, pp. 421–4; Schuyl, *De gonorrhea*; Hoffmann, *Consultationum*, pp. 493f. ('de pollutione nocturna'); for the 19th century see Lallemand, *Des pertes seminales*.
497. 'Pollution nocturne', in: *Encyclopédie*, vol. 12 (1765), pp. 922–4.
498. BCU Lausanne, FT, history of Monsieur Larrei's disease, in the third person but clearly written by himself, 5 January 1784.
499. *Ibid.*, letter from Monsieur Gringet, 4 January 1784; he also masturbated.
500. *Ibid.*, letter from the patient's wife, 29 October 1784.
501. *Ibid.*, letter from Monsieur Chillaud, 4 May 1790.
502. Hoffmann, *Consultationum*, pp. 493f.; Wedel, *De pollutione*; 'Pollution nocturne', in: *Encyclopédie*, vol. 12 (1765), pp. 922–4.
503. UB Leiden, Mss. Marchand 3, letter from Jacob Baldwein von Zweybruck, 30 June 1593 (in Latin).
504. BCU Lausanne, FT, letter from Monsieur Matthis, 1 October 1773.
505. BB Bern, HC, A 184, letter from an unidentified 28-year-old patient, ca. 1775.
506. BCU Lausanne, FT, letter from an unidentified patient, ca. 1775/76.
507. *Ibid.*, letter from Chevalier de Kergas, 17 September 1793; *ibid.*, letter from Prof. Millner, Cambridge, summer 1785.
508. *Ibid.*, series of letters from F. L. Gauteron, 1792; in view of the patient's marked tendendcy towards hypochondria, Tissot advised him to read Montaigne.
509. Cf. for example 'Pollution nocturne' in: *Encyclopédie*, vol. 12 (1765), 922–4.

510. Even though they were accompanied by sexual dreams, J. B. von Zweybruck called his nocturnal emissions a 'kind of gonorrhea' (UB Leiden, Mss. Marchand 3, 30 June 1593, in Latin).
511. Wirsung, *Artzney Buch*, p. 256; Castro, *Universa mulierum medicina*, pp. 95–108; Ettmüller, *Opera*, vol. II, pp. 421–4; Schuyl, *De gonorrhea*.
512. Stupanus, *De chlorosi*; Vieussens, *Histoire*, vol. III, pp. 140–5; Varandaeus, 'De morbis mulierum', in: idem, *Opera*, pp. 483–91; Dahmen, *De Chlorosi*; cf. Irvine Loudon, 'The diseases called chlorosis', *Psychological medicine* 14 (1984), 27–36; Helen King, *The disease of virgins: green sickness, chlorosis, and the problems of puberty*. London/New York 2004.

III Dominant Discourse and the Experience of Disease

1. See Erwin H. Ackerknecht, 'Primitive medicine and culture pattern', *Bulletin of the history of medicine* 12 (1942), 545–74, here p. 546: 'It is an almost hopeless task to try to understand and evaluate the medicine of one primitive tribe while disregarding its cultural background.'
2. Cf. Michael McVaugh, *Medicine before the plague. Practitioners and their patients at the Aragonese court*. Cambridge 1993.
3. For a succinct survey see Ludmilla Jordanova, 'The social construction of medical knowledge', *Social history of medicine* 8 (1995), 361–81.
4. Cf. Arthur Kleinman, *Patients and healers in the context of culture. An exploration of the borderland between anthropology, medicine, and psychiatry*. London 1980, pp. 75–7.
5. For this reason, a radical, outright rejection of any attempt at retrospective diagnosis fails to convince me (see for example Karl Heinz Leven, 'Krankheiten – historische Deutung versus retrospektive Diagnose', in: Norbert Paul/ Thomas Schlich (eds), *Perspektiven der Patientengeschichtsschreibung*. Frankfurt/New York 1998, pp. 153–85). Undoubtedly, a lot of nonsense has been written in this area. Frequently authors are not sufficiently aware of the changing meanings of medical terms and do not take into account that prevailing disease concepts decisively shape the way patients and physicians describe symptoms. Nevertheless, especially in cases when many patients with the same (contemporary) diagnosis suffered from similar complaints, some cautious probabilistic conclusions can be drawn in modern terms. At times they are indeed indispensable, for example when the effect of social status or geography on the dissemination of an epidemic like 19th-century 'cholera' is analyzed or when we want to find out to what degree late medieval fears of 'leprosy' reflected the dissemination of what we would consider a bacterial disease and to what degree they were the result of deeply rooted fears of skin lesions in general.
6. Paul U. Unschuld, 'Die konzeptuelle Überformung der individuellen und kollektiven Erfahrung von Kranksein', in: idem et al. (eds), *Krankheit, Heilkunst, Heilung*. Freiburg/Munich 1978, pp. 491–516, here p. 497f.
7. The concept originated from the notion of 'culture-bound psychosis' (Pong Meng Yap, 'Words and things in comparative psychiatry, with special reference to the exotic psychoses', *Acta psychiatrica scandinavica* 38 (1962), 163–9; cf. Ronald C. Simons/Charles C. Hughes, *The culture-bound*

syndromes. Folk illnesses of psychiatric and anthropological interest. Dordrecht 1985; see also the thematic issue 'New approaches to culture-bound mental disorders', *Social science and medicine* 21 (1985), 163–228.

8. Wolfgang G. Jilek/Louise Jilek-Aall, 'The metamorphosis of "culture-bound" syndromes', *Social science and medicine* 21 (1985), 205–10.

9. Simons/Hughes, *Culture-bound syndromes* [see note 8]; on Amok see John E. Carr, 'Ethno-behaviorism and the culture-bound syndromes. The case of Amok', in: *ibid*, pp. 199–223; on susto see also Arthur J. Rubel et al., *Susto. A folk illness.* Berkeley 1984.

10. The notion has remained contested; cf. Robert A. Hahn, 'Culture-bound syndromes unbound', *Social science and medicine* 21 (1985), 165–71; Ivan Karp, 'Deconstructing culture-bound syndromes', *ibid.*, 221–8.

11. Jilek/Jilek Aall, *Metamorphosis* [see note 8].

12. Raymond Prince, 'The concept of culture-bound syndromes. Anorexia nervosa and brain-fag', *Social science and medicine* 21 (1985), 197–203.

13. Cf. Stanley W. Jackson, 'Acedia the sin and its relationship to sorrow and melancholia', in: Arthur Kleinman/Byron Good (eds), *Culture and depression. Studies in the anthropology and cross-cultural psychiatry of affect and disorder.* Berkeley 1985, pp. 43–62.

14. Caroline Giles Banks, '"Culture" in culture-bound syndromes. The case of anorexia nervosa', *Social science and medicine* 34 (1992), 867–84.

15. Cheyne, *English malady.*

16. Caraccioli, *Dictionnaire*, vol. II, p. 45; in contrast, by 1785, Andry, *Recherches*, p. 1, had already described melancholy as 'the most frequent and most neglected disease', though this clearly also served to underline the importance of his treatise on the matter.

17. Marcard, *Beschreibung*, p. 15 and p. 116.

18. Hoffmann, *Medicina consultatoria*, vol. 5, pp. 93–8, letter from an unidentified 31-year-old patient.

19. Joachim Radkau, *Das Zeitalter der Nervosität. Deutschland zwischen Bismarck und Hitler.* Munich/Vienna 1998.

20. Viridet, *Dissertation*, p. 27; similarly Hunauld, *Vapeurs*, p. 14.

21. Hunauld, *Vapeurs*, p. 73.

22. Cf. for example Sévigné, *Lettres*, pp. 96f.; Mme Sévigné suffered from vapors herself, as did Mme de Graffigny.

23. Ménétra, *Journal*, pp. 45f.

24. Chastelain, *Traité.*

25. Dumoulin, *Nouveau traité*, p. 150.

26. Lorry, *Melancholie*, p. 213.

27. BIM Paris, Ms. 5245. foll. 199r–200v, letter concerning a 52-year-old 'Titius', possibly written by a physician, no date given.

28. BIM Paris, Ms. 5242, foll. 108r–109f., undated letter, written in the third person but probably by the patient himself.

29. Dumoulin, *Nouveau traité* , p. 155.

30. Graffigny, *Correspondance*, vol. III, p. 416.

31. BCU Lausanne, FT, letter from a female patient, ca. 26 years old, no date given.

32. Dumoulin, *Nouveau traité*, pp. 155f.

33. BIM Paris, Ms. 5241, foll. 3r–4v, letter concerning an unidentified patient, probably written by a relative.
34. Lange, *Traité des vapeurs*. Paris 1689; Vieussens, *Histoire*, vol. III, pp. 44–121; similarly Viridet, *Dissertation*, p. 32; see also G. Abricosoff, *L'hystérie aux XVIIᵉ et XVIIIᵉ siècles (étude historique et bibliographique)*. Paris 1897.
35. Bressy (*Recherches*), in 1789, based his account of the 'vapeurs' largely on the notion of hypochondria.
36. Cf. Greek 'hypo' = under; 'chondros' = (rib)cartilage.
37. Cf. for example the epistolary consultation on 'hypochondriacal vapors' by Monsieur la Roquette in Pomme, *Traité*, pp. 203–7.
38. Cf. Massimo Riva, Saturno e le Grazie. Malinconici e ipocondriaci nella letteratura italiana del Settecento. Palermo 1992, pp. 39–65; Stanley W. Jackson, *Melancholia and depression. From Hippocratic times to modern times*. New Haven/London 1986, pp. 274f.
39. BB Bern, HC, de l'Espinasse, 4 October 1762.
40. Hoffmann, *Medicina consultatoria*, vol. 1, pp. 287–91, I. E. F., Leipzig, 5 October 1720, cit. pp. 290f.
41. *Ibid.*, vol. 11, pp. 120–4, letter from Graf von N., 13 June ca. 1735.
42. *Ibid.*, pp. 127–9, additional message, written the same day.
43. *Ibid.*, vol. 4, pp. 12–14, letter from a 34-year-old patient who had worked for decades at the court and in the military. Against the opinion of the other physicians, Hoffmann agreed with the patient's self-diagnosis.
44. BCU Lausanne, FT, letter from a good friend of the 52-year-old patient, 16 August 1772; the immediate reason for the consultation was the patient's impotence.
45. 'Hypochondrisches Übel', in: Zedler, *Universal-Lexikon*, vol. 13 (1735), pp. 1479–87; the Galenic tradition distinguished hypochondriacal and flatulent melancholy; cf. Esther Fischer-Homberger, *Hypochondrie. Melancholie bis Neurose. Krankheiten und Zustandsbilder*. Bern 1970, pp. 15–17; occasionally, French physicians, too, diagnosed a patient as 'hypochondriac' who was out of his mind (BCU Lausanne, FT, letter from a clergyman from the area of Besançon, 12 August 1787, concerning a 26-year-old 'séminariste').
46. Dubois (*Histoire philosophique*, p. 20f.) had already defined hypochondria as a disease which affected only the mind and was characterized by excessive worries about often bizarre and imaginary diseases.
47. Fischer-Homberger, *Hypochondrie* [see note 45], p. 66.
48. BIM Paris, Ms. 5242, foll. 104r–105v, letter, probably written by a physician, summer 1729; her disease was diagnosed as hypochondriacal melancoly.
49. Bressy, *Recherches*, p. 3.
50. Cf. Fischer-Homberger, *Hypochondrie* [see note 45], p. 56.
51. A rare exception was the lawyer Breyand (BCU Lausanne, FT, undated letter), who was also very frank on other matters such as sexual habits, however, and who called himself a 'model of that machine which you describe in your "Onanism"'.
52. *Ibid.*, letter from the Chevalier de Belfontaine, 25 November 1772.
53. The history of 'hysteria' has been studied by numerous scholars but their focus has been mostly on the 19th century; for a historiographical survey see Mark S. Micale, *Approaching hysteria. Disease and its interpretations*.

Princeton 1995; for medical theories of hysteria in the 17th and 18th centuries see Paul Hoffmann, *La femme dans la pensée des lumières*. Paris 1977, pp. 175–99; Abricosoff, *L'hystérie* [see note 34]; Étienne Trillat, *Histoire de l'hystérie*. Paris 1986, pp. 53–77.

54. BIM Paris, Ms. 5242, foll. 71r–72r, letter from an unidentified female patient, 2 March 1730, in the third person but probably written or dictated in part by the patient herself.

55. BCU Lausanne, FT, letter from 34-year-old Mme Disse, 13 June 1774.

56. *Ibid.*, letter from the Marquise de St Innocent, 1785.

57. According to R. Vieussens, some women could feel the process of fermentation physically, perceiving a warming sensation and something moving inside their bodies (Vieussens, *Histoire*, vol. III, p. 46).

58. Fries, *Spiegl*, p. 161.

59. Sennert, *De mulierum*; similarly Fernel, *Universa medicina*, pp. 590f.; cf. Hoffmann, *La femme* [see note 53], p. 177.

60. Baglivi, *Opera*, p. 202.

61. Dumoulin, *Nouveau traité*, p. 152; Pomme, *Abhandlung*, pp. 2f.; BCU Lausanne, FT, letter concerning a 17-year-old girl, probably written by a physician, no date given.

62. One exception is Dumoulin, *Nouveau traité*.

63. As Geoffroy wrote to Dr Dubois in Rennes in 1728, giving gallium was very much 'in fashion' against this type of 'vapeurs' (BIM Paris, Ms. 5241, foll. 166r–167v); see also Hunauld, *Vapeurs*, pp. 62–4.

64. Beauchêne, *Influence*, pp. 40f.; Hunauld, *Vapeurs*, pp. 57–9; Pomme, *Traité*, pp. 1–3.

65. 'Hysterical' women were sometimes also said to have an exaggerated fear of disease (Dumoulin, *Nouveau traité*, pp. 150f.).

66. Jungken, *Medicus*, p. 806.

67. Sydenham, *Epistolary dissertation*, p. 51 (January 1682).

68. For example Platter, *Praxeos*, 567–74, 'De salacitate'.

69. Preface to Whytt, *Vapeurs*, vol. 1, p. III; Jungken (*Medicus*, pp. 806–23) subsumed 'hypochondria' and 'hysteria' under the rubric 'spleen diseases' but assumed that they were caused by pathological changes in the 'spirits' or 'nerves'.

70. Lepois, *Selectiorum observationum*; cf. Abricosoff, *L'hystérie* [see note 34], pp. 24–30; George S. Rousseau, 'Cultural history in a new key. Towards a semiotics of the nerve', in: Joan H. Pittock/Andrew Wear (eds), *Interpretation and cultural history*. Basingstoke/London 1991, pp. 25–81.

71. Viridet, *Dissertation*, p. 24.

72. Pomme, *Abhandlung*, pp. 2–9.

73. 'Vapeurs', in: *Encyclopédie*, vol. 16, pp. 836f.

74. Vandermonde, *Dictionnaire*, p. 403; an identical definition appeared 1772 in Goulin, *Médecin*, pp. 251f.

75. Deffand, *Lettres*, vol. I, p. 352.

76. Pomme, *Abhandlung*; similarly before him Fitzgerald, *Traité*, pp. 186–90; on Pomme see also Paul Juquelier/Jean Vinchon, 'Les vapeurs, les vaporeux et le Dr Pierre Pomme', *Annales médico-psychologiques* 71 (1913), 641–56.

77. 'Vapeurs' in: *Encyclopédie*, vol. 16, pp. 836f.

78. In 1703, Dumoulin (*Nouveau traité*, p. 131) for example wrote about the convulsions 'which the people attribute to supernatural powers'.

79. See also Radkau, *Zeitalter* [see note 19].
80. Cf. Lynn Payer, *Medicine and culture. Notions of health and sickness in Britain, the U.S., France and West Germany.* London 1989; since the liver was, for centuries, considered as the site where food was digested (or 'concocted') into blood, it is tempting to attribute this preference for liver diseases to the importance of cooking and eating in French culture.
81. Jungken, *Medicus*, p. 806.
82. StA Bamberg, K3FIII 1481, Pottenstein 1828; Höchstadt 1828.
83. This may also be a major reason why 'hysteria' in the process of its 'democratization' in the 19th century tended to be associated with much more dramatic, impressive symptoms: only if they exhibited such symptoms could patients expect their suffering to be taken seriously.
84. See also Hermann Faller, 'Subjektive Krankheitstheorien als Forschungsgegenstand von Volkskunde und medizinischer Psychologie', *Curare* 6 (1983), 163–80.
85. BIM Paris, Ms. 5242, foll. 48r–58(bis)v, letters from a patient and a relative of the same name, probably the husband, 1724.
86. BIM Paris, Ms. 5242, foll. 108r–109v, letter concerning an unidentified patient, written in the third person but probably by the patient himself, no date given.
87. BIM Paris, Ms. 5241, foll. 3r–4v, letter from an unidentified patient, written in the third person; the unconventional spelling suggests that the author was a layperson.
88. The first steam engine – 'machine à vapeur' in French – was presented to the public in 1698.
89. Mandeville, *Treatise*, p. 95.
90. BIM Paris, Ms. 5242 48r–58(bis)v, letters from a patient and a relative of the same name, probably the husband, 1724.
91. BIM Paris, Ms. 5241, foll. 5r–7r, letter from an unidentified patient, written in the third person; the unconventional spelling suggests that the author was a layperson.
92. BCU Lausanne, FT, letter from the patient's husband, 28 February 1790.
93. *Ibid.*, letter from an unidentified male patient, no date given.
94. *Ibid.*, letter from Monsieur Gayot, Paris, 14 September 1779.
95. *Ibid.*, letter from the surgeon Rouvière, 26 May 1783.
96. *Ibid.*, account of the disease of Comtesse de Ricci in Lyon, written by herself, ca. 1797; *ibid.*, letter from Mme Herrmann about a lady from Vienne, 21 May 1790; usually terms like 'crispations' and 'tiraillements' were used.
97. *Ibid.*, letter from Mme de Chastenay, 8 November 1784.
98. *Ibid.*, letter from Mme Niels, 15 August 1773.
99. *Ibid.*, letter from Monsieur de Leune, no date given.
100. See also Heinrich Walther Bucher, *Tissot und sein Traité des nerfs. Ein Beitrag zur Medizingeschichte der schweizerischen Aufklärung.* Diss. med. Zürich 1958.
101. BCU Lausanne, FT, letter from Mme Mieg in Mulhouse about her daughter, who was suffering from trembling limbs, 9 May 1790.
102. Boerhaave, *Correspondence*, pp. 138–40, letter from a 36-year-old female patient, June 1736.
103. BCU Lausanne, FT, letter from Monsieur Chillaud, 4 May 1790.

104. *Ibid.*, letter, probably written by an English patient himself.
105. *Ibid.*, letter from Mme de Konauw, 26 January 1773.
106. *Ibid.*, letter from sick Mme Dollfus' husband, 28 February 1790; Tissot agreed.
107. *Ibid.*, letter concerning Isaak Iselin's sick daughter, probably written by an acquaintance, April 1790; the physicians attributed her nervous disease to her births.
108. Radkau. *Zeitalter* [see note 19]; Marijke Gijswijt-Hofstra/Roy Porter (eds), *Cultures of neurasthenia from Beard to the First World War.* Amsterdam 2001.
109. *Recueil des pièces*, p. 34, testimony of a ropemaker's wife, Marie Anne Noël, about her son's injury; *Abregé des miracles*, pp. 64f., on the case of a nun.
110. BCU Lausanne, FT, letter probably written by the patient's parents, no date given.
111. *Ibid.*, letter from the vicar Pétion on behalf of a poor patient, about 30 years of age, 21 April 1775.
112. UB Erlangen, Trew collection, letter from Christian Streubel, 14 December 1754.
113. Cf. for example Vaublanc's account of the literally gripping performance of a German pianist in Paris: 'A lady was suddenly seized by a strange nervous attack; a second lady, and still more, immediately were in the same state.' (Vaublanc, *Mémoires*, p. 150).
114. Pomme, *Traité*, p. 9 and pp. 578–82; Whytt, *Vapeurs*, preface pp. VI/VII; Revillon, *Recherches*, p. 48; Hunauld, *Vapeurs*, p. XXVII; Beauchêne, *Influence*, pp. 32–40.
115. Pomme, *Traité*, p. 10.
116. *Ibid.*, pp. 10f. and p. 580; Louyer-Villermay, *Recherches*, p. 1.
117. Cf. Johanna Bleker, 'Die Stadt als Krankheitsfaktor. Eine Analyse ärztlicher Auffassungen im 19. Jahrhundert', *Medizinhistorisches Journal* 18 (1983), 118–36; Michael Stolberg, *Ein Recht auf saubere Luft? Umweltkonflikte am Beginn des Industriezeitalters.* Erlangen 1994.
118. Michael Stolberg, 'The monthly malady. A history of pre-menstrual suffering', *Medical history* 44 (2000), 301–22.
119. Pomme, *Traité*, p. 9; Beauchêne, *Influence*, esp. pp. 11–18 and pp. 130–8; see also François Azouvi, 'La femme comme modèle de la pathologie au XVIIIème siècle', *Diogenes* 115 (1971), 25–40; Radkau, *Zeitalter* [see note 19], pp. 27f., claims that this association of nervous diseases with women lost importance in the course of the 19th century.
120. Hunauld, *Vapeurs*, p. 14.
121. Cf. Michael Stolberg, 'A woman down to her bones. The anatomy of sexual difference in early modern medicine', *Isis* 94 (2003), 274–99.
122. Cf. Thomas Laqueur, *Making sex. Body and gender from the Greeks to Freud.* Cambridge, MA/London 1990; Londa Schiebinger, 'Skeletons in the closet. The first illustrations of the female skeleton in eighteenth-century anatomy', *Representations* 14 (1986), 42–82; see also eadem, *Nature's body: gender in the making of modern science.* New Brunswick, NJ 2004 (orig. 1993); Claudia Honegger, *Die Ordnung der Geschlechter. Die Wissenschaften vom Menschen und das Weib 1750–1850.* Frankfurt/New York 1991; Ludmilla Jordanova, *Sexual visions: Images of gender in science and medicine between the eighteenth and twentieth centuries.* Madison 1989.

123. Beauchêne, *Influence*, p. 42.
124. Dena Goodman, *The republic of letters. A cultural history of the French Enlightenment*. Ithaca/London 1994; Marguerite Glotz/Madeleine Maire, *Salons du XVIIIᵉ siècle*. Paris 1949.
125. Ian Maclean, *The Renaissance notion of woman. A study in the fortunes of scholasticism and medical science in European intellectual life*. Cambridge 1980.
126. 'Vapeurs', in: *Encyclopédie*, vol. 16 (1765), pp. 836f.
127. Cf. Stolberg, *Monthly malady* [see note 118]; a tension between the medical critique of civilization on the one hand and notions of female nature on the other is also pointed out by Maurice Bloch/Jean H. Bloch, 'Women and the dialectics of nature in eighteenth-century French thought', in: Carol P. MacCormack/Marilyn Strathern (eds), *Nature, culture and gender*. Cambridge 1980, pp. 25–41, here p. 32.
128. Anne C. Vila, *Enlightenment and patholgy. Sensibility in the literature and medicine of eighteenth-century France*. Baltimore/London 1998; Georges Gusdorf, *Les sciences humaines et la pensée occidentale*, vol 7: *Naissance de la conscience romantique au siècle des lumières*. Paris 1976; André Monglond, *Le préromantisme français*, 2 vols. Paris 1965/66.
129. G. J. Barker-Benfield, *The culture of sensibility. Sex and society in eighteenth-century Britain*. Chicago/London 1992.
130. Vila, *Enlightenment* [see note 128].
131. See Stephen Greenblatt, *Renaissance self fashioning*. Chicago 1980.
132. Cf. George S. Rousseau. '"A strange pathology". Hysteria in the early modern world, 1500–1800', in Sander L. Gilman et al. (eds), *Hysteria beyond Freud*. Berkeley 1993, pp. 91–221, here p. 163.
133. According to Stanley Jackson, melancholia played a similar role in England (Jackson, *Melancholia* [see note 38], p. 138).
134. Caraccioli, *Dictionnaire*, vol. II, pp. 11f.
135. *Ibid.*, p. 45.
136. BCU Lausanne, FT, letter from Mme de Chastenay, 8 November 1784.
137. Chastenay, *Mémoires*, vol. I, pp. 43f.; in spite of her harsh words about their incompetence, Mme de Graffigny's relationship with her physicians also played a major role in her life, and she even gave them presents.
138. BCU Lausanne, FT, letter from Mme de Montcharlé, no date given.
139. Bressy, *Recherches*, pp. 1f.; cf. Edmond de Goncourt/Jules de Goncourt, *La femme au XVIIIᵉ siècle*. Paris 1862 (repr. 1982), pp. 272f.
140. Mandeville [see note 89], p. 270.
141. Genlis, *Dictionnaire*, vol. 2, pp. 367f.
142. Paumerelle, *Philosophie*.
143. Caraccioli, *Dictionnaire*, vol. I, p. 155f.
144. Rousseau, *'A strange pathology'* [see note 132], pp. 164f.
145. Dona Lee Davis/Richard G. Whitten, 'Medical and popular traditions of nerves', *Social science and medicine* 26 (1988), 1209–22.
146. Peter J. Guarnaccia/Pablo Farias, 'The social meanings of nervios. A case study of a Central American woman', *Social science and medicine* 26 (1988), 1123–31; Dona Lee Davis/Setha M. Low (eds), *Gender, health, and illness. The case of nerves*. New York 1989; Setha M. Low, 'Culturally interpreted symptoms or culture-bound syndromes. A cross-cultural review of nerves', *Social science*

and medicine 21 (1985), 187–96; eadem, 'Embodied metaphors: nerves as lived experience', in: Thomas J. Csordas (ed.), *Embodiment and experience. The existential ground of culture and self.* Cambridge 1994, pp. 139–62.

147. Kaja Finkler, 'The universality of nerves', in: Davis/Low, *Gender* [see note 147], pp. 169–79, here p. 174; see also Craig R. Janes, 'Imagined lives, suffering, and the work of culture. The embodied discourses of conflict in modern Tibet', *Medical anthropology quarterly* 13 (1999), 391–412.

148. Cf. the words of warning in Arthur Kleinman/Joan Kleinman, 'Somatization: The interconnections in Chinese society among culture, depressive experiences, and meanings of pain', in: Kleinman/Good, *Culture* [see note 13], pp. 429–90.

149. Dona Lee Davis/Richard G. Whitten, 'Medical and popular traditions of nerves', *Social science and medicine* 26 (1988), 1209–22.

150. Edward Shorter, *The making of the modern family.* New York 1975.

151. BCU Lausanne, FT, letter from the Comtesse de Non, 23 April 1785.

152. *Ibid.*, letter concerning Mme de Fayod, probably written by a physician, 2 August 1790.

153. *Ibid.*, letter from Louise Martin Vionnet, 30 September 1793.

154. *Ibid.*, letter from Mlle Kilchberger, 31 May 1790.

155. *Ibid.*, letter from Mme Vivaux, ca. 1797.

156. BIM Paris, Ms. 5241, foll. 71r–72r, letter concerning an unidentified patient, probably written by a relative; Geoffroy advised bloodletting in order to relieve the head and belly.

157. BCU Lausanne, FT, letter from Monsieur Claret, 7 April 1790.

158. *Ibid.*, letter from the Comte d'Albarey, concerning the Comtesse de Mouroux, 13 February 1790.

159. Adrienne Rogers, 'Women and the law', in: Samia I. Spencer (ed.), *French women and the age of Enlightenment.* Bloomington 1984, pp. 33–48.

160. BCU Lausanne, FT, letter from Mme de Ruys, 26 May 1777.

161. Daniel Teysseire, 'Mort du roi et troubles féminins: le premier valet de chambre de Louis XV consulte Tissot pour sa jeune femme (mai 1776)', in: Helmut Holzhey/Urs Boschung (eds), *Gesundheit und Krankheit im 18. Jahrhundert.* Amsterdam/Atlanta 1995, pp. 49–56; BCU Lausanne, FT, letter concerning Monsieur Mazet, possibly written by his physician, 23 August 1796.

162. BCU Lausanne, FT, letter concerning Field Marshal Grivel, written by his wife or another relative of the same name, 24 October 1791.

163. *Ibid.*, letter from Monsieur Dubois, Marseille, 29 January 1792.

164. *Ibid.*, letter from Mme Develay, 21 May 1791.

165. Genlis, *Dictionnaire*, vol. 1, pp. 367f.

166. BCU Lausanne, FT, letter from an unidentified middle-aged woman, 17 December 1783.

167. IGM Stuttgart, HA, Best. B 321014, letter from the Comte de las Cases, 29 July 1832.

168. BCU Lausanne, FT, letter from Monsieur Matthis, 1 October 1773.

169. *Ibid.*, letter from Pierre Toussaint, 3 February 1788.

170. *Ibid.*, letter from an unidentified 60-year-old female patient.

171. *Ibid.*, history of Monsieur Larrei's disease, written in the third person but clearly by himself, 5 January 1784.

172. Kleinman/Kleinman. 'Somatization' [see note 145]; cf. Arthur Kleinman, *Social origins of distress and disease. Depression, neurasthenia, and pain in modern China.* New Haven/London 1986.
173. Heinz Kohut, *Narzißmus. Eine Theorie der psychoanalytischen Behandlung narzißtischer Persönlichkeitsstörungen.* Frankfurt 1976.
174. Deffand, *Lettres*; Meilhan, *Considérations.*
175. Goncourt/Goncourt, *La femme* [see note 139], p. 272 and pp. 287f.
176. BCU Lausanne, FT, letter from Monsieur Gringet, 4 January 1784; in retrospect, one may wonder about a possible connection between the proverbial 'ennui' of the 18th-century French upper classes, especially among women, and their almost total lack of physical work or exercise, which could have promoted a more intense experience of their bodies and themselves. Some patients suggest an awareness of this lack. Julie de l'Espinasse, for example, asked von Haller explicitly to also prescribe her physical exercise, in order to reduce her 'melancholy vapors' with the help of 'gaiety' ('gaieté') (BB Bern, HC, De l'Espinasse, 4 October 1762).
177. Longrois, *Conseils*, pp. 10f.
178. *Ibid.*; Venel, *Sur la santé*, p. 42; cf. Philippe Perrot, *Le travail des apparences ou les transformations du corps féminin XVIIIᵉ-XIXᵉ siècle.* Paris 1984, p. 81.
179. Cf. Goncourt/Goncourt, *La femme* [see note 139], p. 267.
180. Caraccioli, *Dictionnaire*, vol. II, pp. 4f.
181. Sénac de Meilhan, *Considérations*, p. 251.
182. René Pillorget, *La tige et le rameau. Familles anglaise et française*, XVIᵉ-XVIIIᵉ siècle. Paris 1979, p. 276.
183. Cissie Fairchilds, 'Women and family', in: Spencer, *French women* [see note 158], pp. 97–110, here p. 100.
184. BCU Lausanne, FT, letter from Mme de Chastenay, 8 November 1784.
185. *Ibid.*, letter from Monsieur Thomassin, a 20-year-old artillery officer, May 1775.
186. *Ibid.*, second letter from Monsieur Thomassin, 28 May 1776; apparently Tissot had not replied to his first letter.
187. For historical surveys see Jean Stengers/Anne Van Neck, *Histoire d'une grande peur: la masturbation.* Brussels 1984; Franz X. Eder, *Kultur der Begierde. Eine Geschichte der Sexualität.* Munich 2002; Thomas W. Laqueur, *Solitary sex. A cultural history of masturbation.* New York 2003.
188. Gerson, *De confessione*; Benedicti, *Somme*, pp. 132f.; a good survey of the 18th-century debate can be found in Karl-Heinz Bloch, *Die Bekämpfung der Jugendmasturbation im 18. Jahrhundert. Ursachen – Verlauf – Nachwirkungen.* Frankfurt 1998, pp. 73–110; see also idem, *Masturbation und Sexualerziehung in Vergangenheit und Gegenwart. Ein kritischer Literaturbericht.* Frankfurt 1989; P. Hurteau, 'Catholic moral discourse on male sodomy and masturbation in the seventeenth and eighteenth centuries', *Journal of the history of sexuality* 4 (1993/94), 1–26.
189. *Eronania*, part 2, p. 3.
190. *Letters of advice*, p. 26.
191. Capel, *Tentations*, p. 354.
192. Cf. Brandius, *Querela*; *The crime of Onan*; the term 'Onanian' can be found on p. 31; *Onanism display'd.*

193. The term was sometimes used to refer to male homosexuality.
194. The spelling of the term varies, as do the images associated with it; 'masturpatio' suggests the Latin word 'turpis' ('infamous'), 'manustupratio' the Latin word 'stuprum' ('fornication'); cf. Castro, *Universa mulierum medicina*, p. 97; Schurig, *Spermatologia*, pp. 244f.
195. Genesis 38.9–10.
196. Cf. Michael Stolberg, 'The crime of Onan and the laws of Nature. Religious and medical discourses on masturbation in the late 17th and early 18th centuries', *Paedagogica historica* 2003, 707–17.
197. Da Castro, *Universa mulierum medicina*, p. 97; Schuyl, *De gonorrhea*.
198. Ettmüller, *Opera*, vol. II, p. 422.
199. Beverwijk, *Schat*, p. 174; similarly Timaeus von Güldenklee, *Responsa*, pp. 191–3.
200. Capel, *Tentations* (1655 edn), pp. 210f.
201. *Letters* of *advice*, pp. 16f.
202. The original date of publication was for a long time a matter of dispute but it has now been established beyond doubt; see Michael Stolberg, 'Self-pollution, moral reform, and the venereal trade. Notes on the sources and historical context of the "Onania" (1716)', *Journal of the history of sexuality* 9 (2000), 37–61.
203. *Onania*, p. 21.
204. *Ibid.*, p. 19.
205. *Ibid.*, p. 28.
206. *Onania*, 3rd edn (1717), pp. 76–82.
207. *Eronania*; *Crime of Onan*; Philo-Castitatis, *'Onania' examined*; *Practical scheme*, p. 19.
208. *Crime of Onan*, pp. 30f.
209. Stolberg, *Self-pollution* [see note 201].
210. Marten, *Treatise*, pp. 352–429; *ibid.*, 7th edn, London 1711, pp. 106–55. The 4th edn, under the title *A true and succinct account of the venereal disease* (London 1706), mentions masturbation only briefly on p. 226.
211. Rothos, *A whip*; cf. Laqueur, *Solitary sex* [see note 186], footnote p. 425.
212. *London Gazette*, 28–30 July 1709.
213. Schurig, *Spermatologia*, pp. 244f.; Paullini, *Observationes*, p. 513.
214. G. L., preface 'Vom weißen Fluß', in: Stahl, *Abhandlung*, p. 24.
215. Hoffmann, *Consultationum*, pp. 496–8.
216. *De groote zonden van vuile zelfs-bevleckinge, door jonge en oude, mans en vrouws persoonen*. Rotterdam 1730; *Onania, oder die erschreckliche Sünde der Selbst-Befleckung, mit allen ihren entsetzlichen Folgen, so dieselbe bey beyderley Geschlecht nach sich zu ziehen pfleget*. Leipzig 1736.
217. *Neue woleingerichtete Frauenzimmerapotheke*, pp. 3–27; the text is based on: *The ladies' dispensatory*.
218. Zedler, *Universal-Lexikon*, vol. 25 (1740) already has an entry on 'Onanie' but it refers to the entry on 'Selbstbefleckung' ('self-pollution') in: *ibid.*, vol. 36 (1743), pp. 1586–90.
219. Gordon, *Letter to John Hunter*; the full title of Gordon's treatise reads: A letter to John Hunter [...] respecting his treatise on the venereal disease, shewing him to be highly erroneous in his observations on impotence,

and more particularly pointing out the absurdity and immorality of his doctrine in favour of onanism or masturbation.

220. Tissot, *Onanisme*, preface to the 8th edn, Lausanne 1785, pp. XVIf.
221. BCU Lausanne, FT, letter from J. P. Gay, 21 January 1773; he reported the experiences of two clergymen.
222. Fabricius, *Opera*, p. 1030.
223. Beverwijk, *Schat*, p. 174.
224. *Onania*, pp. 105–21 and pp. 125–57.
225. Tauvry, *Nouvelle anatomie*, pp. 155f.
226. Schmieder, *Observatio*.
227. For the more recent past see Lesley A. Hall, 'Forbidden by God, despised by men: masturbation, medical warnings, moral panic, and manhood in Great Britain, 1850–1950', *Journal of the history of sexuality* 2 (1991/92), 365–87; Laqueur, *Solitary sex* [see note 186] pp. 359–420, on 20th-century accounts.
228. Philippe Lejeune, 'Le "dangereux supplément". Lecture d'un aveu de Rousseau', *Annales ESC* 29 (1974), 1009–22; Paule Adamy, *Le corps de Jean-Jacques Rousseau*. Paris 1997.
229. Tim Hitchcock, for example, mentions a passage from John Cannon's memoirs where he confesses having masturbated, in the early 18th century, in front of illustrations from a midwifery book, until his mother took it away from him (T. Hitchcock, *English sexualitites*. Basingstoke/London 1997, p. 14).
230. Since the style and content of these letters are very similar to those of other contemporary patient letters they may well have been authentic, however.
231. He crossed out the more concrete term 'attouchements' and wrote 'libertés' instead.
232. BIM Paris, Ms. 5244, foll. 167r–168r, letter from Monsieur Scellier, 27 July 1727.
233. BCU Lausanne, FT, letter from an unidentified patient, born in 1736, ca. 1775, apparently a member of the army.
234. *Ibid.*, letter from the Chevalier de Belfontaine, 25 November 1772.
235. *Ibid.*, letter from Mme de Chastenay, 8 November 1784.
236. BB Bern, HC, A 184, letter from an unidentified 28-year-old patient, ca. 1775.
237. BCU Lausanne, FT, letter concerning the Baron von Hohenfeld, no date given.
238. BIM Paris, Ms. 5241, foll. 279r–281r, letter concerning an unidentified patient, probably written by a layperson, no date given.
239. BCU Lausanne, FT, letter from the Chevalier de Belfontaine, 25 November 1772.
240. *Ibid.*, letter from an unidentified 25-year-old patient, 27 May 1768.
241. *Ibid.*, letter from an unidentified patient, in English, no date given.
242. BIM Paris, Ms. 5241, foll. 279r–281r.
243. BCU Lausanne, FT, letter from Monsieur de Lavan, 14 July 1793.
244. *Ibid.*, undated letter from an unidentified patient.
245. BCU Lausanne, FT, letter from Monsieur Reichert, 5 March 1793; *ibid.*, letter from Monsieur Chillaud, 4 May 1790.

246. *Ibid.*, letter from Monsieur Chillaud, 4 May 1790.
247. *Ibid.*, letter from Monsieur Roussany, 10 June 1774.
248. *Ibid.*, letter from Monsieur Krizler, 12 November 1791; *ibid.*, letter from Monsieur Barbaroux, 21 August 1793.
249. *Ibid.*, letter from a law student, in English, no date given; *ibid.*, letter from Monsieur Reichert, 5 March 1793; *ibid.*, letter from Monsieur Chillaud, 4 May 1790; *ibid.*, letter from Monsieur Krizler, 12 November 1791.
250. *Ibid.*, letter from Claude Jouple about an unidentified patient, no date given.
251. *Ibid.*, letter from Monsieur Reichert, 5 March 1793; *ibid.*, letter from Monsieur Barbaroux, 21 August 1793; *ibid.*, letter from Monsieur Krizler, 12 November 1791.
252. *Ibid.*, letter from Monsieur Chillaud, 4 May 1790.
253. For example BB Bern, HC, A 184, letter from an unidentified 28-year-old patient, ca. 1775, who called his dry cough a 'first chastisement'.
254. BB Bern, HC, A 184, letter from an unidentified 28-year-old patient, ca. 1775.
255. BCU Lausanne, FT, letter from Monsieur Lavan, 14 July 1793.
256. *Ibid.*, letter from an unidentified patient, no date given.
257. See also my analysis in Michael Stolberg, 'An unmanly vice: Self-pollution, anxiety, and the body in the eighteenth century', *Social history of medicine* 13 (2000), 1–21.
258. Ostervald, *Nature*; cf. Karl-Felix Jacobs, *Die Entstehung der Onanie-Literatur im 17. und 18. Jahrhundert*. Diss. med. Munich 1963.
259. Thus already Benedicti, *Somme*, p. 141.
260. Ostervald, *Nature*, p. XIV and p. XXIII, preface, probably by the translator.
261. Beverland, *De fornicatione*, p. 22.
262. *Letters of advice*, p. 18.
263. *Onania*, 8th edn, pp. 23f.
264. Capel, *Tentations*, pp. 354–6.
265. *Eronania*, pp. 3f.; cf. Isabel V. Hull, *Sexuality, state, and civil society in Germany, 1700–1815*. Ithaca/London 1996, pp. 258–80.
266. *Eronania*, p. 2.
267. See also *Onania*, p. 22.
268. Edward Baynard called it bluntly the 'cursed school wickedness' (Baynard, *Letter*, p. 278).
269. Théodore Tarczylo, *Sexe et liberté au siècle des Lumières*. Paris 1983.
270. Beverland, *De fornicatione*, p. 22.
271. C. Durston, *The family in the English Revolution*. Oxford/New York 1989, pp. 124–30; R. A. Houlbrooke, *The English family 1450–1700*. London/New York 1984, pp. 127–65; cf. Beverland, *De peccato*.
272. Robert Sumser, '"Erziehung", the family, and the regulation of sexuality in late Enlightenment Germany', *German studies review* 15 (1992), 455–74.
273. Michel Foucault, *The care of the self*. New York 1986.

Conclusion: A New Bourgeois Habitus

1. Norbert Elias, *The civilizing process*. New York 1978; see also Barbara Duden, *The woman beneath the skin: A doctor's patients in eighteenth-century Germany*. Cambridge, MA 1991.

2. Hufeland, *Makrobiotik*.

3. *Verhandlungen der Kammer der Abgeordneten der Ständeversammlung des Königreichs Bayern* 15 (1837), p. 315; cf. Michael Stolberg, *Die Homöopathie im Königreich Bayern (1800–1914)*. Heidelberg 1999.

4. *Verhandlungen* [see note 3] 15 (1837), p. 315; similarly *ibid.* 6 (1849/50), p. 495, speech by Dr Sepp.

5. *Verhandlungen der Kammer der Reichsräthe des Königreichs Bayern* 8 (1850), p. 391.

6. *Verhandlungen* [see note 3] 18 (1843), p. 34.

7. SB München, Cgm 6874/111.

8. Cf. Bressy, *Recherches*, p. 3: 'In its revolution, the soul no longer finds obedience in the body it rules.'

9. Dorinda Outram, *The body and the French revolution. Sex, class and political culture*. New Haven/London 1989.

10. Pierre Bourdieu, *Outline of a theory of practice*. Cambridge 1977.

11. Barthez, *Oratio*.

12. N. D. Jewson, 'Medical knowledge and the patronage system in 18th-century England', *Sociology* 8 (1974), 369–85.

Manuscript Sources

Amsterdam, Universiteitsbibliotheek (UB)
  A q 3, 14 CD 1 & 2; Gd. 39, Gd. 40; Y 81; Y 165a, consultations and natal charts by J. Chr. Ludeman
HS IV A II, medical consultations, ca. 1775
  Arnhem, Rijksarchief van Gelderland (RA)
Archive of the Counts of Coulembourg, Ms. 403, letters in times of sickness
  Avignon, Bibliothèque municipale (BM)
Ms. 3192, consultations and medical reports
Ms. 3997 practice journal of Jean-Claude Pancin
  Bamberg, Staatsarchiv (StA)
KIIIF3 1481, medical topographies, ca. 1830
  Basel, Universitätsbibliothek (UB)
  Ms. Fr. Gr. I 6, medical correspondence
  Berlin, Staatsbibliothek (SB)
Ms.germ.fol. 99, 420a, 420b, 421a, 421b, 422a, 422b, 423a, 423b, 424, 425, 426, correspondence of Leonhard Thurneisser
Francke-Nachlaß
  Bern, Burgerbibliothek (BB)
correspondence of Albrecht von Haller (HC)
  Bologna, Archivio di Stato (AS)
Studio 339-51
  Bremen, Staatsarchiv (StA)
2 S.7.a.12 'Quacksalber und Kurpfuscher 1647–1864'
  Dresden, Staatliche Landes- und Universitätsbibliothek (SLUB)
C 466, consilia and consultations, collected by Lorenz Heister 1705/06
  Erlangen, Universitätsbibliothek (UB)
Ms. 948
Ms. 1029
Trew-collection
  Frankfurt, Senckenbergische Bibliothek (SB)
Mss. 332–6, medical correspondence
  Genève, Bibliothèque Universitaire et Publique (BUP)
Ms. Suppl. 647
Mss. Suppl. 1908–09
  Hamburg, Staats- und Universitätsbibliothek (SUB)
Sup. ep. 4 30, letters to Joseph Duchesne
Sup. ep. 109, correspondence of Paul Marquard Schlegel
  Köln, Stadtarchiv (StA)
2/B 846-848 Testamente
Universität A 459, Streitsachen
  Lausanne-Dorigny, Bibliothèque Cantonale et Universitaire (BCU)
IS3784/II, Fonds Tissot (FT)
  Leiden, Universiteitsbibliotheek (UB)

Mss. Marchand 3, letters to Johann and Otto van Heurne
  London, British Library (BL)
Sloane Mss. 4036; 4039; 4075–8.
  London, Library of the Wellcome Institute for the History of Medicine (WL)
Western Manuscripts, Ms. 6114, patient correspondence of Verdeil
Western Manuscripts, Ms. 6873, papers of Robert Whytt
  München, Bayerische Staatsbibliothek (SB)
Cgm 6141, note book of a Coburg physician, 1739
Cgm 6874, medical topographies, ca. 1860
  München, Hauptstaatsarchiv (HStA)
MInn 61717, district physicians' annual reports
  München, Stadtarchiv (StA)
Historischer Verein, Mss. 401/1–40, medical topographies, ca. 1860
  Nürnberg, Stadtarchiv (StA)
B 14 II 9, 12, 15, 18, court records (Stadt-, Bauern- and Untergericht)
D1 201, Stadtalmosenamt
  Paris, Bibliothèque Interuniversitaire de Médecine (BIM)
Ms. 2075, collection of medical consultations
Mss. 5241–5, patient correspondence, mostly of É.-F. Geoffroy
  Regensburg, Fürstlich Thurn und Taxissches Zentralarchiv (FTTZ)
Haus- und Familiensachen, Ms. 3786
  Schwerin, Landeshauptarchiv (LHA)
2.12-1/22, private letters
2.12-2/3, laws and ordinances
  St Gallen, Kantonsbibliothek (KB)
Ms. 94, correspondence of Sebastian and Bartholomäus Schobinger
  Stuttgart, Hauptstaatsarchiv (HStA)
Bestand A 213, Büschel 6734
Bestand A 209, Büschel 547, 1668 and 1921
  Stuttgart, Institut für Geschichte der Medizin der Robert Bosch Stiftung (IGM)
Hahnemann Archiv (HA)
Bestand B, (mostly) German patient letters
Bestand C, (mostly) French patient letters
  Utrecht, Rijksarchief (RA)
Ms. 2200 and 2201, German medical recipe books, 1604 and 1634
Family archive Van Geuns, Ms. 46
  Weimar, Thüringisches Hauptstaatsarchiv (HStA)
Stadt Weimar B 7607a, 'Polizeisachen'
  Wiesbaden, Hessisches Hauptstaatsarchiv (HStA)
Abt. 369, witchcraft trials

Printed Sources

Abrégé des miracles, des graces et merveilles, avenus à l'intercession de la glorieuse Vierge Marie. Brussels 1664.

Ailhaud, Jean Gaspard, *Médecine universelle prouvée par le raisonnement, démontrée par l'expérience.* Carpentras 1760.

Alberti, Michael, *Introductio in universam medicinam tam theoreticam quam practicam.* Halle/Magdeburg 1718.

Andreae, Jakob, *Leben des Jakob Andreae, Doktor der Theologie, von ihm selbst mit großer Treue und Aufrichtigkeit beschrieben, bis auf das Jahr Christi 1562.* Ed. and transl. by Hermann Ehmer. Stuttgart 1991.

Andreä, Johann Valentin, *Johann Valentin Andreä – ein schwäbischer Pfarrer im Dreißigjährigen Krieg.* Ed. by Paul Antony. Heidenheim 1970.

Anhart, Elias, Consilium podagricum, daß ist, wie man sich vor dem Podagra hüten, oder in Zeit dieser Kranckheit curiren und trösten soll, allen Layen so podagrisch seyn zu Gutem gestellet. Ingolstadt 1585.

Armand, P., *Considérations sur l'âge critique des femmes, et sur les moyens de prévenir les maladies qui peuvent survenir à cette période de la vie.* Thèse. Paris 1820.

Astruc, Jean, *A treatise on all the diseases incident to women.* London 1743.

Baglivi, Giorgio, *Opera omnia medico-practica et anatomica.* Lyon 1704.

Becane, B., *Observations sur les effets du virus cancéreux.* Toulouse 1778.

Benedicti, Jean, *La somme des pechez.* Paris 1595.

Betrand, -, *Relation historique de la peste de Marseille en 1720.* Cologne 1721.

Blancardus, Stephanus, *Accurate Abhandlung von dem Podagra und der lauffenden Gicht.* Sine loco 1692.

Barthez, Paul-Joseph, *Oratio academica de principio vitali hominis.* Montpelier 1773.

E. Baynard, 'A letter to Sir John Floyer, Kt. in Litchfield, concerning cold immersion', in: John Floyer, *The ancient psychrologeia revived: or, an essay to prove cold bathing both safe and useful.* London 1702, pp. 207–317.

Beauchêne, [Edme Pierre] de, *De l'influence des affections de l'âme dans les maladies nerveuses des femmes.* Montpelier 1781.

Beverland, Adriaan, *De fornicatione cavenda admonitio, sive adhortatio ad pudicitiam et castitatem.* Sine loco 1698.

Beverland, Adriaan, *De peccato originali.* Sine loco 1679.

Beverwijk, Johan van, *Schat der ongesontheydt, ofte genees-konste van de siechten.* Amsterdam 1672.

Bismarck, Christoph von, 'Das Tagebuch des Christoph von Bismarck aus den Jahren 1625–1640'. Ed. by Georg Schmidt, *Thüringisch-sächsische Zeitschrift* 5 (1915), 67–98.

Boë, François de la, *Idea praxeos medicae in tres libros divisae.* Frankfurt 1671.

Boerhaave, Herman, *Abhandlung vom Krebs und Kranckheiten der Knochen.* Frankfurt 1756.

Boerhaave, Herman, *Boerhaave's medical correspondence; containing various symptoms or chronical distempers.* London 1745.

Bösch, Alexander, *Liber familiarum personalium, das ist, Verzeichnuß waß sich mit mir, und der meinigen in meiner haußhaltung, sonderliches begeben und zugetragen hatt*. Ed. by Lorenz Heiligensetzer. Basel 2000.

Bövingh, Johann Georg, 'Die Lebensbeschreibung des Johann Georg Bövingh (1676–1728)'. Ed. by Elfriede Bachmann, *Rotenburger Schriften* 48/49 (1978), 92–181.

Boscius, Joannis L., *Kurtzer Bericht von dem Podagra und andern Gliedsuchten*. Ingolstadt 1585.

Bozenhart, P. Johannes, 'Schicksale des Klosters Elchingen und seiner Umgebung in der Zeit des dreissigjährigen Krieges (1629–1645). Aus dem Tagebuche des P. Johannes Bozenhart'. Ed. by P. L. Brunner, *Zeitschrift des Historischen Vereins für Schwaben und Neuburg* 1876, 157–282.

Brandis, Henning, *Diarium. Hildesheimische Geschichten aus den Jahren 1471–1528*. Ed. by Ludwig Hänselmann. Hildesheim 1896.

Brandius, Johannes, 'Querela super peccato ononitico enormissimo', in: Beverland, *De fornicatione*, pp. 87–106.

Brenner-Schaeffer, Wilhelm, *Darstellung der sanitätlichen Volkssitten und des medizinischen Volks-Aberglaubens im nordöstlichen Theile der Oberpfalz*. Amberg 1861.

Bressy, Joseph, *Recherches sur les vapeurs*. London 1789.

Brokes, Henrich, 'Aus dem Tagebuche des Lübeckischen Bürgermeisters Henrich Brokes'. Ed. by Dr Pauli, *Zeitschrift des Vereins für Lübeckische Geschichte und Altertumskunde* 1 (1855), 79–92, 173–83 and 281–347; *ibid.*, 2 (1858), 1–37, 254–96 and 367–465.

Brunner, Wolfgang M., *Das nicht erkante,offt turbirte und selten curirte Podagra*. Regensburg sine anno.

Buchan, William, *Domestic medicine*. London 1765.

Buchoz, Pierre-Joseph, *Traité de la phthisie pulmonaire*. Paris 1769.

Burney, Fanny, *The journals and letters of Fanny Burney (Madame d'Arblay)*. Ed. by Joyce Hemlow, vol. 6. Oxford 1975.

Butzbach, Johannes, *Odeporicon. Eine Autobiographie aus dem Jahre 1506*. Ed. and transl. by Andreas Beriger. Weinheim 1991.

Capel, Richard, *Tentations: Their nature, danger, cure*. London 1633.

Caraccioli, Luigi Antonio, *Dictionnaire critique, pittoresque et sentencieux propre à faire connoître les usages du siècle, ainsi que ses bisarreries*, 3 vols. Lyon 1768.

Castro, Roderigo da, *Universa mulierum medicina, pars 2, sive praxis*. Hamburg 1662.

Caylus, Madame de, *Les souvenirs*. Amsterdam 1770.

Chastelain, Jean, *Traité des convulsions et des mouvemens convulsifs, qu'on appelle à présent vapeurs*. Paris 1691.

Chastenay, Victorine de, *Mémoires 1771–1815*, 2 vols. Ed. by A. Roserot. Paris 1896.

Cheyne, George, *The English malady, or a treatise of nervous diseases of all kinds, as spleen, vapours, lowness of spirits, hypochondriacal, and hysterical distempers*, 2nd edn. London 1734.

Clacius, Georg, *Consilia medica über gantz besondere, schwere, auch rare Casus*. Frankfurt/Leipzig 1739.

Conversations-Lexicon für die gebildeten Stände, 10 vols. Leipzig/Altenberg 1814–19.

Cornaro, Luigi/ Lessius, Leonardus, *Hygiasticon, or, the right course of preserving life and health unto extream old age: together with soundnesse and integritie of the senses, judgement, and memorie [...]. A treatise of temperance and sobrietie.* Cambridge 1634.

Dahmen, Jacobus Lambertus, *De Chlorosi, vulgo von der Jungfern-Krankheit.* Strasbourg 1747.

Deffand, Marie de, *Lettres à Horace Walpole, écrites dans les années 1766 à 1780; auxquelles sont jointes des lettres de Madame du Deffand à Voltaire, écrites dans les années 1759 à 1775,* 4 vols. Paris 1824.

Diesbach, Ludwig von, *Die autobiographischen Aufzeichnungen Ludwig von Diesbachs.* Ed. by Urs Martin Zahnd. Bern 1986.

Dreytwein, Dionysius, *Esslingische Chronik. 1548–1564.* Ed. by Adolf Diehl. Tübingen 1901.

Dryander, Ioannes, *Der gantzen Artzenei gemeyner Inhalt, wes einem Artzt bede in der Theorie vnd Practic zusteht.* Frankfurt 1542.

Dubois, E. Frédéric, *Histoire philosophique de l'hypochondrie.* Paris 1837.

Dumoulin, M., *Nouveau traité du rhumatisme et des vapeurs.* Paris 1703.

Encyclopédie ou dictionnaire raisonné des sciences, des arts et des métiers. Neufchastel 1765.

Eronania. On the misusing of the marriage bed by Er and Onan. London 1724.

Essich, Johann Gottfried, *Gesundheitswörterbuch für das Landvolk und den gemeinen Mann.* Augsburg 1789.

Ettmüller, Michael, *Opera omnia theoretica et practica.* Lyon 1685.

Fabricius, Theodor: 'Lebensbeschreibung des ersten anhaltischen Superintendenten'. Ed. by Dr Münnich, *Zerbster Jahrbuch* 16 (1931), 37–94.

Fabricius, Wilhelm, *Deß weitberühmten Guilhelmi Fabricii Hildani Wund-Artzney.* Frankfurt 1652.

Fabricius, Wilhelm, *Opera omnia quae extant.* Frankfurt 1656.

Fernel, Jean, *Universa medicina.* Geneva 1644.

Ferrières, Charles-Élie Marquis de, *Mémoires.* Ed. by Berville Barrière, 2nd edn. Paris 1822.

Ficino, Marsilio, *Three books on life. A critical edition and translation with introduction and notes.* Ed. by Carol V. Kaske and John R. Clark. Binghamton, NY 1989.

Fischer, Johann A., *Consilia medica.* Frankfurt 1705.

Fitzgerald, Gerald, *Traité des maladies de femmes.* Paris 1758.

Flamant, -, *Die Kunst sein eigener Medicus zu seyn.* Franckenhausen 1721.

Fleck, Peter, 'Aus einer oberhessischen Familienchronik', *Aschaffenburger Jahrbuch* 15 (1992), 217–22.

Flügel, Dr, *Volksmedizin und Aberglaube im Frankenwalde.* Munich 1863.

Foreest, Pieter van, *Observationum et curationum medicinalium libri duo: nempe decimussextus De pectoris pulmonisque vitiis ac morbis: Et decimusseptimus De cordis ac quibusdam mammillarum affectibus.* Leiden 1603.

Fothergill, John, *Conseils pour les femmes de quarante-cinq à cinquante ans. Ou conduite à tenir lors de la cessation des règles.* London 1788.

Freind, John, *Emmenologia.* London 1729.

Fries, Lorenz, *Spiegl der Artzny.* Strasbourg 1519.

Fröschel, Hieronymus, 'Der Augsburger Jurist Dr Hieronymus Fröschel und seine Hauschronik von 1528–1600'. Ed. by Friedrich Roth, *Zeitschrift des Historischen Vereins für Schwaben und Neuburg* 38 (1912), 1–82.

Fuchs, Leonhart, *De curandis totius humani corporis morbis*. Frankfurt 1567.

Galen, *Opera omnia*. Ed. by C. G. Kühn. Leipzig 1822 (repr. Hildesheim 1964).

Garnier, Joseph-François, *Considérations sur l'âge critique, et sur l'hygiène des femmes à cette époque*, vol. 4. Paris 1820.

Geizkofler, Lucas, *Lucas Geizkofler und seine Selbstbiographie. 1550–1620*. Ed. by Adam Wolf. Vienna 1873.

Gendron, Deshais, *Recherches sur la nature et la guérison des cancers*. Paris 1700.

Genlis, Stéphanie-Félicité Comtesse du Crest de St Aubin, *Souvenirs de Félicie L**. Paris 1804.

Genlis, Stéphanie-Félicité Comtesse du Crest de St Aubin, *Dictionnaire critique et raisonné des étiquettes de la cour, des usages du monde, des amusemens, des modes, des mœurs, etc., des françois, depuis la mort de Louis XIII jusqu'à nos jours*, 2 vols. Paris 1818.

Gerson, Johannes, 'De confessione mollitiei', in: idem, *Œuvres complètes*, vol. 8. Paris 1971, pp. 71–5.

Glückel von Hameln, *Denkwürdigkeiten der Glückel von Hameln*. Ed. by Alfred Feilchenfeld. Frankfurt 1987.

Gockel, Eberhard, *Consilia medicinales: decades sex*. Augsburg 1683.

Gordon, Duncan, *A letter to John Hunter [...] respecting his treatise on the venereal disease*. London 1786.

[Goulin, Jean], *Le médecin des dames, ou l'art de les conserver en santé*. Paris 1772.

Graffigny, Françoise de, *Correspondance*, 12 vols. Oxford 1985–2008.

Guyon, Jeanne Marie Bouvier de la Motte, *La vie de Madame Guyon écrite par elle-même*. Paris 1983.

Hartung, Gangolf, 'Die chronikalischen Aufzeichnungen des Fuldaer Bürgers Gangolf Hartung (1607–1666)'. Ed. by Th. Haas, *Fuldaer Geschichtsblätter* 9 (1910), 49–176.

Harvey, William, *Exercitatio anatomica de motu cordis et sanguinis in animalibus*. Frankfurt 1628.

Helmont, Johann Baptist van, *Aufgang der Artzney-Kunst*. Sulzbach 1683.

Helvetius, Jean Adrien, *Traité des maladies les plus fréquentes et des remedes propres à les guérir*. Paris 1739.

Hirsch, Caspar, 'Familienaufzeichnungen'. Ed. by Ferdinand Mencik, *Jahrbuch der Gesellschaft für die Geschichte des Protestantismus in Österreich* 22 (1901), 18–52.

Hoffmann, Friedrich, *Consultationum et responsorum medicinalium centuria secunda complectens morbos abdominis*, vol. 2. Amsterdam 1735.

Hoffmann, Friedrich, *Medicina consultatoria*, 12 vols. Halle 1721–39.

Holsten, Hieronymus Christian von, *Kriegsabenteuer des Rittmeisters Hieronymus Christian von Holsten 1655–1666*. Ed. by Helmut Lahrkamp. Wiesbaden 1971.

Houppeville, M. de, *La guérison du cancer du sein*. Rouen 1693.

Hufeland, Christoph Wilhelm, *Geschichte der Gesundheit nebst einer physischen Karakteristik des jetzigen Zeitalters. Eine Skizze*, 2nd edn. Berlin 1813.

Hufeland, Christoph Wilhelm, *Makrobiotik oder die Kunst das Leben zu verlängern*, 5th edn. Berlin 1823.

Hummel, Johann Heinrich, 'Histori des Lebens Johannis Henrici Hummelii. Eine Autobiographie aus dem 17. Jahrhundert'. Ed. by Chr. Erni, *Berner Zeitschrift für Geschichte und Heimatkunde* 1950, 24–57.

Hunauld, Pierre, *Dissertation sur les vapeurs et les pertes de sang.* Paris 1756.

Hunter, John, *A treatise on the venereal disease.* London 1786.

Hutten, Ulrich von, 'Des Ritters Ulrich von Hutten Brief [vom 25 October 1518] an den Nürnberger Patrizier Willibald Pirckheimer, in dem er über sein Leben Rechenschaft ablegt', in: idem, *Deutsche Schriften.* Ed. by Peter Ukena. Munich 1970, pp. 317–40.

Jacquin, Armand Pierre, *Abhandlung von der Gesundheit.* Augsburg 1764.

Jeannet de Longrois, [Jean-Baptiste-Claude], *Conseils aux femmes de quarante ans.* Paris 1787.

Jungken, Johann Helferich, *Wohlunterrichtender sorgfältiger Medicus.* Nuremberg 1725.

Jurin, James, *The correspondence of James Jurin (1684–1750). Physician and secretary to the Royal Society.* Ed. by Andrea Rusnock. Amsterdam/Atlanta 1996.

Khevenhüller, Hans: *Hans Khevenhüller kaiserlicher Botschafter bei Philipp II. Geheimes Tagebuch 1548–1605.* Ed. by Georg Khevenhüller-Metsch. Graz 1971.

Kiechel, Samuel, *Die Reisen des Samuel Kiechel, aus drei Handschriften.* Ed. K. D. Hassler. Stuttgart 1866.

Krafft, Hans Ulrich, *Reisen und Gefangenschaft Hans Ulrich Kraffts.* Ed. by Konrad Dietrich Hassler. Stuttgart 1861.

Lallemand, François-Claude, *Des pertes seminales involontaires*, 3 vols. Paris 1836–42.

La Tour du Pin, Henriette, *Journal d'une femme de cinquante ans 1778–1815*, 7th edn, vol. 2. Paris 1913.

Ledel, Johann Samuel, *Breviarium epistolicum medico-consultatorium sive brevis manuductio ad epistolas medico-consultatorias conscribendas neo-medicis admodum utilis cui accesserunt observationum medico-physicarum decuriae tres.* Sorau 1734.

Lepois, Charles, *Selectiorum observationum et consiliorum [...] liber singularis.* Pont-à-Mousson 1618.

Letters of advice from two reverend divines to a young gentleman, about a weighty case of conscience. London 1676.

Levis, Pierre Marc Gaston de, *Souvenirs et portraits, 1780–1789.* Paris 1813.

Lind, James, *A treatise of the scurvy, in three parts. Containing an inquiry into the nature, causes, and cure, of that disease.* Edinburgh 1753.

Lorry, Anne-Charles, *Von der Melancholie und den melancholischen Krankheiten*, vol. 1. Frankfurt/Leipzig 1770.

Louyer-Villermay, J. B., *Recherches historiques et médicales sur l'hypocondrie.* Paris 1802.

Malboissière, Geneviève de, *Lettres de Geneviève de Malboissière à Adélaïde Méliand 1761–1766.* Ed. by Albert de Luppé. Paris 1924.

Mandeville, Bernard, *A treatise on the hypocondriack and hysterick diseases in three dialogues*, 3rd edn. London 1730.

Maraise, Marie-Catherine-Renée de, *Une femme d'affaires au XVIIIe siècle. La correspondance de Mme de Maraise, collaboratrice d'Oberkampf.* Ed. by Serge Chassagne. Toulouse 1981.

Marcard, Heinrich Matthias, *Beschreibung von Pyrmont*, vol. 2. Leipzig 1785.

Marmontel, Jean-François, *Mémoires*. Ed. by John Renwick. Clermont-Ferrand 1972.

Marten, John, *A treatise of the venereal disease*, 6th edn. London 1708.

Massia, Edouard de, *Âge critique chez la femme*. Thèse. Montpelier 1851.

Melanchthon, Philipp, *Liber de anima*. Wittenberg 1552.

Ménétra, Jacques-Louis, *Journal de ma vie*. Ed. by Daniel Roche. Paris 1982.

Mettrie, Julien Offray de la, *Traité du vertige*. Paris 1738.

Milow, Margarethe E., *Ich will aber nicht murren*. Ed. by Rita Bake and Birgit Kiupel. Hamburg 1993.

Montaigne, Michel de, *Die Essais*. Ed. by Arthur Franz. Leipzig 1953.

Musitano, Carlo, *Opera omnia*, vol. 2. Geneva 1716.

Neue woleingerichtete Frauenzimmerapotheke. Hamburg 1741.

Oldecop, Johan, *Chronik des Johan Oldecop*. Ed. by Karl Euling. Tübingen 1891.

Onania, or the heinous sin of self pollution, 8th edn. London 1723 (repr. New York/ London 1986).

Onanism display'd, 2nd edn. London 1719.

Orléans, Elisabeth Charlotte von, *Leben und Charakter der Elisabeth Charlotte, Herzogin von Orléans nebst einem Auszuge des Denkwürdigsten aus ihren Briefen. Ein Beitrag zur Charakteristik des französischen Hofes Ludwigs XIV*. Ed. by Prof. Schütz. Leipzig 1820.

Ostervald, Jean-Frédéric, *The nature of uncleanliness consider'd*. London 1708.

Pansa, Martin, *Köstlicher und heilsamer Extract der gantzen Artzneykunst*. Leipzig 1618.

Paré, Ambroise, *Œuvres complètes*. Paris 1840.

Passer, Justus Eberhard, 'Berichte des Hessen-Darmstädtischen Gesandten Justus Eberh. Passer an die Landgräfin Elisabeth Dorothea über die Vorgänge am kaiserlichen Hofe und in Wien von 1680–1683'. Ed. by Ludwig Bauer, *Archiv für Österreichische Geschichte* 37 (1867), 271–409.

Pauli, Friedrich, *Die in der Pfalz und den angrenzenden Gegenden üblichen Volksheilmittel*. Landau 1842.

Paullini, Christian Franz, *Observationes medico-physicae rarae, selectae et curiosae*. Leipzig 1706.

Paumerelle, C. J. de B. de, *La philosophie des vapeurs, ou lettres raisonnées d'une jolie femme, sur l'usage des symptomes vaporeux*. Lausanne 1774.

Pellikan, Konrad, *Die Hauschronik Konrad Pellikans von Rufach. Ein Lebensbild aus der Reformationszeit*. Transl. by Th. Vulpinus. Strasbourg 1892.

Peyrilhe, Bernard, *Sur le cancer*. Paris 1776.

Philo-Castitatis, *'Onania' examined and detected*, 2nd edn. London 1724.

Pictorius, Georg, *LeibsArtzney. Ein sehr nutzliches Handtbüchlein vom Grund vnd Innhalt der gantzen Artzney*. Frankfurt 1566.

Pirckheimer, Willibald, 'Autobiographie', in: W. P. Eckert/Christoph von Imhoff (eds), *Willibald Pirckheimer. Dürers Freund im Spiegel seines Lebens, seiner Werke und seiner Umwelt*. Cologne 1971, pp. 125–34.

Platter, Felix, *Praxeos medicae opus, quinque libris adornatum*, 3rd edn. Basel 1666.

Platter, Felix, *Tagebuch (Lebensbeschreibung) 1536–1567*. Ed. by Valentin Lötscher. Basel/Stutttgart 1976.

Pomme, Pierre, *Abhandlung von den hysterisch- und hypochondrischen Nerven-Krankheiten beider Geschlechter, oder den Vapeurs*. Breslau/Leipzig 1775.

Pomme, Pierre, *Traité des affections vaporeuses des deux sexes*. Lyon 1763.

Practical scheme for the secret disease and broken constitutions, 24th edn. London 1719.

Questelius, Caspar, *Dissertatio academica de pulvinari morientibus non subtrahendo*. Jena 1678.

Ranzovius, Heinrich, *Exempla quibus astrologiae scientiae certitudo, doctissimorum cum veterum, tum recentiorum auctoritate astruitur*. Cologne 1585.

Recueil des miracles opérés au tombeau de M. de Paris Diacre. [Paris] 1732.

Recueil des pieces [...] sur les miracles opérés par l'intercession de feu Messire Felix Vialard, Évêque Comte de Chalons Pair de France, mort en odeur de sainteté dans le mois de juin 1680. Nancy 1735.

Relation de la maladie et de la guérison miraculeuse de Mlle Dumoulin opérée par l'intercession de M. de Paris. Sine loco 1735.

Rem, Lucas, *Tagebuch des Lucas Rem aus den Jahren 1494–1541. Ein Beitrag zur Handelsgeschichte der Stadt Augsburg*. Ed. by Bernd Greiff. Augsburg 1861.

Rétif de la Bretonne, Nicolas-Edme, *Monsieur Nicolas ou le coeur humain dévoilé*, 2 vols. Ed. by Pierre Testud. Paris 1989.

Revillon, Claude, *Recherches sur la cause des affections, appelées communément vapeurs*. Paris 1779.

Rivière, Lazare, *Opera medica universa*. Lyon 1663.

Rothos, Math., *A whip for the quack: or, some remarks on M---N's Supplement to his Onania*. London 1727.

Roussel, Pierre, *Système physique et moral de la femme*. Paris 1775.

Ryff, Andreas, 'Selbstbiographie des Andreas Ryff (bis 1574)', *Beiträge zur vaterländischen Geschichte* 9 (1870), 37–121.

Santorio, Santorio, *De statica medicina*. The Hague 1657.

Schad, Hans, 'Das Memorial- und Reisebuch des Hans Schad. Ein Beitrag zur Geschichte Ulms im 17. Jahrhundert'. Ed. by Prof. Greiner, *Württembergische Vierteljahreshefte für Landesgeschichte* N.F. 17 (1908), 334–420.

Schertlin, Sebastian, *Leben und Taten des weiland wohledeln Ritters Sebastian Schertlin von Burtenbach*. Ed. by Engelbert Hegaur. Munich 1910.

Schmieder, L. S[alomon], 'Observatio de seminis regressu ad massam sanguineam', in: *Actorum eruditorum quae Lipsiae publicantur supplementa*. Leipzig 1713, pp. 408–14.

Schmitt, August, 'Ueber die Sitten, Gebräuche und Volksmittel in der Rhön', *Aerztliches Intelligenz-Blatt. Münchener Medicinische Wochenschrift* 27 (1880), 361–4.

Schurig, Martin, *Parthenologia historico-medica*. Dresden/Leipzig 1729.

Schurig, Martin, *Spermatologia historico-medica h. e. seminis humani consideratio physico-medico-legalis*. Frankfurt 1720.

Schuyl, Florentius, *De gonorrhea*. Exp. Hermannus Grube. Leiden 1666.

Scultetus, Abraham, *Die Selbstbiographie des Heidelberger Theologen und Hofpredigers Abraham Scultetus (1566–1624)*. Ed. by Gustav A. Benrath. Karlsruhe 1966.

Second recueil des miracles opérés par l'intercession de M. de Paris. [Paris] 1732.

Sénac de Meilhan, Gabriel, *Considérations sur l'esprit et les moeurs*. London 1778.

Sennert, Daniel, 'De mulierum et infantium morbis ac symptomatibus', in: idem, *Practica medicina*, book 4. 2nd edn. Wittenberg 1649, pp. 230–51.

Sennert, Daniel, *Opera omnia*, 2 vols. Lyon 1656.

Sévigné, Marie de, *Lettres de Madame de Sévigné*. Paris 1976.

Spener, Philipp Jacob, 'D. Phil. Jacob Speners eigenhaendig aufgesetzter Lebens-Lauf', in: Heinrich Anshelm von Ziegler und Kliphausen (ed.), *Continuirter Historischer Schau-Platz und Labyrinth der Zeit*. Leipzig 1718, pp. 856–64.

Staal Delaunay, Marguerite Jeanne de, *Mémoires*. Ed. by François Barrière. Paris 1846.

Staël-Holstein, Anne Louise Germaine de, *Mémoires sur la vie privée de mon père*. Paris 1818.

Stahl, Georg Ernst, *Ausführliche Abhandlung von den Zufällen und Kranckheiten des Frauenzimmers*. Ed. by G. L. Leipzig 1724.

Staiger, Klara, *Tagebuch. Aufzeichnungen während des Dreißigjährigen Krieges im Kloster Mariastein bei Eichstätt*. Ed. by Ortrun Fina. Regensburg 1981.

Stampfer, Maria Elisabeth, *Das Hausbüchl der Stampferin (1679–1699)*. Klagenfurt 1982.

Storch, Johann, *Von Kranckheiten der Weiber* (later under the title 'Von Weiberkranckheiten'), 7 vols. Gotha 1748–51.

Stupanus, Emmanuel, *De chlorosi seu morbo virgineo*. Basel 1619.

Sydenham, Thomas, 'Epistolary dissertation to Dr. Cole', in: idem, *The whole works*, 6th edn. London 1715, vol. 2, pp. 51–118.

Tauvry, Daniel, *Nouvelle anatomie raisonnée, ou les usages de la structure du corps de l'homme*. Paris 1690.

The crime of Onan [...] or the hainous vice of self-defilement with all its dismal consequences, vol. 1. London [1723?].

The ladies' dispensatory, or every woman her own physician. London [?] ca. 1739.

Timaeus von Güldenklee, Baldassar, *Responsa medica, et diaeteticon opus posthumum*. Leipzig 1668.

Tissot, Samuel Auguste, *Advice to people in general with respect to their health*, 2 vols. Edinburgh 1766.

Tissot, Samuel Auguste, *An essay on the disorders of people of fashion*. Edinburgh [1771?].

Tissot, Samuel Auguste, *Anleitung für das Landvolk in Absicht auf seine Gesundheit*. Zürich 1767.

Tissot, Samuel Auguste, *Avis au peuple sur sa santé*. Lausanne 1761.

Tissot, Samuel Auguste, *De la santé des gens de lettres*, 3rd rev. edn. Lausanne 1775.

Tissot, Samuel Auguste, *L'onanisme ou dissertation physique sur les maladies produites par la masturbation*. Lausanne 1760.

Vandermonde, Charles Augustin, *Dictionnaire portatif de santé*, 2nd edn. Paris 1760.

Varandaeus, Ioannis, *Opera omnia*. Lyon 1658.

Vaublanc, Vincent, *Mémoires*. Ed. by François Barrière. Paris 1857.

Venel, Jean-André, *Essai sur la santé et sur l'éducation médicinale des filles destinées au mariage*. Yverdon 1776.

Venel, Jean-André, *Observations on the diseases of people of fashion*. London 1815.

Vetter, Anna, 'Von den Gesichten Annae Vetterin. Ihr lebenslauff, den sie auf begehren aufgeschrieben, und sonst mündlich zum öffteren erzehlet', in: Marianne Beyer-Fröhlich (ed.): *Selbstzeugnisse aus dem Dreißigjährigen Krieg und dem Barock*. Leipzig 1930, pp. 72–80.

Vieussens, Raymond de, *Histoire des maladies internes*, 3 vols. Toulouse 1775.

Vincentz, Wolfgang, *Die Goldschmiede-Chronik. Die Erlebnisse der ehrbaren Goldschmiede-Ältesten Martin und Wolfgang, auch Mag. Peters Vincentz*. Hanover [1918].

Viridet, -, *Dissertation sur les vapeurs*. Yverdon 1726.

Wedel, Georg Wolffgang, *De pollutione nocturna*. Jena 1667.

Weinsberg, Herrmann von, *Das Buch Weinsberg. Aus dem Leben eines Kölner Ratsherrn*. Ed. by Johann Jakob Hässlin. Stuttgart 1961.

Weinsberg, Hermann von, *Das Buch Weinsberg. Kölner Denkwürdigkeiten aus dem 16. Jahrhundert*, 5 vols. Düsseldorf 2000 (reprint).

Whytt, Robert, *Les vapeurs et maladies nerveuses, hypocondriaques ou hystériques reconnues et traitées dans les deux sexes*. Paris 1767.

Wierus, Ioannis, *Medicarum observationum rararum liber I*. Basel 1567.

Wirsung, Christophorus, *Artzney Buch*. Heidelberg 1568.

Wittich, Johannes, *Praeservator sanitatis. Ein nützlicher Bericht von den sechs unvormeidlichen Dingen, zur Gesundheit gantz ersprießlichen, wie man sich in denselben beydes zu Hause, und auch uber Land verhalten sol*. Leipzig [1590].

Wolf, Hieronymus, *Commentariolus de vita sua*. Ed. by Helmut Zäh. Diss. phil. Munich 1998.

Zedler, Johann Heinrich, Grosses vollständiges Universal-Lexikon aller Wissenschafften und Künste, 64 vols. Leipzig/Halle 1732–4.

Zimmern von [Graf Froben Christoph], *Die Chronik der Grafen von Zimmern*. Ed. by Hansmartin Decker-Hauff with the collaboration of Rudolf Seigel, 3 vols. Konstanz/Stuttgart 1964 and 1967, Sigmaringen 1972.

Zwinger, Theodor, *Theatrum praxeos medicae*. Basel 1710.

Index